PEARSON

Math Makes Sense

Preparation and Practice Book

Robert Berglind

Delcy Rolheiser

David Sufrin

Publisher
Mike Czukar

Research and Communications Manager
Barbara Vogt

Publishing Team
Claire Burnett
Enid Haley
Cristina Getson
Alison Rieger
Ioana Gagea
Nirmala Nutakki
Ellen Davidson
Alison Dale
Jane Schell
Karen Alley
Judy Wilson

Design
David Cheung

Composition
Lapiz Digital Services, India

Illustrations (pp. 345, 356, 362, 367, 376)
© 2009 Jupiterimages Corporation

ISBN-10 0-321-52418-7
ISBN-13 978-0-321-52418-8

Printed and bound in Canada

3 2019

Contents

1 Square Roots and Surface Area 1

Skill Builder 1.1:
- Side Lengths and Areas of Squares 2
- Whole Number Squares and Square Roots 3
- Perfect Squares 3

1.1 Square Roots of Perfect Squares 4

Skill Builder 1.2:
- Degree of Accuracy 10
- Squares and Square Roots on Number Lines 11
- The Pythagorean Theorem 12

1.2 Square Roots of Non-Perfect Squares 13
- Checkpoint 19

Skill Builder 1.3: Surface Areas of Rectangular Prisms 22

1.3 Surface Areas of Objects Made from Right Rectangular Prisms 24

Skill Builder 1.4:
- Surface Areas of Triangular Prisms 33
- Surface Areas of Cylinders 34

1.4 Surface Areas of Other Composite Objects 36
- Puzzle 44
- Study Guide 45
- Review 46

2 Powers and Exponent Laws 51

Skill Builder 2.1:
- Multiplying Integers 52
- Multiplying More than 2 Integers 53

2.1 What Is a Power? 54

Skill Builder 2.2:
- Patterns and Relationships in Tables 59
- Writing Numbers in Expanded Form 60

2.2 Powers of Ten and the Zero Exponent 61

Skill Builder 2.3:
- Adding Integers 64
- Subtracting Integers 65
- Dividing Integers 66

2.3 Order of Operations with Powers 67
- Checkpoint 72

Skill Builder 2.4: Simplifying Fractions 75

2.4 Exponent Laws I 76

Skill Builder 2.5:
- Grouping Equal Factors 82
- Multiplying Fractions 82

2.5 Exponent Laws II 83
- Puzzle 88
- Study Guide 89
- Review 90

3 Rational Numbers 93

Skill Builder 3.1:
- Equivalent Fractions 94
- Comparing Fractions 95
- Common Denominators 96
- Converting between Fractions and Decimals 97

3.1 What Is a Rational Number? 98

Skill Builder 3.2:
- Adding Fractions 102
- Adding Mixed Numbers 103

3.2 Adding Rational Numbers 104

Skill Builder 3.3: Converting Mixed Numbers to Improper Fractions 109

3.3 Subtracting Rational Numbers 110
- Checkpoint 114

Skill Builder 3.4:
- Writing a Fraction in Simplest Form 117
- Multiplying Proper Fractions 118
- Multiplying Mixed Numbers 119

3.4 Multiplying Rational Numbers 120

Skill Builder 3.5: Dividing Fractions 125

3.5 Dividing Rational Numbers 126

3.6 Order of Operations with Rational Numbers 130
- Puzzle 135
- Study Guide 136
- Review 137

4 Linear Relations 141

Skill Builder 4.1:
- Algebraic Expressions 142
- Relationships in Patterns 142

4.1 Writing Equations to Describe Patterns 144

Skill Builder 4.2:
- The Coordinate Grid 149
- Graphing Relations 150

4.2 Linear Relations 151

Skill Builder 4.3: Solving Equations 157

4.3 Another Form of the Equation for a Linear Relation 158
- Checkpoint 163

4.4 Matching Equations and Graphs 166

4.5 Using Graphs to Estimate Values 170
- Puzzle 173
- Study Guide 174
- Review 175

5 Polynomials **179**

Skill Builder 5.1: Modelling Expressions 180

5.1 Modelling Polynomials **181**

Skill Builder 5.2: Modelling Integers 186

5.2 Like Terms and Unlike Terms **187**

Skill Builder 5.3: Adding Integers 192

5.3 Adding Polynomials **193**

Skill Builder 5.4: Subtracting Integers Symbolically 198

5.4 Subtracting Polynomials **199**

Checkpoint 203

Skill Builder 5.5:

The Distributive Property 206

Multiplying and Dividing Integers 207

5.5 Multiplying and Dividing a Polynomial by a Constant **208**

Skill Builder 5.6: Multiplying Monomials 214

5.6 Multiplying and Dividing a Polynomial by a Monomial **215**

Puzzle 220

Study Guide 221

Review 222

6 Linear Equations and Inequalities **225**

Skill Builder 6.1:

Order of Operations 226

The Distributive Property 227

6.1 Solving Equations by Using Inverse Operations **228**

Skill Builder 6.2: Solving Equations Using Models 233

6.2 Solving Equations by Using Balance Strategies **234**

Checkpoint 240

6.3 Introduction to Linear Inequalities **242**

6.4 Solving Linear Inequalities by Using Addition and Subtraction **246**

6.5 Solving Linear Inequalities by Using Multiplication and Division **250**

Puzzle 255

Study Guide 256

Review 257

7 Similarity and Transformations **261**

Skill Builder 7.1: Converting between Metric Units of Length 262

7.1 Scale Diagrams and Enlargements **263**

7.2 Scale Diagrams and Reductions **267**

Skill Builder 7.3: Polygons 271

7.3 Similar Polygons **272**

Skill Builder 7.4: Sum of the Angles in a Triangle 278

7.4 Similar Triangles **279**

Checkpoint 286

Skill Builder 7.5:

Lines of Symmetry in Quadrilaterals 289

Reflections 290

7.5 Reflections and Line Symmetry **291**

Skill Builder 7.6: Rotations 297

7.6 Rotations and Rotational Symmetry **298**

Skill Builder 7.7: Translations 304

7.7 Identifying Types of Symmetry on the Cartesian Plane **305**

Puzzle 311

Study Guide 312

Review 313

8 Circle Geometry **317**

Skill Builder 8.1: Solving for Unknown Measures in Triangles 318

8.1 Properties of Tangents to a Circle **319**

8.2 Properties of Chords in a Circle **323**

Checkpoint 329

8.3 Properties of Angles in a Circle **332**

Puzzle 338

Study Guide 339

Review 340

9 Probability and Statistics **343**

Skill Builder 9.1:

Relating Fractions, Decimals, and Percents 344

Theoretical Probability 345

Experimental Probability 346

9.1 Probability in Society **347**

Skill Builder 9.2: Writing a Questionnaire 352

9.2 Potential Problems with Collecting Data **353**

9.3 Using Samples and Populations to Collect Data **358**

Checkpoint 363

9.4 Selecting a Sample **365**

9.5 Designing a Project Plan **371**

Puzzle 374

Study Guide 375

Review 376

About *Math Makes Sense 9* Preparation and Practice Book

Welcome to *Math Makes Sense 9*.
These pages describe how this *Preparation and Practice Book* can help to support your success this year.

For the main lessons of *Math Makes Sense 9*, this workbook gives you:

- ***Skill Builder pages***
  - Each topic identifies a prerequisite skill or concept you will need in that lesson.
  - Each topic starts with a quick review and a short example.
  - General rules are highlighted to help you find them more easily.
  - Try the ***Check*** questions to see whether you're ready to move on.
- ***Matching lessons for the ones in your Student Book***
  - General rules are highlighted to help you find them more easily.
  - ***Examples*** show you what you need to know and be able to do.
  - For each Example, ***Check*** lets you try a question on your own before moving on.
  - ***Practice*** questions help you put it all together during the lesson.
- ***Step-by-step support***
  - Every question supports your work by breaking down the steps you need to follow to arrive at a correct solution.

For each unit of *Math Makes Sense 9*, you'll find:

- an opening page that tells you ***What You'll Learn*** and ***Why It's Important***, as well as an expanded list of ***Key Words***
- ***Checkpoint*** pages with mid-unit review
- a ***Unit Puzzle*** for fun and reinforcement
- a ***Unit Study Guide*** to summarize key ideas
- ***Unit Review*** pages to help you see what you remember

UNIT 1

Square Roots and Surface Area

What You'll Learn

- Find square roots of fractions and decimals that are perfect squares.
- Approximate the square roots of fractions and decimals that are not perfect squares.
- Find the surface areas of composite objects.

Why It's Important

Square roots are used by

- police officers, to estimate the speed of a vehicle when it crashed
- vets, to calculate drug dosages

Surface area is used by

- painters, to find the number of cans of paint needed to paint a room
- farmers, to find the amount of fertilizer needed for a field

Key Words

square
square root
perfect square
non-perfect square
terminating decimal
repeating decimal
non-terminating, non-repeating decimal
surface area
composite object

1.1 Skill Builder

Side Lengths and Areas of Squares

The side length and area of a square are related.

- The area is the **square** of the side length.

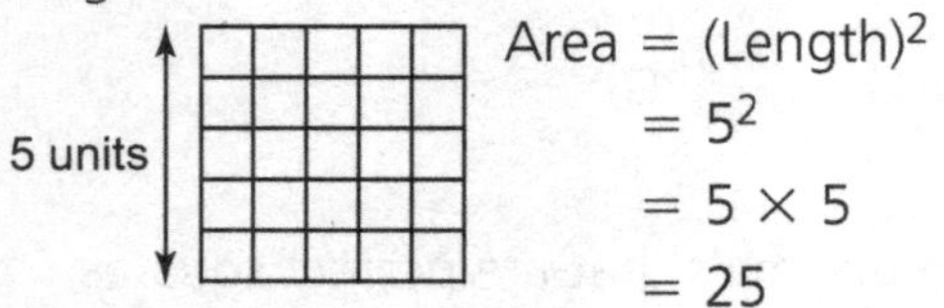

$$\text{Area} = (\text{Length})^2 = 5^2 = 5 \times 5 = 25$$

The area is 25 square units.

- The side length is the **square root** of the area.

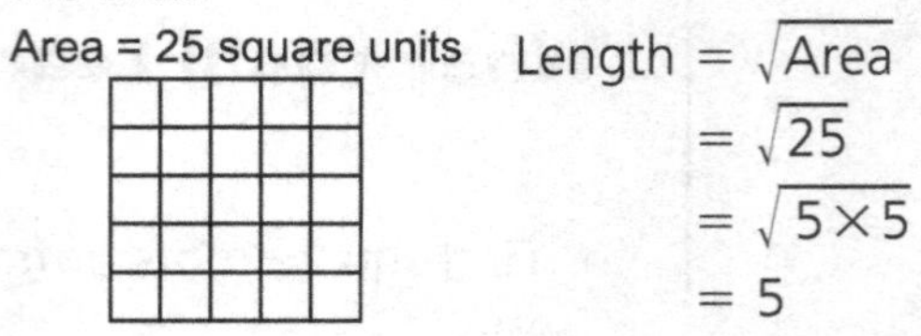

$$\text{Length} = \sqrt{\text{Area}} = \sqrt{25} = \sqrt{5 \times 5} = 5$$

The side length is 5 units.

Check

1. Which square and square root are modelled by each diagram?

Diagram	Square Modelled	Square Root Modelled
a) Area = 49 square units; 7 units	$(\text{Length})^2 = \text{Area}$ $7^2 =$ ______ The area is 49 square units.	$\sqrt{\text{Area}} = \text{Length}$ $\sqrt{49} =$ ______ The side length is 7 units.
b) 4 units	______ = ______ The area is ______ square units.	$\sqrt{\_\_\_\_\_\_}$ = ______ The side length is ______ units.
c) 8 units	______ = ______ The area is ______ square units.	$\sqrt{\_\_\_\_\_\_}$ = ______ The side length is ______ units.
d) 11 units	______ = ______ The area is ______ square units.	$\sqrt{\_\_\_\_\_\_}$ = ______ The side length is ______ units.

Whole Number Squares and Square Roots

- The square of a number is the number multiplied by itself.
- A square root of a number is one of 2 equal factors of the number.
- Squaring and taking a square root are inverse operations.

$5^2 = 5 \times 5$
$= 25$
$\sqrt{25} = \sqrt{5 \times 5}$
$= 5$
$5^2 = 25$ and $\sqrt{25} = 5$

Check

1. Complete each sentence.

a) $4^2 = 16$, so $\sqrt{16} =$ ____ **b)** $12^2 =$ ____, so $\sqrt{\_\_\_\_} =$ ____

c) $\sqrt{25} =$ ____, since ____ $= 25$ **d)** $\sqrt{100} =$ ____, since ____ $=$ ____

Perfect Squares

A number is a **perfect square** if it is the product of 2 equal factors.
25 is a perfect square because $25 = 5 \times 5$.
24 is a **non-perfect square.** It is not the product of 2 equal factors.

Check

1. Complete each sentence.

First 12 Whole-Number Perfect Squares			
Perfect Square	**Square Root**	**Perfect Square**	**Square Root**
$1^2 = 1 \times 1 = 1$	$\sqrt{1} = 1$	$7^2 =$ ___ × ___ = ___	$\sqrt{\_\_\_}$ = ___
$2^2 = 2 \times 2 = 4$	$\sqrt{4} = 2$	$8^2 =$ ___ × ___ = ___	$\sqrt{\_\_\_}$ = ___
$3^2 =$ ___ × ___ = ___	$\sqrt{\_\_\_}$ = ___	$9^2 =$ ___ × ___ = ___	$\sqrt{\_\_\_}$ = ___
$4^2 =$ ___ × ___ = ___	$\sqrt{\_\_\_}$ = ___	$10^2 =$ ___ × ___ = ___	$\sqrt{\_\_\_}$ = ___
$5^2 =$ ___ × ___ = ___	$\sqrt{\_\_\_}$ = ___	$11^2 =$ ___ × ___ = ___	$\sqrt{\_\_\_}$ = ___
$6^2 =$ ___ × ___ = ___	$\sqrt{\_\_\_}$ = ___	$12^2 =$ ___ × ___ = ___	$\sqrt{\_\_\_}$ = ___

1.1 Square Roots of Perfect Squares

FOCUS **Find the square roots of decimals and fractions that are perfect squares.**

The square of a fraction or decimal is the number multiplied by itself.

$$\left(\frac{2}{3}\right)^2 = \frac{2}{3} \times \frac{2}{3} = \frac{2 \times 2}{3 \times 3} = \frac{4}{9}$$

$$(1.5)^2 = 1.5 \times 1.5 = 2.25$$

$\frac{4}{9}$ and 2.25 are perfect squares because they are the product of 2 equal factors.

$\frac{2}{3} \times \frac{2}{3} = \frac{4}{9}$, so

$\frac{2}{3}$ is a square root of $\frac{4}{9}$.

We write: $\sqrt{\frac{4}{9}} = \frac{2}{3}$

$2.25 = 1.5 \times 1.5$, so

1.5 is a square root of 2.25.

We write: $\sqrt{2.25} = 1.5$

Each equal factor is a square root of the perfect square.

Example 1 Finding a Perfect Square Given Its Square Root

Calculate the number whose square root is:

a) $\frac{5}{8}$

b) 1.2

Solution

A square root of a number is one of two equal factors of the number.

a) $\frac{5}{8}$

$$\frac{5}{8} \times \frac{5}{8} = \frac{5 \times 5}{8 \times 8} = \frac{25}{64}$$

So, $\frac{5}{8}$ is a square root of $\frac{25}{64}$.

b) 1.2

Use a calculator.

$1.2 \times 1.2 = 1.44$

So, 1.2 is a square root of 1.44.

Check

1. Calculate the perfect square with the given square root.

a) $\frac{3}{8}$

$\frac{3}{8} \times \frac{3}{8} = \frac{3 \times 3}{8 \times 8}$

$= \frac{9}{64}$

$\frac{3}{8}$ is a square root of $\frac{9}{64}$.

b) $\frac{3}{2}$

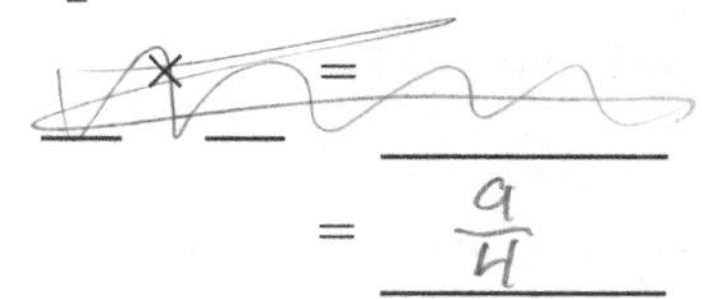

$= \frac{9}{4}$

$\frac{3}{2}$ is a square root of $\frac{9}{4}$.

c) 0.5

$0.5 \times 0.5 =$ 0.25

0.5 is a square root of 0.25.

d) 2.5

$2.5 \times 2.5 =$ 6.25

2.5 is a square root of 6.25.

Example 2 Identifying Fractions that Are Perfect Squares

Is each fraction a perfect square? If so, find its square root.

a) $\frac{16}{25}$

b) $\frac{9}{20}$

Solution

Check if the numerator and denominator are perfect squares.

a) $\frac{16}{25}$

$16 = 4 \times 4$, so 16 is a perfect square.

$25 = 5 \times 5$, so 25 is a perfect square.

So, $\frac{16}{25}$ is a perfect square.

b) $\frac{9}{20}$

$9 = 3 \times 3$, so 9 is a perfect square.

20 is not a perfect square.

So, $\frac{9}{20}$ is not a perfect square.

Check

1. Determine whether the fraction is or is not a perfect square. How do you know?

a) $\frac{9}{49}$

9 ________ a perfect square because ________________________________.

49 ________ a perfect square because ________________________________.

So, $\frac{9}{49}$ ________ a perfect square.

b) $\frac{25}{13}$

25 ________ a perfect square because ________________________________.

13 ________ a perfect square because ________________________________.

So, $\frac{25}{13}$ ________ a perfect square.

c) $\frac{64}{81}$ 64 is a perfect square because ______.

81 is a perfect square because ______.

So, $\frac{64}{81}$ is a perfect square.

2. Find the value of each square root.

a) $\sqrt{\frac{9}{4}} = \sqrt{\frac{\_\_\_ \times \_\_\_}{\_\_\_ \times \_\_\_}} = \frac{3}{2}$

b) $\sqrt{\frac{16}{81}} = \sqrt{\frac{\_\_\_ \times \_\_\_}{\_\_\_ \times \_\_\_}} = \frac{4}{9}$

A **terminating decimal** ends after a certain number of decimal places.
A **repeating decimal** has a repeating pattern of digits in the decimal expansion.
The bar shows the digits that repeat.

Terminating	Repeating	Non-terminating and non-repeating
0.5 0.28	0.333 333 … = $0.\overline{3}$	1.414 213 56 … 7.071 067 812 …
	0.191 919 … = $0.\overline{19}$	

You can use a calculator to find out if a decimal is a perfect square.
The square root of a perfect square decimal is either a terminating decimal or a repeating decimal.

Example 3 Identifying Decimals that Are Perfect Squares

Is each decimal a perfect square? How do you know?

a) 1.69 **b)** 3.5

Solution

Use a calculator to find the square root of each number.

a) $\sqrt{1.69} = 1.3$

The square root is the terminating decimal 1.3.
So, 1.69 is a perfect square.

b) $\sqrt{3.5} \doteq 1.870\ 828\ 693$

The square root appears to be a decimal that neither repeats nor terminates.
So, 3.5 is not a perfect square.

The symbol ≐ means "approximately equal to".

Check

1. Complete the table to find whether each decimal is a perfect square.
The first one is done for you.

	Decimal	Value of square root	Type of decimal	Is decimal a perfect square?
a)	70.5	8.396 427 811 ...	Non-repeating Non-terminating	No
b)	5.76	________	________ ________	________
c)	0.25	________	________ ________	________
d)	2.5	________	________ ________	________

Practice

1. Calculate the number whose square root is:

a) $\frac{1}{4}$

$\frac{1}{4} \times \frac{1}{4} = \frac{\_\_\_ \times \_\_\_}{\_\_\_ \times \_\_\_}$

$= \frac{\_\_\_}{\_\_\_}$

$\frac{1}{4}$ is a square root of ____.

b) $\frac{2}{7}$

$\frac{\_\_\_}{\_\_\_} \times \frac{\_\_\_}{\_\_\_} = \frac{\_\_\_ \times \_\_\_}{\_\_\_ \times \_\_\_}$

$= \frac{\_\_\_}{\_\_\_}$

$\frac{2}{7}$ is a square root of ____.

c) 0.6

_____ × _____ = _____

0.6 is a square root of _____.

d) 1.1

_____ × _____ = _____

1.1 is a square root of _____.

2. Identify the fractions that are perfect squares. The first one has been done for you.

	Fraction	Is numerator a perfect square?	Is denominator a perfect square?	Is fraction a perfect square?
a)	$\frac{81}{125}$	Yes; 9 × 9 = 81	No	No
b)	$\frac{25}{49}$	________	________	________
c)	$\frac{36}{121}$	________	________	________
d)	$\frac{17}{25}$	________	________	________
e)	$\frac{9}{100}$	________	________	________

3. Find each square root.

a) $\sqrt{\frac{49}{100}} = \sqrt{\frac{\_\_\_ \times \_\_\_}{\_\_\_ \times \_\_\_}}$

$= \_\_\_$

b) $\sqrt{\frac{25}{144}} = \sqrt{\frac{\_\_\_ \times \_\_\_}{\_\_\_ \times \_\_\_}}$

$= \_\_\_$

c) $\sqrt{\frac{1}{16}} = \sqrt{\frac{\_\_\_ \times \_\_\_}{\_\_\_ \times \_\_\_}}$

$= \_\_\_$

d) $\sqrt{\frac{9}{400}} = \sqrt{\frac{\_\_\_ \times \_\_\_}{\_\_\_ \times \_\_\_}}$

$= \_\_\_$

4. Use a calculator. Find each square root.

a) $\sqrt{8.41} = \_\_\_$ **b)** $\sqrt{0.0676} = \_\_\_$ **c)** $\sqrt{51.125} = \_\_\_$ **d)** $\sqrt{6.25} = \_\_\_$

5. Which decimals are perfect squares?

a) 1.44

$\sqrt{1.44} = \_\_\_\_\_\_$

The square root is a decimal that ____________.

So, 1.44 ____ a perfect square.

b) 30.25

$\sqrt{30.25} = \_\_\_\_\_\_$

The square root is a decimal that ____________.

So, 30.25 ____ a perfect square.

c) 8.5

$\sqrt{8.5} \doteq \_\_\_\_\_\_$

The square root is a decimal that ____________.

So, 8.5 ____ a perfect square.

d) 0.0256

$\sqrt{0.0256} = \_\_\_\_\_\_$

The square root is a decimal that ____________.

So, 0.0256 ____ a perfect square.

6. Find the area of each square.

a)

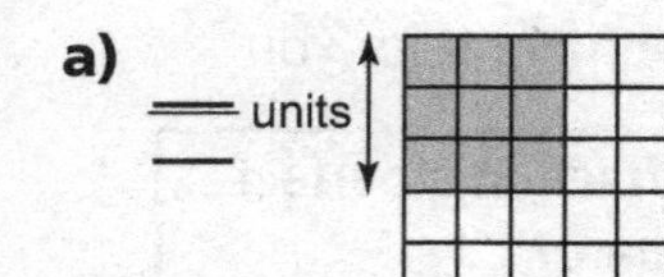

Area = ____

= ____

b)

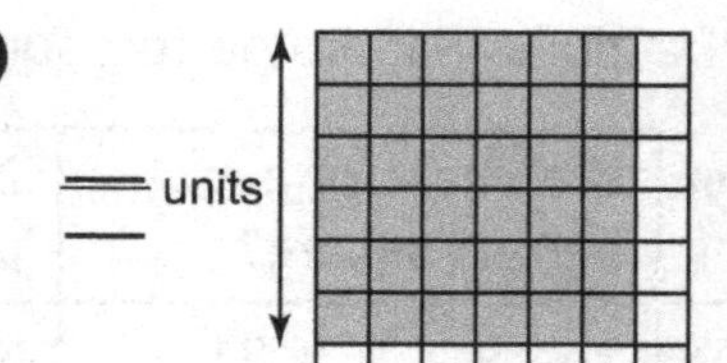

Area = ____

= ____

Area = $(\text{Length})^2$

The area is ________ ________

c) 5.4 units

Area = _____

= _____ × _____

= ________

____

d) 2.1 units

Area = _____

= _____ × _____

= ________

____

7. Find the side length of each square.

a) Area = $\frac{9}{100}$ square units

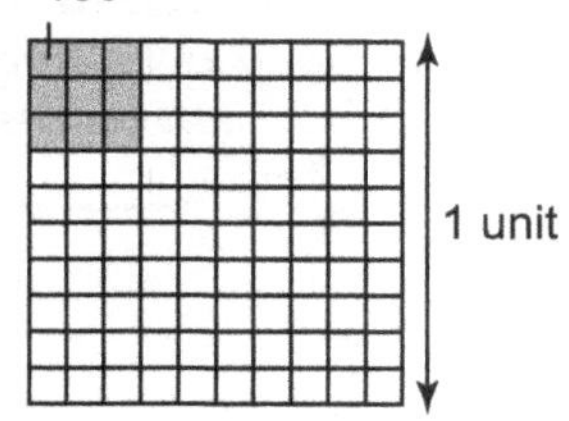

Side Length = $\sqrt{}$

= ____________

Length = $\sqrt{\text{Area}}$

b) The side length is ___ units.

Area = $\frac{25}{36}$ square units

1 unit

Length = $\sqrt{}$

= ____________

____

c) Area = 0.01 square units

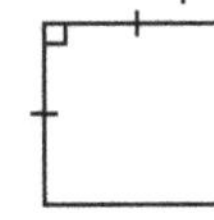

Length = $\sqrt{}$

= ____________

____

d) Area = 46.24 square units

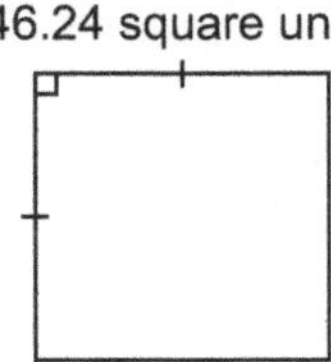

Length = $\sqrt{}$

= ____________

____

1.2 Skill Builder

Degree of Accuracy

We are often asked to write an answer to a given decimal place.
To do this, we can use a number line.

To write 7.3 to the nearest whole number:
Place 7.3 on a number line in tenths.

3 is the last digit. It is in the tenths position. So, use a number line in tenths.

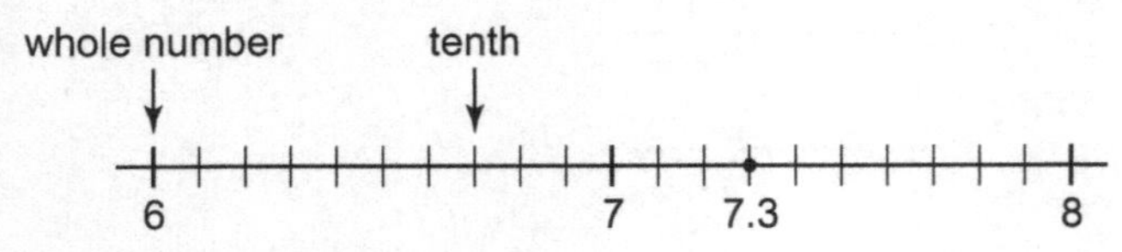

7.3 is closer to 7 than to 8.
So, 7.3 to the nearest whole number is: 7

To write 3.67 to the nearest tenth:
Place 3.67 on a number line in hundredths.

7 is the last digit. It is in the hundredths position. So, use a number line in hundredths.

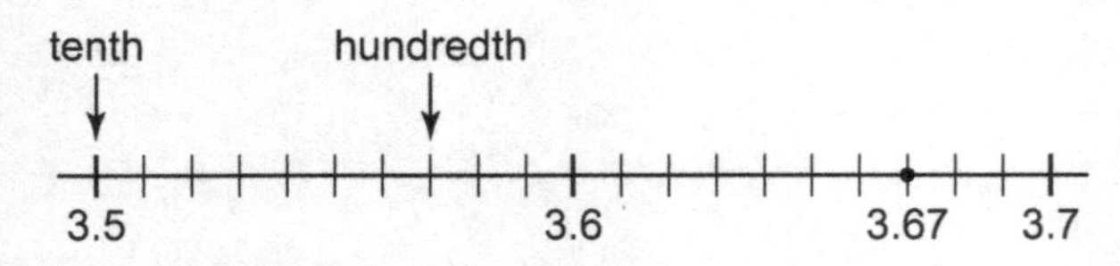

3.67 is closer to 3.7 than to 3.6.
So, 3.67 to the nearest tenth is: 3.7

Check

1. Write each number to the nearest whole number.
Mark it on the number line.

a) 5.3 _____ **b)** 6.8 _____ **c)** 7.1 _____ **d)** 6.4 _____

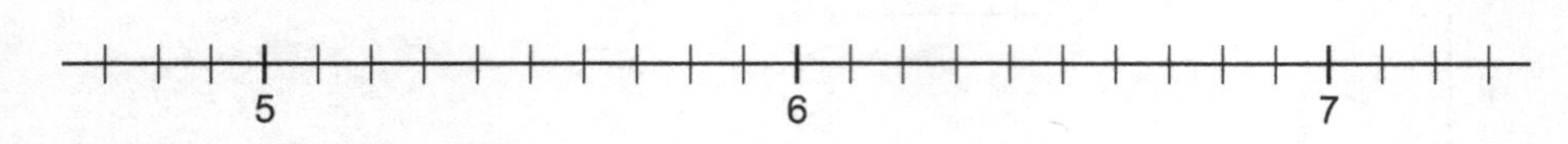

2. Write each number to the nearest tenth.
Mark it on the number line.

a) 2.53 _____ **b)** 2.64 _____ **c)** 2.58 _____ **d)** 2.66 _____

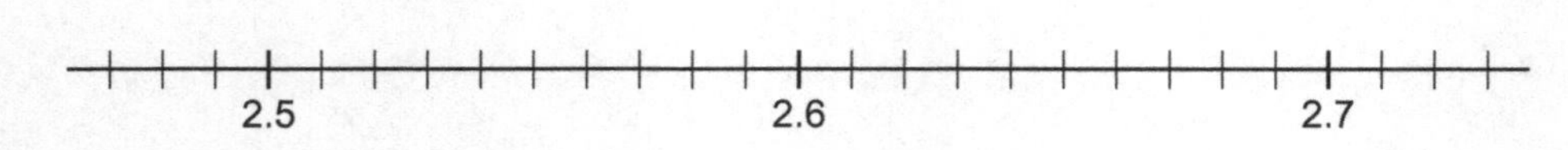

Squares and Square Roots on Number Lines

Most numbers are not perfect squares.
You can use number lines to estimate the square roots of these numbers.

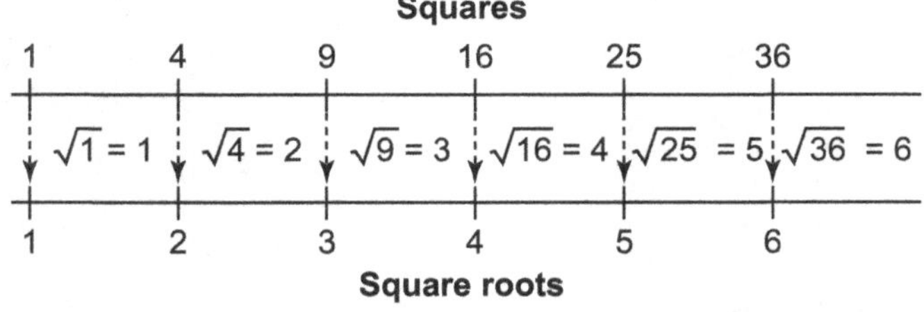

10 is between the perfect squares 9 and 16.
So, $\sqrt{10}$ is between $\sqrt{9}$ and $\sqrt{16}$.
$\sqrt{9} = 3$ and $\sqrt{16} = 4$
So, $\sqrt{10}$ is between 3 and 4.

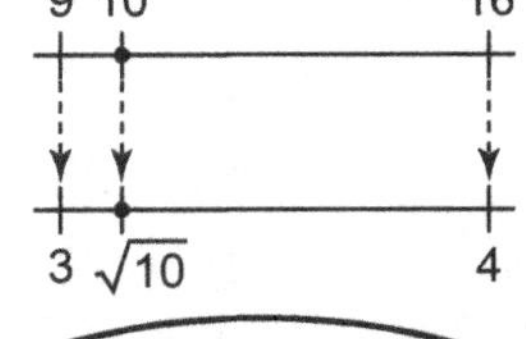

10 is closer to 9 than 16, so $\sqrt{10}$ is closer to 3 than 4.

Check with a calculator.
$\sqrt{10} \doteq 3.2$, which is between 3 and 4.

Check

1. Between which 2 consecutive whole numbers is each square root? Explain.

Refer to the squares and square roots number lines.

a) $\sqrt{22}$

22 is between the perfect squares 16 and 25.
So, $\sqrt{22}$ is between $\sqrt{\_\_\_\_}$ and $\sqrt{\_\_\_\_}$.
$\sqrt{\_\_\_\_}$ = ____ and $\sqrt{\_\_\_\_}$ = ____
So, $\sqrt{22}$ is between ____ and ____.

b) $\sqrt{6}$

6 is between the perfect squares ____ and ____.
So, $\sqrt{6}$ is between $\sqrt{\_\_\_\_}$ and $\sqrt{\_\_\_\_}$.
$\sqrt{\_\_\_\_}$ = ____ and $\sqrt{\_\_\_\_}$ = ____
So, $\sqrt{6}$ is between ____ and ____.

The Pythagorean Theorem

You can use the Pythagorean Theorem to find unknown lengths in right triangles.

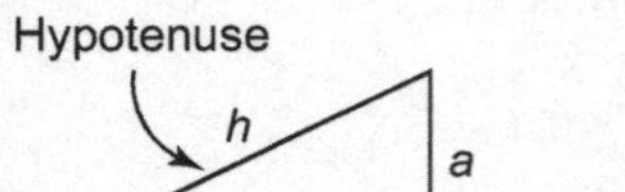
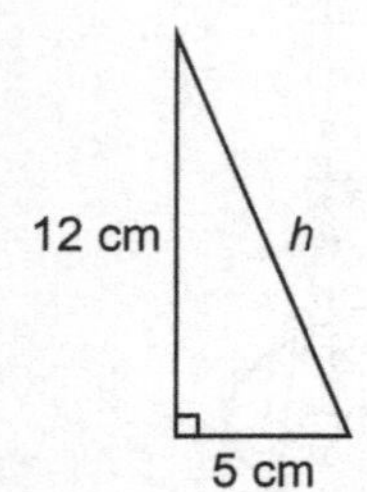

Pythagorean Theorem

$h^2 = a^2 + b^2$

To find the length of the hypotenuse, h, in this triangle:

12 cm h

5 cm

$h^2 = 5^2 + 12^2$

$h^2 = 25 + 144$

$h^2 = 169$

$h = \sqrt{169}$

$h = 13$

The length of the hypotenuse is 13 cm.

Check

1. Use the Pythagorean Theorem to find the length of each hypotenuse, h.

a)

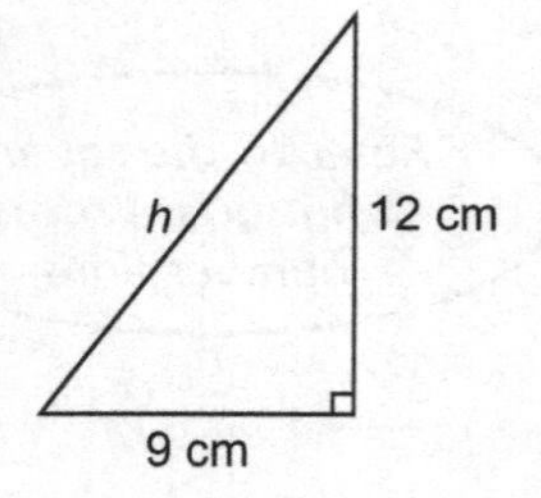

$h^2 =$ _____ + _____

$h^2 =$ _____ + _____

$h^2 =$ _____

$h = \sqrt{\_\_\_\_\_}$

$h =$ _____

The length of the hypotenuse is _____ cm.

b)

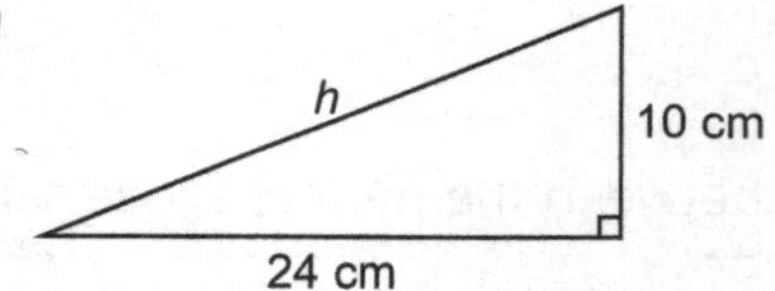

$h^2 =$ _____ + _____

$h^2 =$ _____ + _____

$h^2 =$ _____

$h = \sqrt{\_\_\_\_\_}$

$h =$ _____

The length of the hypotenuse is _____ cm.

1.2 Square Roots of Non-Perfect Squares

FOCUS **Approximate the square roots of decimals and fractions that are not perfect squares.**

The top number line shows all the perfect squares from 1 to 100.

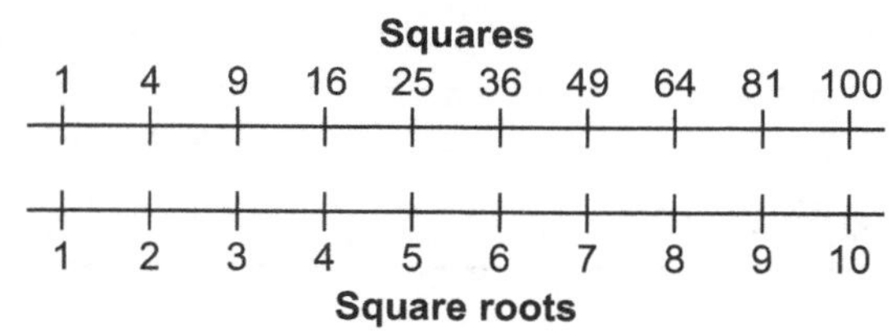

The bottom number line shows the square root of each number in the top line. You can use these lines to estimate the square roots of fractions and decimals that are not perfect squares.

Example 1 Estimating a Square Root of a Decimal

Estimate: $\sqrt{68.5}$

Solution

68.5 is between the perfect squares 64 and 81.
So, $\sqrt{68.5}$ is between $\sqrt{64}$ and $\sqrt{81}$.
That is, $\sqrt{68.5}$ is between 8 and 9.
Since 68.5 is closer to 64 than 81, $\sqrt{68.5}$ is closer to 8 than 9.
So, $\sqrt{68.5}$ is between 8 and 9, and closer to 8.

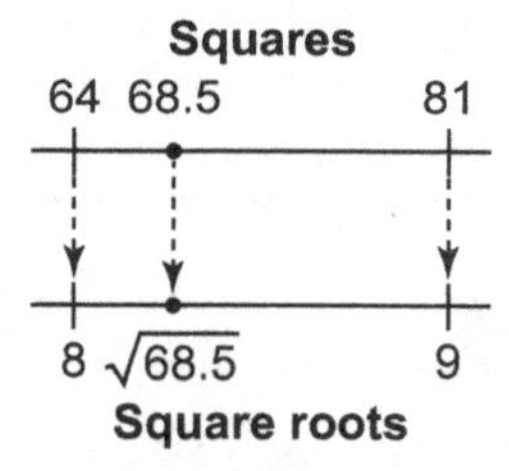

Check

1. Estimate each square root.
Explain your estimate.

a) $\sqrt{13.5}$

13.5 is between the perfect squares ____ and ____.
So, $\sqrt{13.5}$ is between $\sqrt{\_\_\_\_}$ and $\sqrt{\_\_\_\_}$.
That is, $\sqrt{13.5}$ is between ____ and ____.
Since 13.5 is closer to ____ than ____, $\sqrt{13.5}$ is closer to ____ than ____.
So, $\sqrt{13.5}$ is between ____ and ____, and closer to ____.

b) $\sqrt{51.5}$

51.5 is between the perfect squares ______ and ______.

So, $\sqrt{51.5}$ is between $\sqrt{\_\_\_\_}$ and $\sqrt{\_\_\_\_}$.

That is, $\sqrt{51.5}$ is between ____ and ____.

Since 51.5 is closer to ____ than ____, $\sqrt{51.5}$ is closer to ____ than ____.

So, $\sqrt{51.5}$ is between ____ and ____, and closer to ____.

Example 2 Estimating a Square Root of a Fraction

Estimate: $\sqrt{\frac{3}{10}}$

Solution

Find the closest perfect square to the numerator and denominator.

In the fraction $\frac{3}{10}$:

3 is close to the perfect square 4.

10 is close to the perfect square 9.

So, $\sqrt{\frac{3}{10}} \doteq \sqrt{\frac{4}{9}}$ and $\sqrt{\frac{4}{9}} = \frac{2}{3}$

So, $\sqrt{\frac{3}{10}} \doteq \frac{2}{3}$

Check

1. Estimate each square root.

a) $\sqrt{\frac{23}{80}}$

23 is close to the perfect square ____.

80 is close to the perfect square ____.

So, $\sqrt{\frac{23}{80}} \doteq \sqrt{\frac{\_\_\_\_}{\_\_\_\_}}$

$= \frac{\_\_\_\_}{\_\_\_\_}$

So, $\sqrt{\frac{23}{80}} \doteq$ ____

b) $\sqrt{\frac{8}{17}}$

8 is close to the perfect square ____.

17 is close to the perfect square ____.

So, $\sqrt{\frac{8}{17}} \doteq \sqrt{\frac{\_\_\_\_}{\_\_\_\_}}$

$= \frac{\_\_\_\_}{\_\_\_\_}$

So, $\sqrt{\frac{8}{17}} \doteq$ ____

Example 3 Finding a Number with a Square Root between Two Given Numbers

Identify a decimal that has a square root between 5 and 6.

Solution

$5^2 = 25$, so 5 is a square root of 25.
$6^2 = 36$, so 6 is a square root of 36.
So, any decimal between 25 and 36 has a square root between 5 and 6.
Choose 32.5.

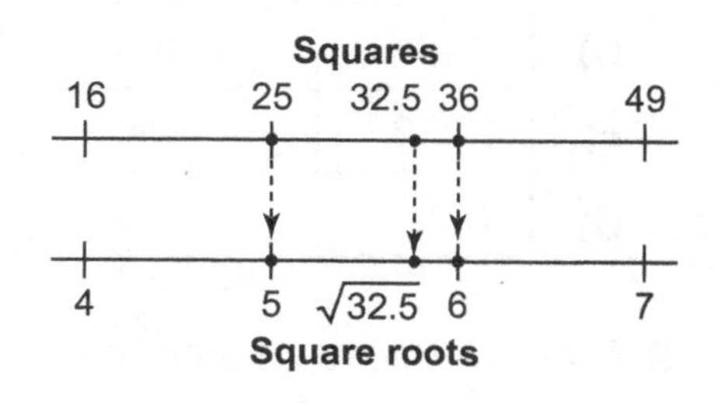

Check the answer by using a calculator.
$\sqrt{32.5} \doteq 5.7$, which is between 5 and 6.
So, the decimal 32.5 is one correct answer.
There are many more correct answers.

Check

1. a) Identify a decimal that has a square root between 7 and 8.
Check the answer.
$7^2 =$ _____ and $8^2 =$ _____
So, any decimal between _____ and _____ has a square root between 7 and 8.
Choose _____.
Check the answer on a calculator.
$\sqrt{\_\_\_\_\_} \doteq$ _____
The decimal _____ is one correct answer.

b) Identify a decimal that has a square root between 11 and 12.
_____ = _____ and _____ = _____
So, any decimal between _____ and _____ has a square root between 11 and 12.
Choose _____.
$\sqrt{\_\_\_\_\_} \doteq$ _____
So, _____ is one correct answer.

Practice

1. For each number, name the 2 closest perfect squares and their square roots.

	Number	Two closest perfect squares	Their square roots
a)	44.4	____ and ____	____ and ____
b)	10.8	____ and ____	____ and ____
c)	125.9	____ and ____	____ and ____
d)	87.5	____ and ____	____ and ____

2. For each fraction, name the closest perfect square and its square root for the numerator and for the denominator.

	Fraction	Closest perfect squares	Their square roots
a)	$\frac{5}{11}$	Numerator: ____; denominator: ____	____ and ____
b)	$\frac{17}{45}$	Numerator: ____; denominator: ____	____ and ____
c)	$\frac{3}{24}$	Numerator: ____; denominator: ____	____ and ____
d)	$\frac{11}{62}$	Numerator: ____; denominator: ____	____ and ____

3. Estimate each square root.
Explain.

a) $\sqrt{1.6}$

1.6 is between ____ and ____.
So, $\sqrt{1.6}$ is between $\sqrt{\_\_\_\_}$ and $\sqrt{\_\_\_\_}$.
That is, $\sqrt{1.6}$ is between ____ and ____.
Since 1.6 is closer to ____ than ____, $\sqrt{1.6}$ is closer to ____ than ____.
So, $\sqrt{1.6}$ is between ____ and ____, and closer to ____.

b) $\sqrt{44.5}$

44.5 is between ____ and ____.
So, $\sqrt{44.5}$ is between $\sqrt{\_\_\_\_}$ and $\sqrt{\_\_\_\_}$.
That is, $\sqrt{44.5}$ is between ____ and ____.
Since 44.5 is closer to ____ than ____, $\sqrt{44.5}$ is closer to ____ than ____.
So, $\sqrt{44.5}$ is between ____ and ____, and closer to ____.

c) $\sqrt{75.8}$

75.8 is between ____ and ____.

So, $\sqrt{75.8}$ is between $\sqrt{\_\_\_\_}$ and $\sqrt{\_\_\_\_}$.

That is, $\sqrt{75.8}$ is between ____ and ____.

Since 75.8 is closer to ____ than ____, $\sqrt{75.8}$ is closer to ____ than ____.

So, $\sqrt{75.8}$ is between ____ and ____, and closer to ____.

4. Estimate each square root. Explain.

a) $\sqrt{\frac{7}{15}}$

7 is close to ____; 15 is close to ____.

So, $\sqrt{\frac{7}{15}} \doteq \sqrt{\frac{\_\_\_\_}{\_\_\_\_}}$

$\doteq$ ____

b) $\sqrt{\frac{2}{7}}$

2 is close to ____; 7 is close to ____.

So, $\sqrt{\frac{2}{7}} \doteq \sqrt{\frac{\_\_\_\_}{\_\_\_\_}}$

$\doteq$ ____

c) $\sqrt{\frac{35}{37}}$

35 is close to ____; 37 is close to ____.

So, $\sqrt{\frac{35}{37}} \doteq \sqrt{\frac{\_\_\_\_}{\_\_\_\_}}$

$\doteq$ ____

d) $\sqrt{\frac{99}{122}}$

99 is close to ____; 122 is close to ____.

So, $\sqrt{\frac{99}{122}} \doteq \sqrt{\frac{\_\_\_\_}{\_\_\_\_}}$

$\doteq$ ____

5. Identify a decimal that has a square root between the two given numbers. Check the answer.

a) 1 and 2

$1^2 =$ _____ and $2^2 =$ _____

So, any number between _____ and _____ has a square root between 1 and 2.

Choose _____.

Check: $\sqrt{\_\_\_\_\_} \doteq$ _____

The decimal _____ is one possible answer.

b) 8 and 9

$8^2 =$ _____ and $9^2 =$ _____

So, any number between _____ and _____ has a square root between 8 and 9.

Choose _____.

Check: $\sqrt{\_\_\_\_\_} \doteq$ _____

The decimal _____ is one possible answer.

c) 2.5 and 3.5

_______ = _______ and _______ = _______

So, any number between _______ and _______ has a square root between 2.5 and 3.5.

Choose _______.

Check: $\sqrt{\_\_\_\_\_\_\_} \doteq$ _______

The decimal _______ is one correct answer.

d) 20 and 21

_______ = _______ and _______ = _______

So, any number between _______ and _______ has a square root between 20 and 21.

Choose _______.

Check: $\sqrt{\_\_\_\_\_\_\_} \doteq$ _______

The decimal _______ is one correct answer.

6. Determine the length of the hypotenuse in each right triangle.
Write each answer to the nearest tenth.

a)

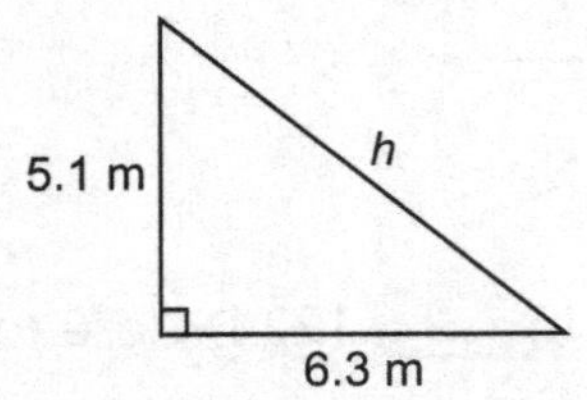

$h^2 = 5.1^2 + 6.3^2$

$h^2 =$ _____ + _____

$h^2 =$ _____

$h = \sqrt{\_\_\_\_\_}$

$h \doteq$ _____

So, h is about _____ m.

b)

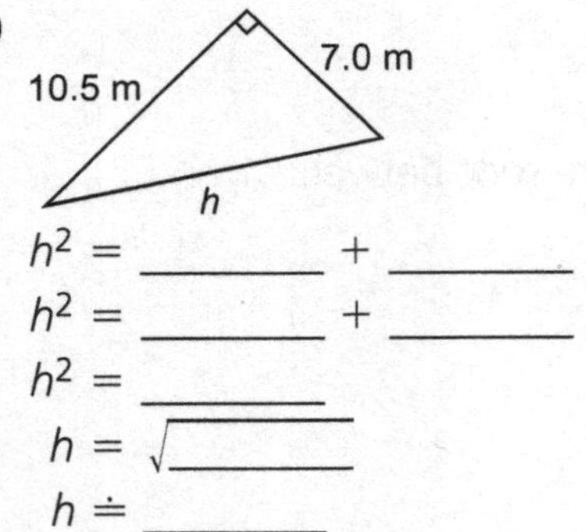

$h^2 =$ _______ + _______

$h^2 =$ _______ + _______

$h^2 =$ _______

$h = \sqrt{\_\_\_\_\_\_\_}$

$h \doteq$ _______

So, h is about _______ m.

Can you ...

- Identify decimals and fractions that are perfect squares?
- Find the square roots of decimals and fractions that are perfect squares?
- Approximate the square roots of decimals and fractions that are not perfect squares?

1.1 **1.** Calculate the number whose square root is:

a) $\frac{2}{7}$

$\frac{2}{7} \times \frac{2}{7} = \_\_\_\_$

$\frac{2}{7}$ is a square root of ____.

b) $\frac{8}{11}$

$\_\_\_ \times \_\_\_ = \_\_\_\_$

$\frac{8}{11}$ is a square root of ____.

c) 0.1

____ × ____ = ____

0.1 is a square root of ____.

d) 1.4

1.4 × 1.4 = ____

1.4 is a square root of ____.

2. Identify the fractions that are perfect squares.
The first one has been done for you.

	Fraction	Is numerator a perfect square?	Is denominator a perfect square?	Is fraction a perfect square?
a)	$\frac{64}{75}$	Yes; 8 × 8 = 64	No	No
b)	$\frac{9}{25}$	____________	____________	__________
c)	$\frac{25}{55}$	____________	____________	__________

3. Find each square root.

a) $\sqrt{\frac{9}{49}} = \sqrt{\frac{\_\_\_ \times \_\_\_}{\_\_\_ \times \_\_\_}}$
$= \_\_\_\_$

b) $\sqrt{\frac{16}{25}} = \sqrt{\frac{\_\_\_ \times \_\_\_}{\_\_\_ \times \_\_\_}}$
$= \_\_\_\_$

c) $\sqrt{\frac{36}{121}} = \sqrt{\frac{\_\_\_ \times \_\_\_}{\_\_\_ \times \_\_\_}}$
$= \_\_\_\_$

4. a) Put a check mark beside each decimal that is a perfect square.

i) 4.84 _____ **ii)** 3.63 _____ **iii)** 98.01 _____ **iv)** 67.24 _____

b) Explain how you identified the perfect squares in part a.

____

____

5. a) Find the area of the shaded square.

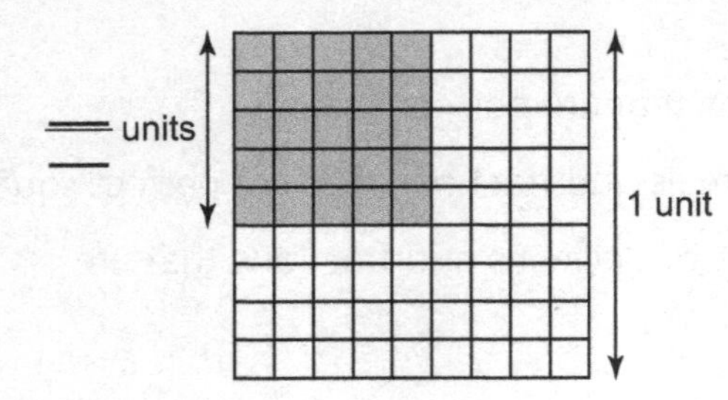

Area = (Length)^2

$= \left(\frac{\_}{\_}\right)^2$

$= \frac{\_}{\_} \times \frac{\_}{\_}$

$= \frac{\_}{\_}$

The area is _____ square units.

b) Find the side length of the shaded square.

Area = $\frac{81}{100}$ square units

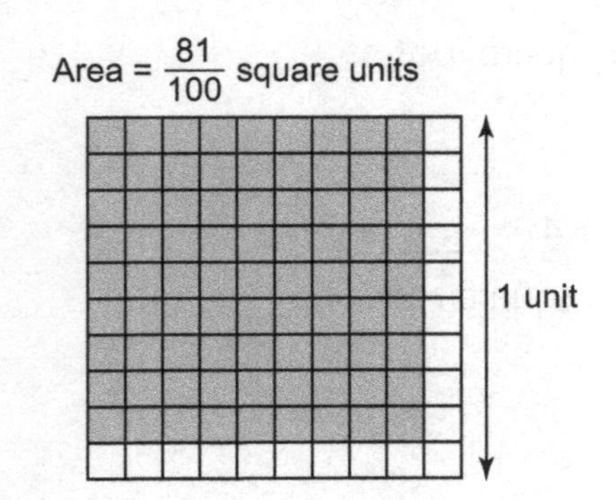

Length = $\sqrt{\text{Area}}$

$= \sqrt{\frac{\_}{\_}}$

$= \sqrt{\frac{\_ \times \_}{\_ \quad \_}}$

$= \frac{\_}{\_}$

The side length is _____ units.

1.2 **6.** Estimate each square root.
Explain.

a) $\sqrt{7.5}$

7.5 is between _____ and _____.
So, $\sqrt{7.5}$ is between $\sqrt{\_\_\_}$ and $\sqrt{\_\_\_}$.
That is, $\sqrt{7.5}$ is between _____ and _____.
Since 7.5 is closer to _____ than _____, $\sqrt{7.5}$ is closer to _____ than _____.
So, $\sqrt{7.5}$ is between _____ and _____, and closer to _____.

b) $\sqrt{66.6}$

66.6 is between _____ and _____.
So, $\sqrt{66.6}$ is between $\sqrt{\_\_\_}$ and $\sqrt{\_\_\_}$.
That is, $\sqrt{66.6}$ is between _____ and _____.
Since 66.6 is closer to _____ than _____, $\sqrt{66.6}$ is closer to _____ than _____.
So, $\sqrt{66.6}$ is between _____ and _____, and closer to _____.

7. Estimate each square root.

a) $\sqrt{\frac{15}{79}}$

15 is close to ____; 79 is close to ____.

So, $\sqrt{\frac{15}{79}} \doteq \sqrt{\frac{\_\_\_\_}{\_\_\_\_}}$

$\doteq \frac{\_\_\_\_}{\_\_\_\_}$

b) $\sqrt{\frac{23}{50}}$

23 is close to ____; 50 is close to ____.

So, $\sqrt{\frac{23}{50}} \doteq \sqrt{\frac{\_\_\_\_}{\_\_\_\_}}$

$\doteq \frac{\_\_\_\_}{\_\_\_\_}$

8. Identify a decimal whose square root is between the given numbers.
Check your answer.

a) 2 and 3

$2^2 =$ _____ and $3^2 =$ _____

So, any number between _____ and _____ has a square root between 2 and 3.

Choose _____.

Check: $\sqrt{\_\_\_\_} \doteq$ _____

The decimal _____ is one correct answer.

b) 6 and 7

$6^2 =$ _____ and $7^2 =$ _____

So, any number between _____ and _____ has a square root between 6 and 7.

Choose _____.

$\sqrt{\_\_\_\_} \doteq$ _____

The decimal _____ is one correct answer.

9. Find the length of each hypotenuse.

a)

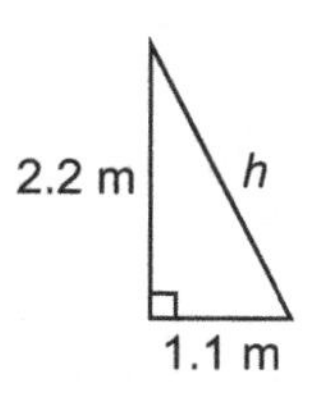

$h^2 =$ _______ + _______

$h^2 =$ _______ + _______

$h^2 =$ _______

$h = \sqrt{\_\_\_\_\_\_\_}$

$h \doteq$ _______

The length of the hypotenuse is about _______ m.

b)

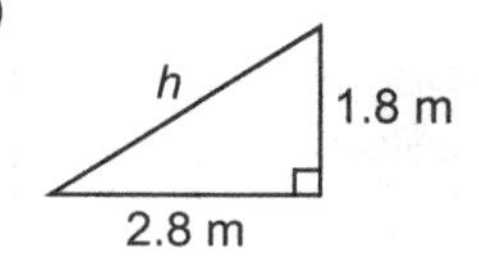

$h^2 =$ _______ + _______

$h^2 =$ _______ + _______

$h^2 =$ _______

$h = \sqrt{\_\_\_\_\_\_\_}$

$h \doteq$ _______

The length of the hypotenuse is about _______ m.

1.3 Skill Builder

Surface Areas of Rectangular Prisms

The **surface area** of a rectangular prism is the sum of the areas of its 6 rectangular faces. Look for matching faces with the same areas.

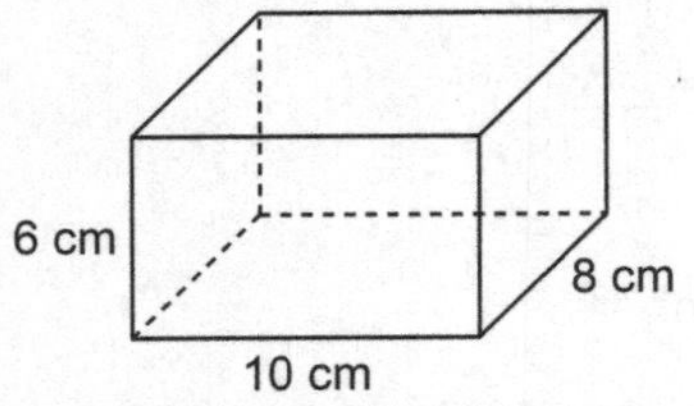

The matching faces in each pair have the same area. We find the area of one face and multiply by 2.

For each rectangular face, area equals its length times its width.

Matching Faces	Diagram	Corresponding Area (cm^2)
Back, Front (6 cm, 10 cm) →	6 cm, 10 cm	$2(10 \times 6) = 120$
Top, Bottom (10 cm, 8 cm) →	10 cm, 8 cm	$2(10 \times 8) = 160$
Left side, Right side (8 cm, 6 cm) →	8 cm, 6 cm	$2(8 \times 6) = 96$
Total		376

The surface area is 376 cm^2.

Check

1. Determine the surface area of each rectangular prism.

a)

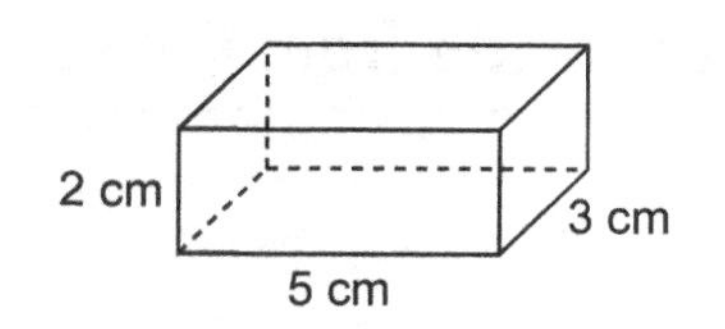

Matching Faces	Diagram	Corresponding Area (cm^2)
Front Back	__ cm × __ cm	2(__ × __) = __
Top Bottom	__ cm × __ cm	2(__ × __) = __
Right Left	__ cm × __ cm	2(__ × __) = __
Total		__

The surface area is ___ cm^2.

b)

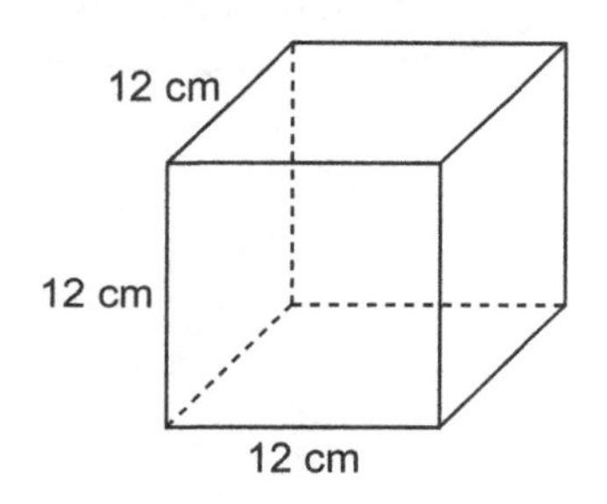

Matching Faces	Diagram	Corresponding Area (cm^2)
Front Back	__ cm × __ cm	2(__ × __) = ___
Top Bottom	__ cm × __ cm	2(__ × __) = ___
Right Left	__ cm × __ cm	2(__ × __) = ___
Total		___

The surface area is ____ cm^2.

1.3 Surface Areas of Objects Made from Right Rectangular Prisms

FOCUS **Find the surface areas of objects made from rectangular prisms.**

Example 1 Finding the Surface Area of an Object Made from Cubes

Make this object with 1-cm cubes.
What is the surface area of the object?

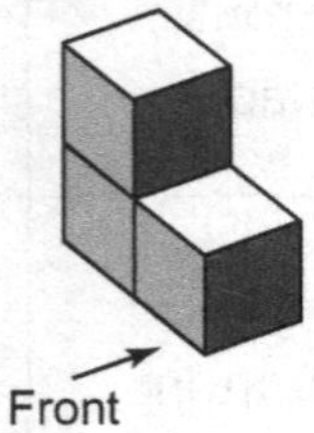

Solution

Think of tracing each face, or "opening" the object.

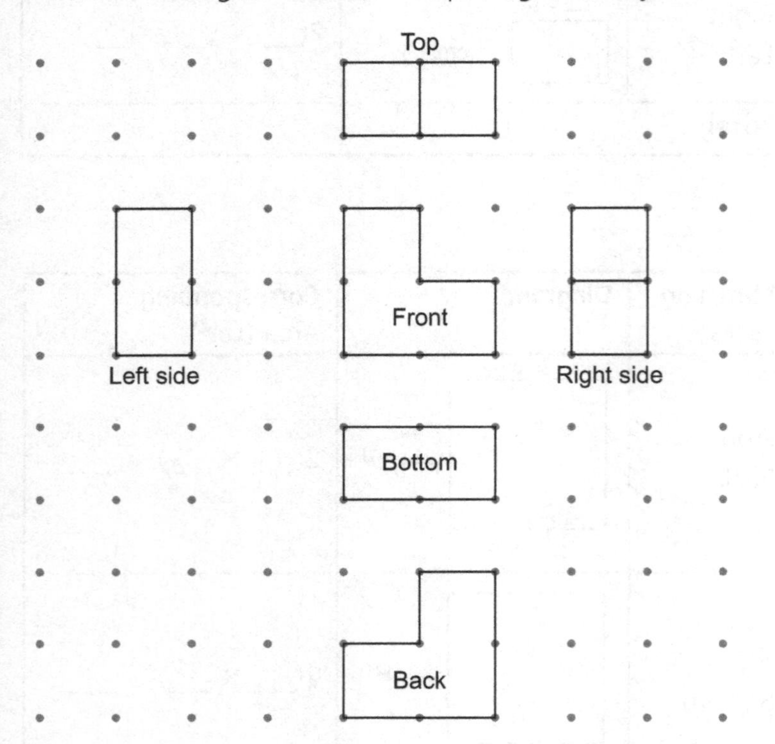

Turn the object to see each view.

Look for matching views.

Matching Views	Corresponding Area (cm^2)
Front / Back	2(3) = 6
Top / Bottom	2(2) = 4
Right / Left	2(2) = 4
Total	14

The surface area is 14 cm^2.

Check

1. Make this object with 1-cm cubes, then find its surface area.

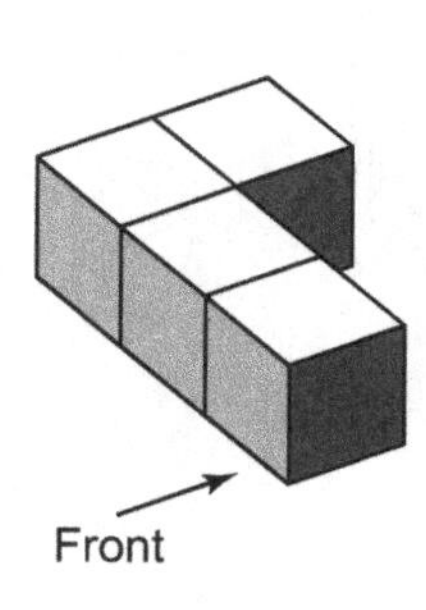

Matching Views	Diagram	Corresponding Area (cm^2)
Front Back		2(__) = __
Top Bottom		2(__) = __
Right Left		2(__) = __
Total		___

The surface area is ____ cm^2.

A **composite object** is made from 2 or more objects.

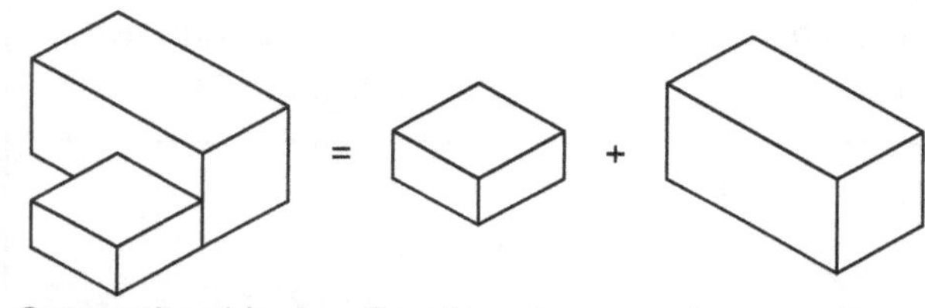

Composite object Smaller prism Larger prism

To find the surface area of a composite object, imagine dipping the object in paint. The surface area is the area of all the faces covered in paint.

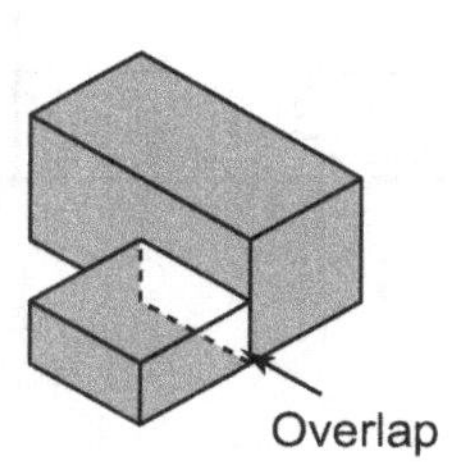

Where objects overlap, there is a hidden surface. The paint doesn't reach the hidden surface.

The overlap is not painted, so it is not part of the surface area.

Example 2 Finding the Surface Area of a Composite Object

Find the surface area of this composite object.

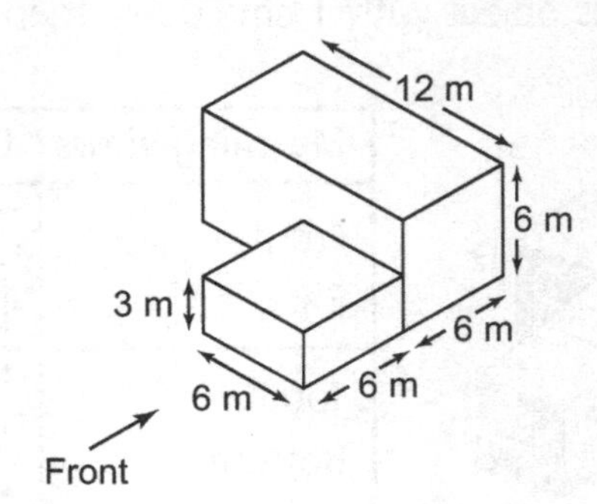

Solution

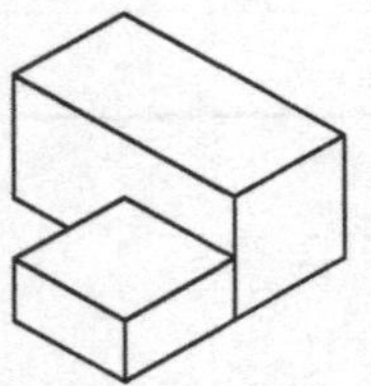 = 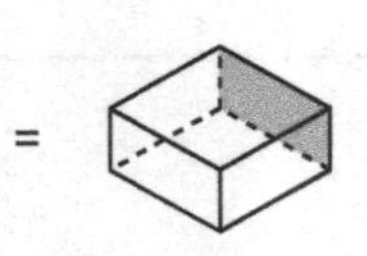 + 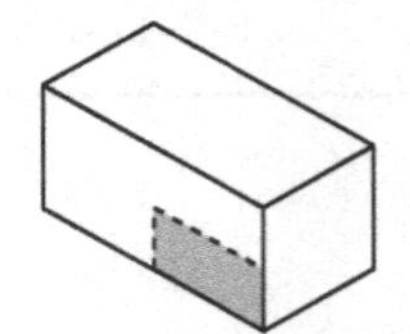 − 2(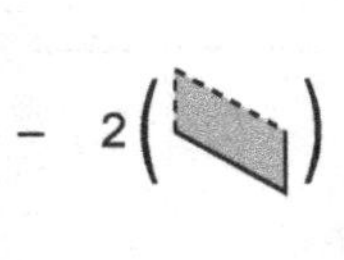)

Surface area of composite object = Surface area of smaller prism + Surface area of larger prism − 2(Area of overlap)

Surface area of smaller prism

Matching Faces	Diagram	Corresponding Area (m²)
Front Back Right Left	6 m, 3 m	$4(6 \times 3) = 72$
Top Bottom	6 m, 6 m	$2(6 \times 6) = 72$
Total		144

The surface area is 144 m^2.

Surface area of larger prism

Matching Faces	Diagram	Corresponding Area (m²)
Front Back Top Bottom	12 m, 6 m	$4(12 \times 6) = 288$
Right Left	6 m, 6 m	$2(6 \times 6) = 72$
Total		360

The surface area is 360 m^2.

Area of overlap

Diagram	Corresponding Area (m²)
6 m, 3 m	$6 \times 3 = 18$

The area of overlap is 18 m^2.

SA of composite object $= 144 + 360 - 2(18) = 468$

The surface area of the composite object is 468 m^2.

SA means surface area.

Check

1. The diagram shows the surface areas of the two prisms that make up a composite object.

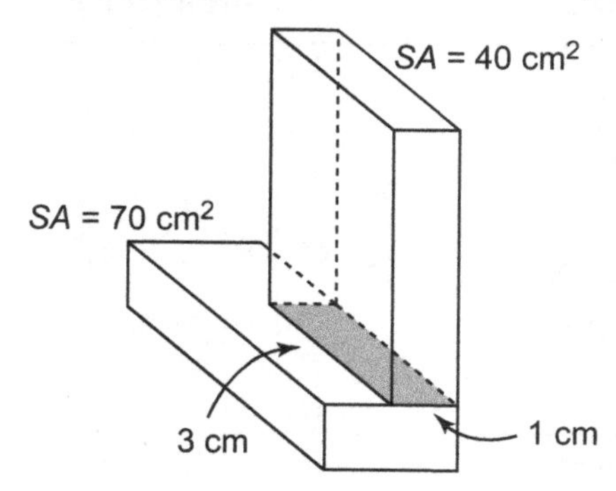

a) What is the area of the overlap?

The overlap is a _____-cm by _____-cm rectangle.

Area of overlap = _____ cm × _____ cm

= _____ cm^2

b) What is the surface area of the composite object?

SA composite object = SA smaller prism + SA larger prism − 2(Area of overlap)

= _____ cm^2 + _____ cm^2 − 2(_____) cm^2

= _____ cm^2

2. Find the surface area of this composite object.

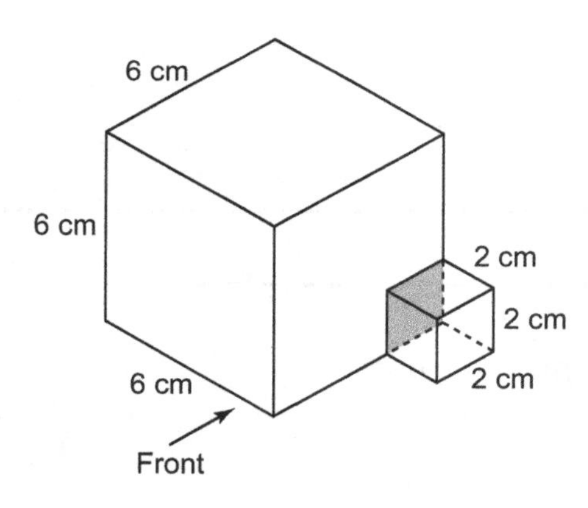

A cube has _____ congruent faces.

Surface area of larger cube

Matching Faces	Diagram	Corresponding Area (cm^2)
Front Back Top Bottom Right Left	__ cm by __ cm	6(__ × __) = ____
Total		____

The surface area is ____ cm^2.

Surface area of smaller cube

Matching Faces	Diagram	Corresponding Area (cm^2)
Front Back Top Bottom Right Left	__ cm by __ cm	6(__ × __) = ____
Total		____

The surface area is ___ cm^2.

Area of overlap

Diagram	Corresponding Area (cm^2)
__ cm by __ cm	__ × __ = __

The area of overlap is __ cm^2.

SA composite object = SA larger cube + ________________ − ______________________

= _____ + _____ − 2(__)

= _____

The surface area of the composite object is ______ cm^2.

Practice

1. The diagram shows the 6 views of an object made from 1-cm cubes.
Identify pairs of matching views in the first column of the table.
Then, find the surface area of the object.

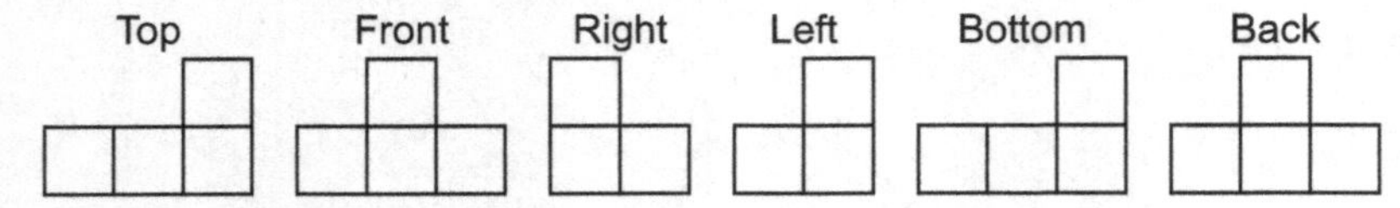

Matching Views	Corresponding Area (cm^2)
Front / _______	____________
Top / _______	____________
Right / _______	____________
Total	____________

The surface area is ___ cm^2.

2. Each object is made with 1-cm cubes. Find the surface area of each object.

a)

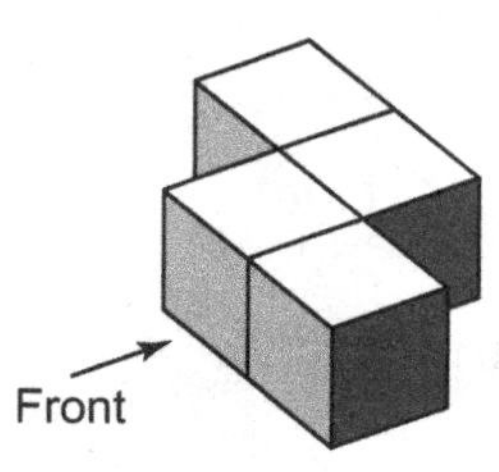

Matching Views	Diagram	Corresponding Area (cm^2)
Front Back		2(__) = __
Top Bottom		2(__) = __
Right Left		2(__) = __
Total		___

The surface area is ___ cm^2.

b)

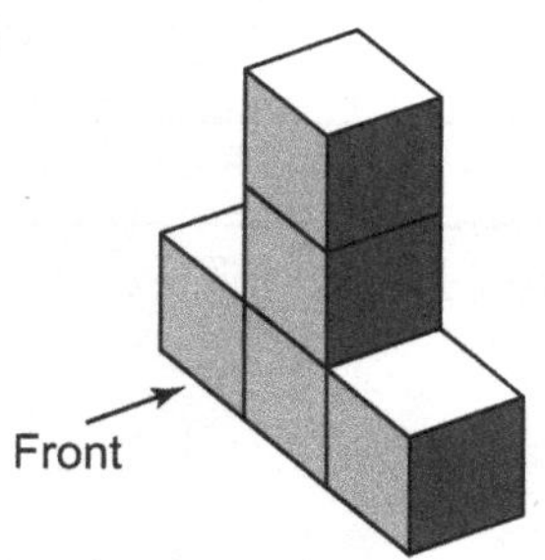

Matching Views	Diagram	Corresponding Area (cm^2)
Front Back		2(__) = ___
Top Bottom		________________
Right Left		________________
Total		___

The surface area is ___ cm^2.

3. Find the surface area of this composite object.

10 m
10 m
5 m
5 m
20 m
15 m
Front

Surface area of larger prism

Matching Faces	Diagram	Corresponding Area (m^2)
Front Back		2(___ × __) = ___
Top Bottom		________________
Right Left		________________
Total		________________

The surface area is ___ m^2.

Surface area of smaller prism

Matching Faces	Diagram	Corresponding Area (m^2)
Front Back		2(__ × ___) = ___
Top Bottom		________________
Right Left		________________
Total		________________

The surface area is ___ m^2.

Area of overlap

Diagram	Corresponding Area (m^2)
	___ × ___ = ___

The area of overlap is ___ m^2.

Surface area of composite object

SA composite object = ____________ + ____________ − ____________

= ___ + ___ − 2(___)

= ____

The surface area of the composite object is ____ m^2.

4. Find the surface area of this composite object.

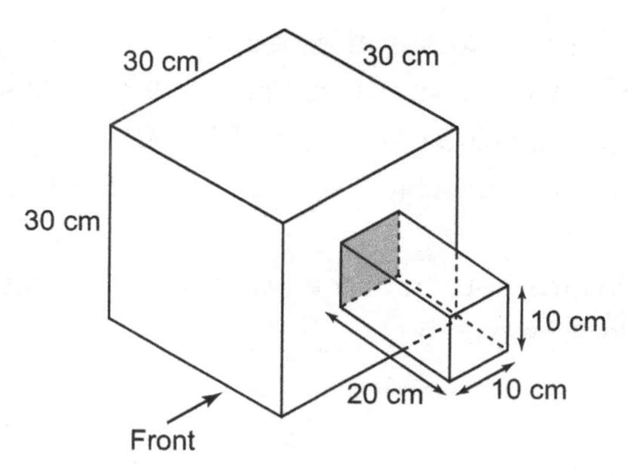

Surface area of cube

Matching Faces	Diagram	Corresponding Area (cm²)
Front / Back Top / Bottom Right / Left	___ cm ___ cm	6(___ × ___) = _____
Total		_____

The surface area is _____ cm².

Surface area of rectangular prism

Matching Faces	Diagram	Corresponding Area (cm²)
Front / Back		2(___ × ___) = ___
Top / Bottom		____________
Right / Left		____________
Total		____________

The surface area is _____ cm².

Area of overlap

Diagram	Corresponding Area (cm²)
	___ × ___ = ___

The area of overlap is ____ cm².

Surface area of composite object

SA composite object = ____________ + ____________ − ____________

= _____ + _____ − _____

= ____________

The surface area of the composite object is _____ cm².

5. A loading dock is attached to one wall of a warehouse. The exterior of the buildings is to be painted at a cost of $\$2.50/m^2$. How much will it cost to paint the buildings?

Will the bottom of the warehouse and loading dock be painted? ________

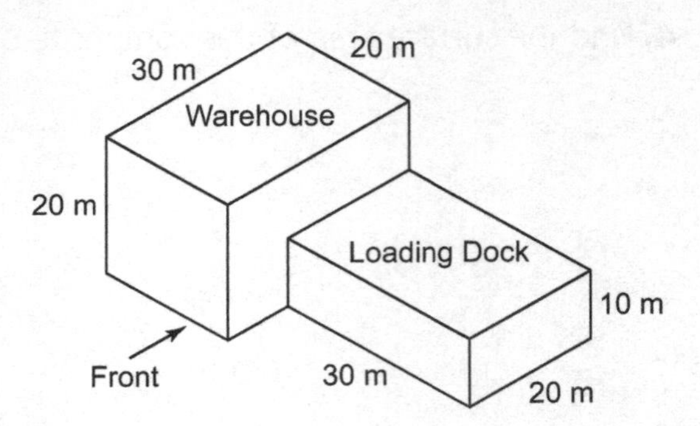

Surface area of warehouse to be painted

Matching Faces	Diagram	Corresponding Area (m^2)
Front Back		2(___ × ___) = ____
Top Sides		3(___ × ___) = ____
Total		______

The surface area of the warehouse to be painted is _____ m^2.

Area of overlap

Diagram	Corresponding Area (m^2)
	___ × ___ = ___

The area of overlap is ____ m^2.

Surface area of loading dock to be painted

Matching Faces	Diagram	Corresponding Area (m^2)
Front Back		2(___ × ___) = ____
Top		___ × ___ = ____
Sides		2(___ × ___) = ____
Total		______

The surface area of the loading dock to be painted is _____ m^2.

Surface area of composite object to be painted

_____ + _____ − _______ = _____

The surface area of the composite object to be painted is _____ m^2.

So, the area to be painted is _____ m^2.

The cost per square metre is: $_____

The cost to paint the buildings is: _____ × $_____ = _____

1.4 Skill Builder

Surface Areas of Triangular Prisms

To find the surface area of a right triangular prism, add the areas of its 5 faces.
Look for matching faces with the same areas.

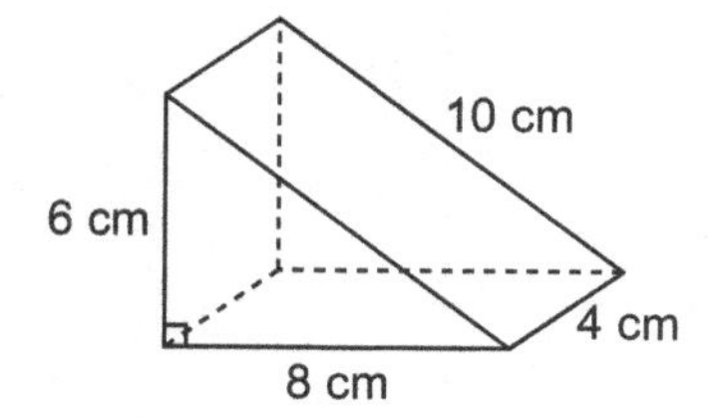

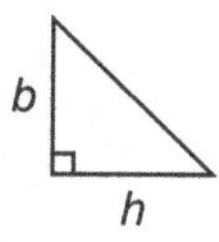

$A = \frac{1}{2}bh$

There are 2 congruent triangular faces. Find the area of one, then multiply it by 2.

Faces	Diagram	Corresponding Area (cm^2)
Triangular	6 cm, 8 cm → 6 cm, 8 cm	$2(\frac{1}{2} \times 6 \times 8) = 48$
Rectangular	10 cm, 4 cm → 10 cm, 4 cm	$10 \times 4 = 40$
	4 cm, 6 cm → 6 cm, 4 cm	$6 \times 4 = 24$
	8 cm, 4 cm → 4 cm, 8 cm	$8 \times 4 = 32$
Total		144

The surface area is 144 cm^2.

Check

1. Find the surface area of the triangular prism.

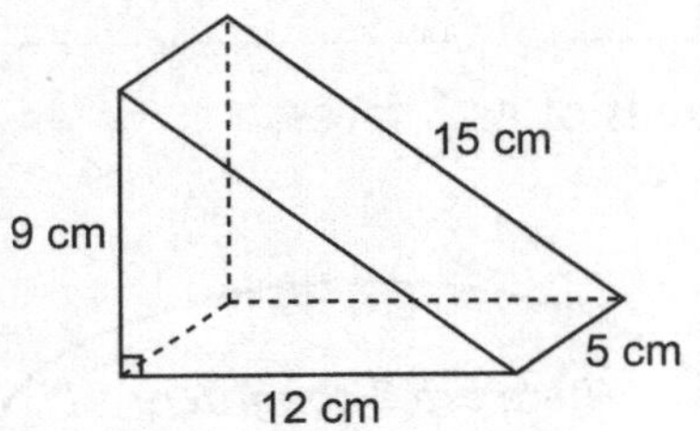

Faces	Diagram	Corresponding Area (cm^2)
Triangular	___ cm; ___ cm	$2(\frac{1}{2} \times$ ___ $\times$ ___$) =$ ___
Rectangular	___ cm; ___ cm	___ $\times$ ___ $=$ ___
	___ cm; ___ cm	___ $\times$ ___ $=$ ___
	___ cm; ___ cm	___ $\times$ ___ $=$ ___
Total		___

The surface area is ___ cm^2.

Surface Areas of Cylinders

To find the surface area of a right cylinder, add the areas of:

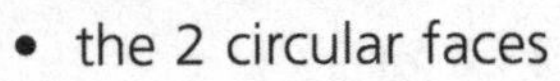

- the 2 circular faces
- the curved surface

Look for matching faces with the same areas.

Faces	Diagram	Corresponding Area
Top Bottom	Top, Bottom, r → $A = \pi r^2$	$2 \times \pi r^2$
Curved surface	$C = 2\pi r$, r, h → $C = 2\pi r$, h → $C = 2\pi r$, h	$2\pi rh$

The side can be unrolled into a rectangle, whose length is the circumference of the circle.

The surface area is: $2\pi r^2 + 2\pi rh$

To calculate the surface area of this cylinder:

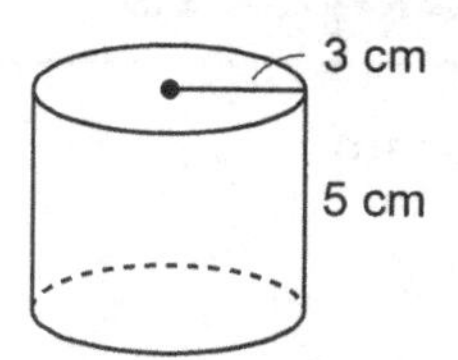

Faces	Diagram	Corresponding Area (cm^2)
Top Bottom	3 cm	$2 \times \pi \times 3^2$ $\doteq 56.55$
Curved surface	$2\pi(3)$ cm 5 cm	$2 \times \pi \times 3 \times 5$ $\doteq 94.25$
Total		150.80

The dimensions of the cylinder are given to the nearest centimetre, so we give the surface area to the nearest square centimetre.

The surface area is about 151 cm^2.

Check

1. Find the surface area of the cylinder.

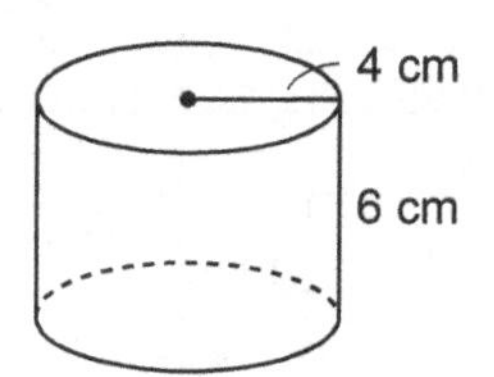

Faces	Diagram	Corresponding Area (cm^2)
Top Bottom	___ cm	___ × ___ × ___ $\doteq$ _______
Curved surface	2π___ cm ___ cm	___ × ___ × ___ × ___ $\doteq$ _______
Total		_______

The surface area is about ____ cm^2.

1.4 Surface Areas of Other Composite Objects

FOCUS **Find the surface areas of composite objects made from right prisms and right cylinders.**

Example 1 Finding the Surface Area of a Composite Object Made from a Rectangular Prism and a Triangular Prism

Find the surface area of this composite object.

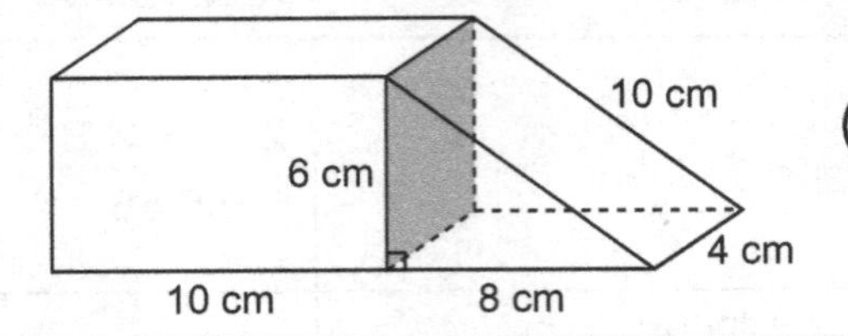

The shaded area is the area of overlap.

Solution

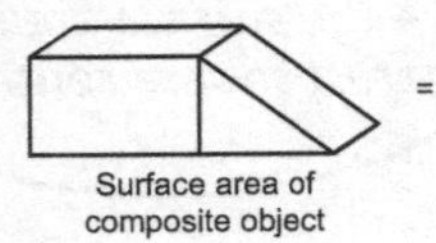

=

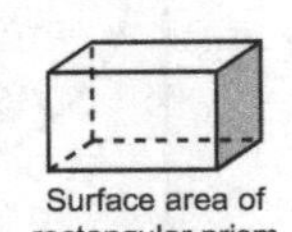

+

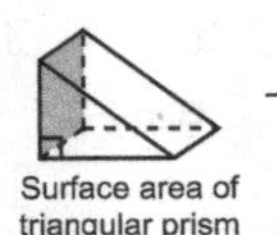

−

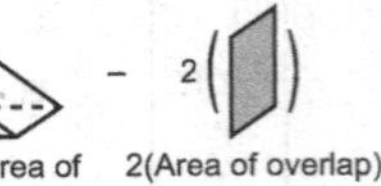

Surface area of rectangular prism

Faces	Diagram	Corresponding Area (cm²)
Front Back	10 cm by 6 cm	$2(6 \times 10)$ $= 120$
Top Bottom	10 cm by 4 cm	$2(10 \times 4)$ $= 80$
Right Left	4 cm by 6 cm	$2(6 \times 4)$ $= 48$
Total		248

The surface area is 248 cm^2.

Surface area of triangular prism

Faces	Diagram	Corresponding Area (cm²)
Triangular	6 cm, 8 cm	$2(\frac{1}{2} \times 6 \times 8)$ $= 48$
Rectangular	10 cm by 4 cm	$10 \times 4 = 40$
	4 cm by 6 cm	$6 \times 4 = 24$
	8 cm by 4 cm	$8 \times 4 = 32$
Total		144

The surface area is 144 cm^2.

Area of overlap

Diagram	Corresponding Area (cm²)
4 cm by 6 cm	$6 \times 4 = 24$

The area of overlap is 24 cm^2.

Surface area of composite object $= 248 + 144 - 2(24) = 344$

The surface area of the composite object is 344 cm^2.

Check

1. The diagram shows the surface area of the two prisms that make up a composite object.

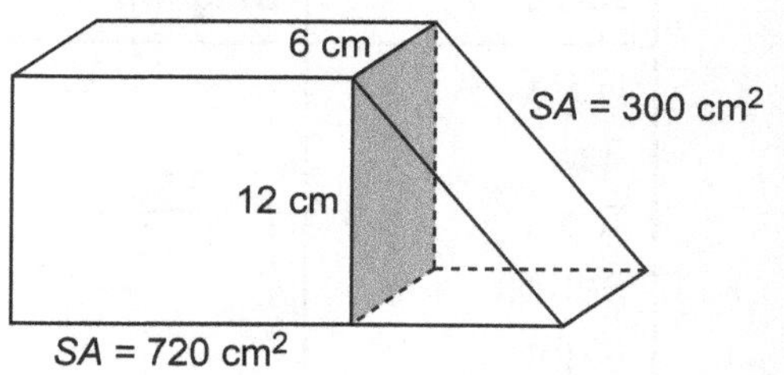

   a) What is the area of the overlap?
      The overlap is a 12 -cm by 6 -cm rectangle.
      Area of overlap = 12 cm × 6 cm = 72 cm²

   b) What is the surface area of the composite object?
      Surface area of composite object = Surface area of 2 prisms − 2(Area of overlap)
      = 300 + 720 − 144 = 876
      The surface area of the composite object is 876 .

2. Find the surface area of this composite object.

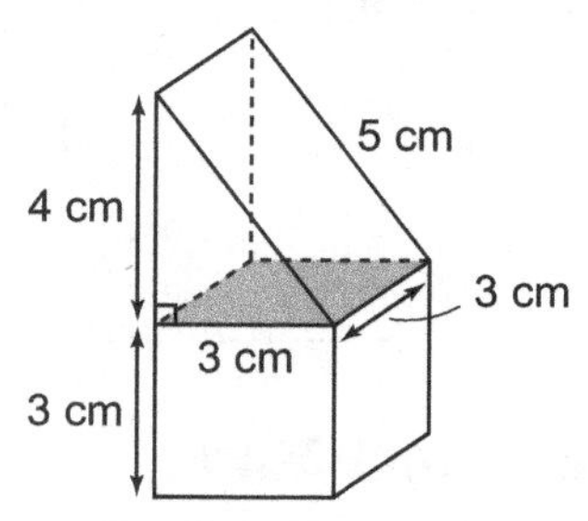

Surface area of triangular prism

Faces	Diagram	Corresponding Area (cm²)
Triangular	4 cm, 3 cm	$2(\frac{1}{2} \times$ 4 × 3) = 12
Rectangular	5 cm, 3 cm	5 × 3 = 15
	4 cm, 3 cm	4 × 3 = 12
	3 cm, 3 cm	3 × 3 = 9
Total		48

The surface area is 48 cm².

Surface area of cube

Faces	Diagram	Corresponding Area (cm^2)
Front Back Top Bottom Right Left	3 cm × 3 cm square	6(3 × 3) = 54
Total		54

The surface area is 54 cm^2.

Area of overlap

Diagram	Corresponding Area (cm^2)
3 cm × 3 cm square	3 × 3 = 9

The area of overlap is 9 cm^2.

Surface area of composite object = Surface area of 2 prisms − 2(Area of overlap)
= 48 + 54 − 18
= 84

The surface area of the composite object is 84 cm^2.

Example 2 Finding the Surface Area of a Composite Object Made from a Rectangular Prism and a Cylinder

Find the surface area of this object.

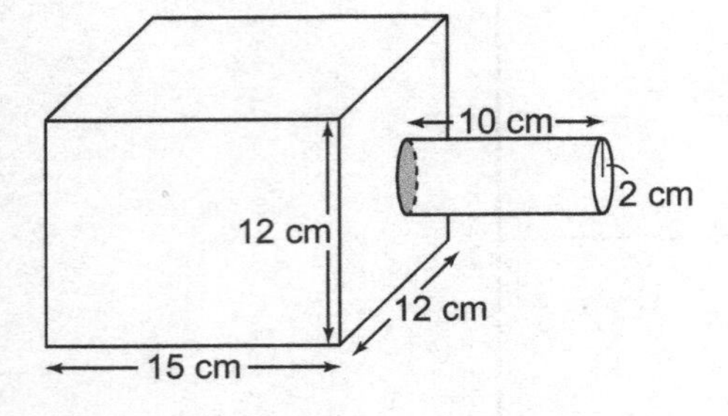

Surface area of rectangular prism

Faces	Diagram	Corresponding Area (cm^2)
Front Back Top Bottom	15 cm × 12 cm rectangle	4(12 × 15) = 720
Right Left	12 cm × 12 cm square	2(12 × 12) = 288
Total		1008

The surface area is 1008 cm^2.

Surface area of cylinder

Faces	Diagram	Corresponding Area (cm^2)
Top Bottom	2 cm	$2 \times \pi \times 2^2 \doteq 25.13$
Curved surface	$2\pi(2)$ cm 10 cm	$2 \times \pi \times 2 \times 10 \doteq 125.67$
Total		150.80

The surface area is about 150.80 cm^2.

Area of overlap

Diagram	Corresponding Area (cm^2)
2 cm	$\pi \times 2^2 \doteq 12.57$

The area of overlap is about 12.57 cm^2.

$$\begin{aligned} \text{SA composite object} &= \text{SA rectangular prism} + \text{SA cylinder} - 2(\text{Area of overlap}) \\ &\doteq 1008 + 150.80 - 2(12.57) \\ &\doteq 1133.66 \end{aligned}$$

The surface area is about 1134 cm^2.

Check

1. The diagram shows the surface area of the rectangular prism and cylinder that make up a composite object.

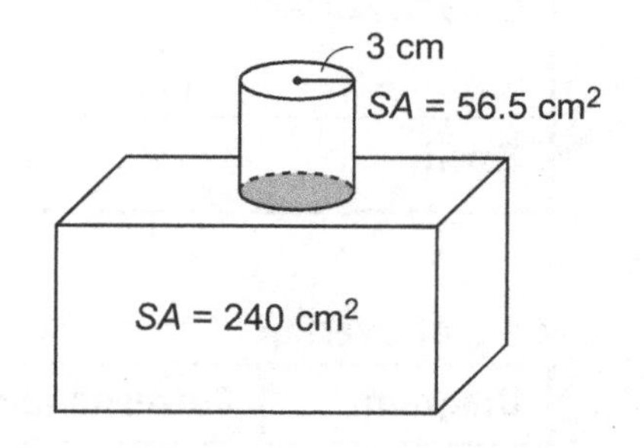

a) What is the area of the overlap?

The overlap is a circle.

Area of overlap = $\pi 3^2$

$\doteq$ 28.27 cm^2

b) What is the surface area of the composite object?

SA composite object = SA 56.5 + SA 240 − 2(28.27)

= 56.5 + 240 − 56.55

= 239.95

The surface area of the composite object is about 240 cm^2.

2. Find the surface area of this composite object.

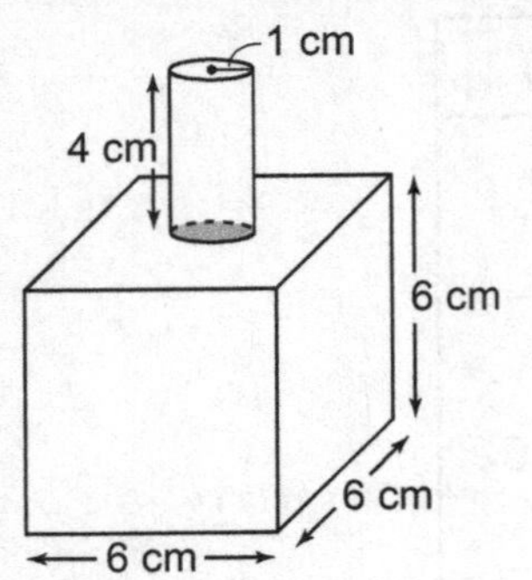

Surface area of cube

Faces	Diagram	Corresponding Area (cm^2)
Front Back Top Bottom Right Left	6 cm (top), 6 cm (side)	6(6 × 6) = 216
Total		216

Surface area of cylinder

Faces	Diagram	Corresponding Area (cm^2)
Top Bottom	1 cm	π × 1 × 1 ≐ 3.14
Curved surface	2π 1 cm, 4 cm	2 × π × 1 × 4 ≐ 25.13
Total		25.13

Area of overlap

Diagram	Corresponding Area (cm^2)
1 cm	____ × ____ ≐ ____

SA composite object = SA ______ + SA __________ − 2(________________)

≐ ________ + ________ − ________

≐ ________

The surface area of the composite object is about ______ cm^2.

Practice

1. Find the surface area of this composite object.

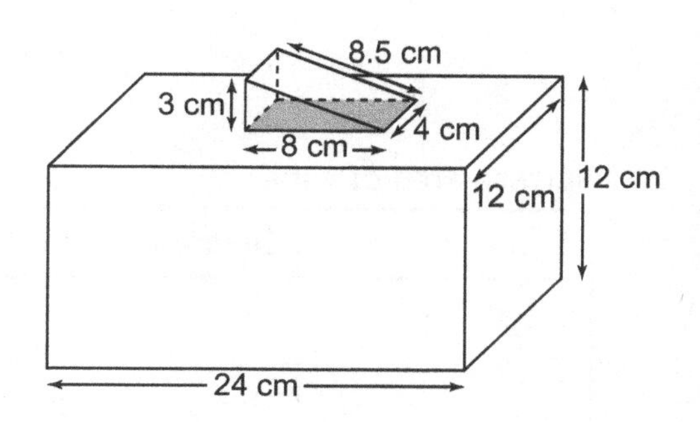

Surface area of rectangular prism

Faces	Diagram	Corresponding Area (cm^2)
Front Back Top Bottom		______ = ______
Right Left		______ = ______
Total		______

The surface area is ______ cm^2.

Surface area of triangular prism

Faces	Diagram	Corresponding Area (cm^2)
Triangular		______
Rectangular		______
		______
		______
Total		______

The surface area is ______ cm^2.

Area of overlap

Diagram	Area (cm^2)
	__ × __ = __

The area of overlap is ______ cm^2.

Surface area of composite object

SA composite object

= ______________

= ______________

The surface area of the composite object is ______ cm^2.

2. Find the surface area of this composite object.

2 cm
2 cm
2 cm
4 cm
6 cm

Surface area of cube

Faces	Diagram	Corresponding Area (cm²)
Front Back Top Bottom		6(___ × ___) = ___
Total		___

The surface area is ___ cm².

Surface area of cylinder

Faces	Diagram	Corresponding Area (cm²)
Top Bottom		___ × ___ × ___ ≐ ___
Curved surface		___ × ___ × ___ × ___ ≐ ___
Total		___

The surface area is about ___ cm².

Area of overlap

Diagram	Corresponding Area (cm²)
	___ × ___ = ___

The area of overlap is ___ cm².

Surface area of composite object

SA composite object ≐ ___ + ___ − ___

≐ ___

The surface area of the composite object is about ___ cm².

3. Calculate the surface area of the cake at the right. Write your answer to the nearest tenth.

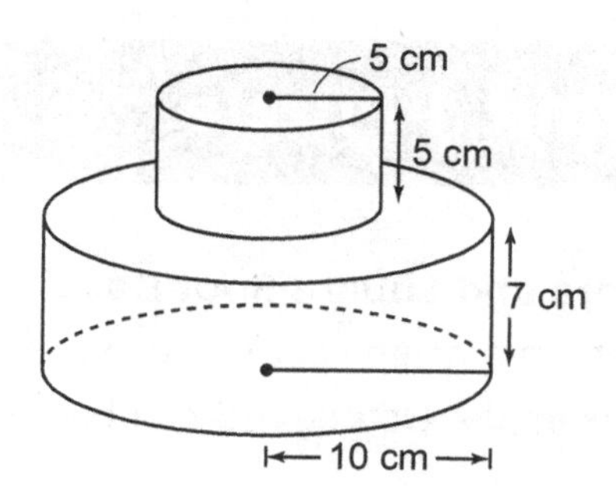

Surface area of smaller cake

Faces	Diagram	Corresponding Area (cm^2)
Top Bottom		___ × ___ × ___ ≐ ______
Curved surface		___ × ___ × ___ × ___ ≐ ______
Total		______

The surface area is about ________ cm^2.

Surface area of larger cake

Faces	Diagram	Corresponding Area (cm^2)
Top Bottom		___ × ___ × ___ ≐ ______
Curved surface		___ × ___ × ___ × ___ ≐ ______
Total		______

The surface area is about ________ cm^2.

Area of overlap

Diagram	Corresponding Area (cm^2)
	___ × ___ ≐ ______

The area of overlap is about ______ cm^2.

Surface area of cake ≐ ________ + ________ − ________

≐ ________

The surface area of the cake is about ________ cm^2.

Unit 1 Puzzle

Square and Square-Root Days

A date in a given year can be written as the month number followed by the day number. For example, October 25 can be written as 10/25.

- In a *square-root day*, the month is the square root of the day.
  For example, March 9 is a square-root day because it is written as 3/9, and 3 is the square root of 9.

  List all the square-root days in a year.

  ____

- In a *square day*, the month is the square of the day.
  For example, April 2 is a square day because it is written as 4/2, and 4 is the square of 2.

  List all the square days in a year.

  ____

- A *square year* is a year which is a perfect square.
  For example, the year 1600 is a square year because $1600 = 40 \times 40$.

  List all the square years from 1000 to the present.

  ____

Unit 1 Study Guide

Skill	Description	Example
Identify fractions that are perfect squares and find their square roots.	A fraction is a perfect square if it can be written as the product of 2 equal fractions. The square root is one of the 2 equal fractions.	$\frac{16}{25} = \frac{4}{5} \times \frac{4}{5}$ $\sqrt{\frac{16}{25}} = \frac{4}{5}$
Identify decimals that are perfect squares.	Use a calculator. The square root is a repeating or terminating decimal.	$\sqrt{1.69} = 1.3$
Estimate square roots of numbers that are not perfect squares.	Find perfect squares close to the number. Use the squares and square roots number lines.	$\sqrt{\frac{3}{10}} \doteq \sqrt{\frac{4}{9}} \doteq \frac{2}{3}$ 3 is close to 4; 10 is close to 9. 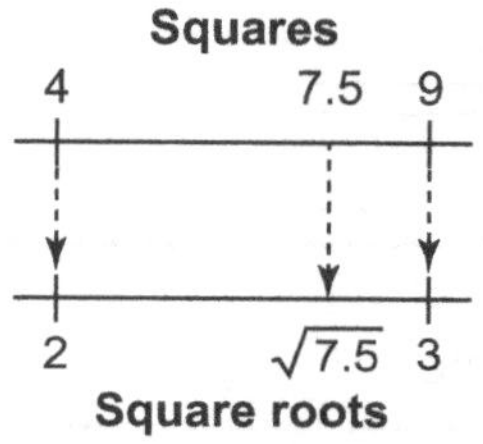 Squares: 4, 7.5, 9 Square roots: 2, $\sqrt{7.5}$, 3
Calculate the surface area of a composite object.	Add the areas of each of the 6 views. Or Add surface areas of the parts, then subtract for the overlap.	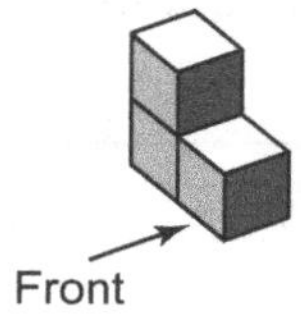 Front Top, Left side, Front, Right side, Bottom, Back The surface area is 14 square units. 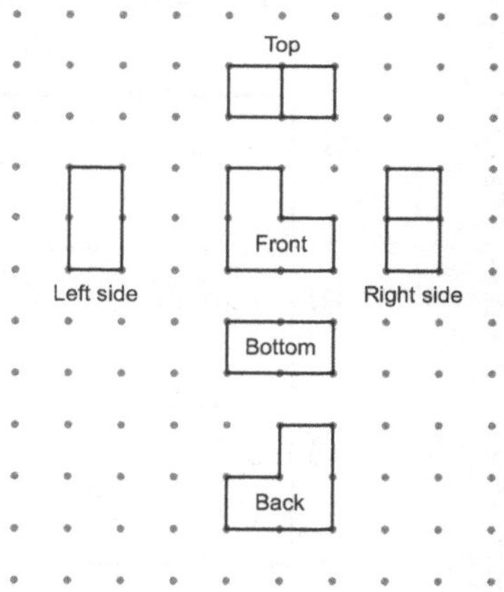 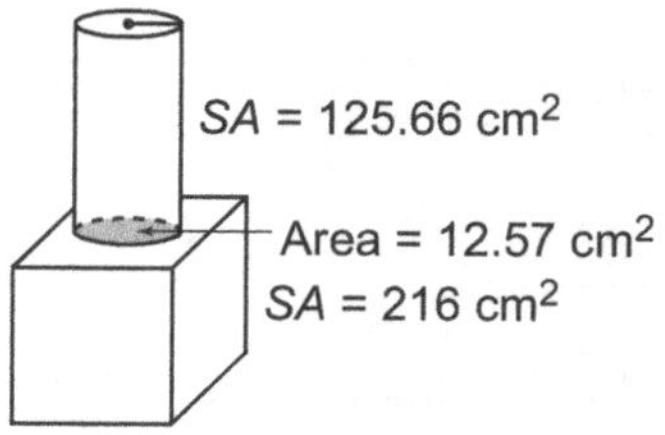 $SA = 125.66\text{ cm}^2$ Area = 12.57 cm^2 $SA = 216\text{ cm}^2$ $SA = 216 + 125.66 - 2(12.57)$ $= 316.52$ The surface area is about 317 cm^2.

Unit 1 Review

1.1 **1.** Calculate the number whose square root is:

a) $\frac{3}{7}$

$\_\_ \times \_\_ = \_\_$

$\frac{3}{7}$ is a square root of ___.

b) 9.9

$9.9 \times 9.9 =$ ______

9.9 is a square root of ______.

2. Complete the table.

	Fraction	Is numerator a perfect square?	Is denominator a perfect square?	Is fraction a perfect square?
a)	$\frac{25}{81}$	________	________	____
b)	$\frac{7}{4}$	________	________	____
c)	$\frac{49}{65}$	________	________	____

3. Complete the table.

	Decimal	Value of Square Root	Type of Decimal	Is decimal a perfect square?
a)	5.29	________	____________ ____________	____
b)	156.25	________	____________ ____________	____
c)	6.4	________	____________ ____________	____

4. Find the square root of each number.

a) $\sqrt{\frac{25}{81}} =$ ____

b) $\sqrt{59.29} =$ ____

1.2 **5.** Estimate $\sqrt{14.5}$. Explain your estimate.

14.5 is between ____ and ____.

So, $\sqrt{14.5}$ is between $\sqrt{\_\_\_\_}$ and $\sqrt{\_\_\_\_}$. That is, $\sqrt{14.5}$ is between ____ and ____.

Since 14.5 is closer to ____ than ____, $\sqrt{14.5}$ is closer to ____ than ____.

So, $\sqrt{14.5}$ is between ____ and ____, and closer to ____.

6. Estimate each square root. Explain.

a) $\sqrt{\frac{2}{13}}$

2 is close to ____; 13 is close to ____.

So, $\sqrt{\frac{2}{13}} \doteq \sqrt{\frac{\_\_\_\_}{\_\_\_\_}}$

$\doteq \frac{\_\_\_\_}{\_\_\_\_}$

b) $\sqrt{\frac{11}{70}}$

11 is close to ____; 70 is close to ____.

So, $\sqrt{\frac{11}{70}} \doteq \sqrt{\frac{\_\_\_\_}{\_\_\_\_}}$

$\doteq \frac{\_\_\_\_}{\_\_\_\_}$

7. Identify a decimal that has a square root between the two given numbers. Check the answer.

a) 2 and 3

$2^2 =$ ____ and $3^2 =$ ____

So, any number between ____ and ____ has a square root between 2 and 3.

Choose ____.

Check: $\sqrt{\_\_\_\_} \doteq$ ____

The decimal ____ is one possible answer.

b) 6.5 and 7.5

_____ = _____ and _____ = _____

So, any number between _____ and _____ has a square root between 6.5 and 7.5.

Choose _____.

Check: $\sqrt{\_\_\_\_\_} \doteq$ _____.

The decimal _____ is one possible answer.

8. Find the length of the hypotenuse of each right triangle.

a)

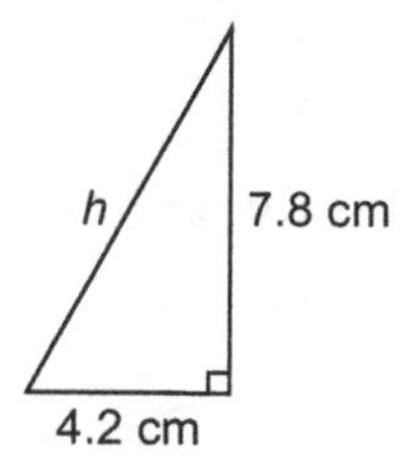

$h^2 =$ _____ + _____

$h^2 =$ _____ + _____

$h^2 =$ _____

$h = \sqrt{\_\_\_\_\_}$

$h \doteq$ _____

The length of the hypotenuse is about _____ cm.

b)

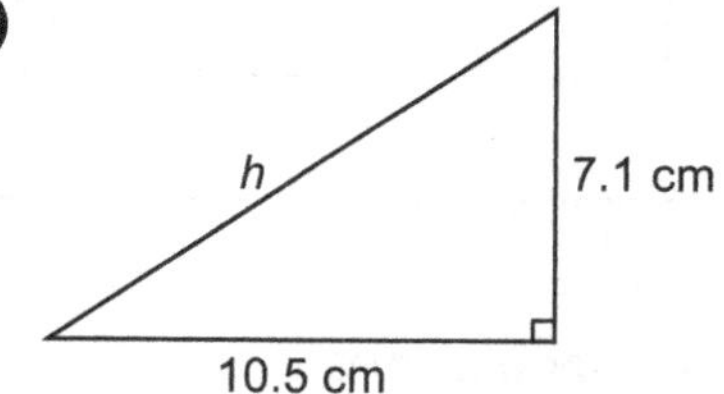

$h^2 =$ _____ + _____

$h^2 =$ _____ + _____

$h^2 =$ _____

$h = \sqrt{\_\_\_\_\_}$

$h \doteq$ _____

____

____

1.3 **9.** This object is made from 1-cm cubes. Find its surface area.

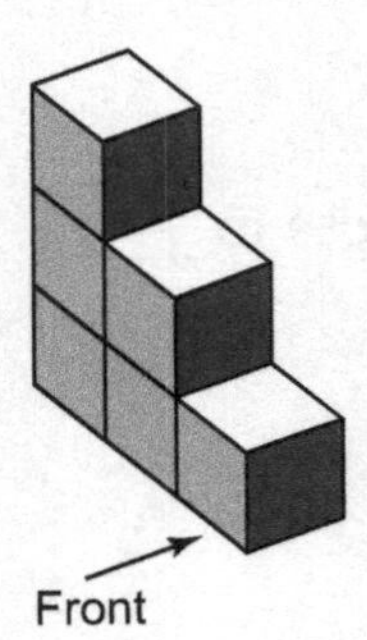

Matching Views	Diagram	Corresponding Area (cm^2)
______ ______		______
______ ______		______
______ ______		______
Total		______

The surface area is ____ cm^2.

10. Calculate the surface area of this composite object.

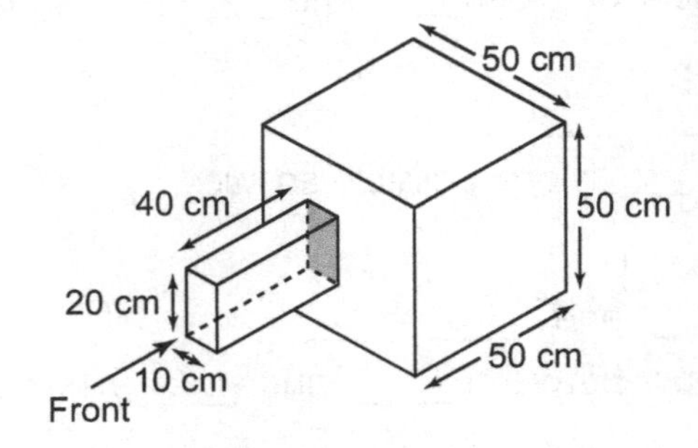

Surface area of cube

Matching Faces	Diagram	Corresponding Area (cm^2)
_____/_____ ____/_______ _______/_____		6(___ × ___) = ______
Total		______

The surface area is _______ cm^2.

Surface area of rectangular prism

Matching Faces	Diagram	Corresponding Area (cm^2)
_____/_____		______________
____/______		______________
_____/_____		______________
Total		______________

The surface area is ______ cm^2.

Area of overlap

Diagram	Corresponding Area (cm²)
	___ × ___ = ___

The area of overlap is ____ cm².

SA composite object = ________________ + ________________ − ________________

= _______ + _______ − _______

= _______

The surface area of the composite object is _______ cm².

1.4 **11.** Find the surface area of this composite object.

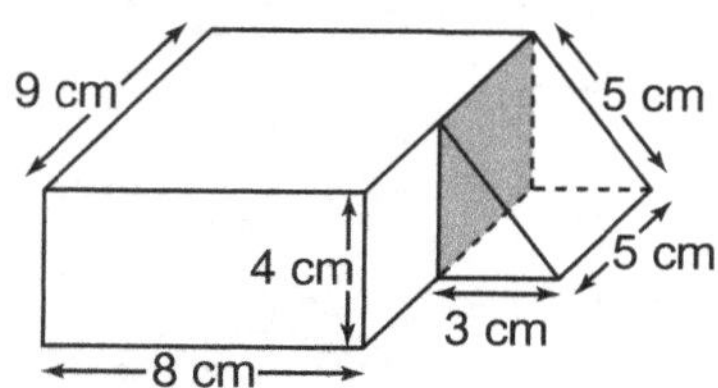

Surface area of rectangular prism

Faces	Diagram	Corresponding Area (cm²)
______ ______		________
______ ______		________
______ ______		________
Total		______

The surface area is _____ cm².

Surface area of triangular prism

Faces	Diagram	Corresponding Area (cm²)
Triangular		________ ________
Rectangular		________
		________
		________
Total		____

The surface area is ____ cm².

Area of overlap

Diagram	Corresponding Area (cm²)
	________

The area of overlap is ____ cm².

SA = ________________ + ________________ − ________________

= _____ + _____ − _____

= _____

The surface area of the composite object is _____ cm².

12. Find the surface area of this composite object.

16 cm
12 cm
12 cm
4 cm

The larger cylinder has diameter ____ cm, so its radius is ___ cm.
The smaller cylinder has diameter ___ cm, so its radius is ___ cm.

Surface area of smaller cylinder

Faces	Diagram	Corresponding Area (cm^2)
Top Bottom		__ × __ × __ ≐ ______
Curved surface		__ × __ × __ × __ ≐ ______
Total		______

The surface area is about _______ cm^2.

Surface area of larger cylinder

Faces	Diagram	Corresponding Area (cm^2)
Top Bottom		__ × __ × __ ≐ ______
Curved surface		__ × __ × __ × __ ≐ ______
Total		______

The surface area is about _______ cm^2.

Area of overlap

Diagram	Corresponding Area (cm^2)
	__ × __ ≐ ______

The area of overlap is about ______ cm^2.

Surface area of the composite object ≐ ______ + ________ − ________
≐ ________

The surface area is about _______ cm^2.

UNIT 2

Powers and Exponent Laws

What You'll Learn

- Use powers to show repeated multiplication.
- Evaluate powers with exponent 0.
- Write numbers using powers of 10.
- Use the order of operations with exponents.
- Use the exponent laws to simplify and evaluate expressions.

Why It's Important

Powers are used by

- lab technicians, when they interpret a patient's test results
- reporters, when they write large numbers in a news story

Key Words

integer
opposite
positive
negative
factor
power
base
exponent
squared
cubed
standard form
product
quotient

2.1 Skill Builder

Multiplying Integers

When multiplying 2 integers, look at the sign of each integer:

- When the integers have the same sign, their product is positive.
- When the integers have different signs, their product is negative.

×	(−)	(+)
(−)	(+)	(−)
(+)	(−)	(+)

$6 \times (-3)$ These 2 integers have different signs, so their product is negative.

$6 \times (-3) = -18$

$(-10) \times (-2)$ These 2 integers have the same sign, so their product is positive.

$(-10) \times (-2) = 20$

When an integer is positive, we do not have to write the + sign in front.

Check

1. Will the product be positive or negative?

a) 7×4 ____________ **b)** $3 \times (-6)$ ____________

c) $(-9) \times 10$ ____________ **d)** $(-5) \times (-9)$ ____________

2. Multiply.

a) $7 \times 4 =$ ______ **b)** $3 \times (-6) =$ ______

c) $(-9) \times 10 =$ ______ **d)** $(-5) \times (-9) =$ ______

e) $(-3) \times (-5) =$ ______ **f)** $2 \times (-5) =$ ______

g) $(-8) \times 2 =$ ______ **h)** $(-4) \times 3 =$ ______

Multiplying More than 2 Integers

We can multiply more than 2 integers.
Multiply pairs of integers, from left to right.

$$\underbrace{(-1) \times (-2)} \times (-3)$$
$$= 2 \times (-3)$$
$$= -6$$

The product of 3 negative factors is negative.

$$\underbrace{(-1) \times (-2)} \times (-3) \times (-4)$$
$$= \underbrace{2 \times (-3)} \times (-4)$$
$$= (-6) \times (-4)$$
$$= 24$$

The product of 4 negative factors is positive.

Multiplying Integers
When the number of negative factors is *even*, the product is positive.
When the number of negative factors is *odd*, the product is negative.

We can show products of integers in different ways:
$(-2) \times (-2) \times 3 \times (-2)$ is the same as $(-2)(-2)(3)(-2)$.

So, $(-2) \times (-2) \times 3 \times (-2) = (-2)(-2)(3)(-2)$
$= -24$

Check

1. Multiply.

a) $(-3) \times (-2) \times (-1) \times 1$ ______

b) $(-2)(-1)(-2)(-2)(2)$ ______

c) $(-2)(-2)(-1)(-2)(-2)$ ______

d) $3 \times 3 \times 2$ ______

Is the answer positive or negative? How can you tell?

2.1 What Is a Power?

FOCUS Show repeated multiplication as a power.

We can use powers to show repeated multiplication.

$2 \times 2 \times 2 \times 2 \times 2 = 2^5$

Repeated multiplication
5 **factors** of 2

Power

2 is the ***base.***
5 is the ***exponent.***
2^5 is a ***power.***

We read 2^5 as "2 to the 5th."
Here are some other powers of 2.

Repeated Multiplication	Power	Read as...
$\underbrace{2}_{\text{1 factor of 2}}$	2^1	2 to the 1st
$\underbrace{2 \times 2}_{\text{2 factors of 2}}$	2^2	2 to the 2nd, or 2 squared
$\underbrace{2 \times 2 \times 2}_{\text{3 factors of 2}}$	2^3	2 to the 3rd, or 2 cubed
$\underbrace{2 \times 2 \times 2 \times 2}_{\text{4 factors of 2}}$	2^4	2 to the 4th

In each case, the exponent in the power is equal to the number of factors in the repeated multiplication.

Example 1 Writing Powers

Write as a power.

a) $4 \times 4 \times 4 \times 4 \times 4 \times 4$

b) 3

Solution

a) The base is 4.

$\underbrace{4 \times 4 \times 4 \times 4 \times 4 \times 4}_{\text{6 factors of 4}} = 4^6$

So, $4 \times 4 \times 4 \times 4 \times 4 \times 4 = 4^6$

b) The base is 3.

$\underbrace{3}_{\text{1 factor of 3}}$

So, $3 = 3^1$

Check

1. Write as a power.

a) $2 \times 2 \times 2 \times 2 \times 2 \times 2 = 2$___

b) $5 \times 5 \times 5 \times 5 = 5$___

c) $(-10)(-10)(-10) =$ _______

d) $4 \times 4 =$ _____

e) $(-7)(-7)(-7)(-7)(-7)(-7)(-7)(-7) =$ _______

2. Complete the table.

	Repeated Multiplication	Power	Read as…
a)	$8 \times 8 \times 8 \times 8$	_______	8 to the 4th
b)	7×7	_______	_______________
c)	$3 \times 3 \times 3 \times 3 \times 3 \times 3$	_______	3 to the 6th
d)	$2 \times 2 \times 2$	_______	_______________

Power	Repeated Multiplication	Standard Form
2^5	$2 \times 2 \times 2 \times 2 \times 2$	32

Example 2 Evaluating Powers

Write as repeated multiplication and in standard form.

a) 2^4

b) 5^3

Solution

a) $2^4 = 2 \times 2 \times 2 \times 2$ — As repeated multiplication
$= 16$ — Standard form

b) $5^3 = 5 \times 5 \times 5$ — As repeated multiplication
$= 125$ — Standard form

Check

1. Complete the table.

Power	Repeated Multiplication	Standard Form
2^3	$2 \times 2 \times 2$	______
6^2	__________________	36
3^4	__________________	______
10^4	__________________	______
8 squared	__________________	______
7 cubed	__________________	______

To evaluate a power that contains negative integers, identify the base of the power. Then, apply the rules for multiplying integers.

Example 3 Evaluating Expressions Involving Negative Signs

Identify the base, then evaluate each power.

a) $(-5)^4$

b) -5^4

Solution

a) $(-5)^4$

The brackets tell us that the base of this power is (-5).

$$(-5)^4 = (-5) \times (-5) \times (-5) \times (-5)$$
$$= 625$$

There is an even number of negative integers, so the product is positive.

b) -5^4

There are no brackets. So, the base of this power is 5. The negative sign applies to the whole expression.

$$-5^4 = -(5 \times 5 \times 5 \times 5)$$
$$= -625$$

Check

1. Identify the base of each power, then evaluate.

a) $(-1)^3$

The base is _____.

$(-1)^3 =$ ____________

$=$ ____________

b) -10^3

The base is _____.

$-10^3 =$ ____________

$=$ ____________

c) $(-7)^2$

The base is _____.

$(-7)^2 =$ ____________

$=$ ____________

d) $-(-5)^4$

The first negative sign applies to the whole expression.

The base is _____.

$-(-5)^4 =$ ____________

$=$ ____________

Practice

1. Write as a power.

a) $\underbrace{8 \times 8 \times 8 \times 8 \times 8 \times 8 \times 8}_{\text{7 factors of 8}}$

The base is 8. There are _____ equal factors, so the exponent is _____.

$8 \times 8 \times 8 \times 8 \times 8 \times 8 \times 8 = 8^{\_\_}$

b) $\underbrace{10 \times 10 \times 10 \times 10 \times 10}_{\text{5 factors of 10}}$

The base is _____. There are _____ equal factors, so the exponent is _____.

So, $10 \times 10 \times 10 \times 10 \times 10 =$ _____

c) $\underbrace{(-2)(-2)(-2)}_{\text{3 factors of \_\_\_\_\_}}$

The base is _____. There are _____ equal factors, so the exponent is _____.

So, $(-2)(-2)(-2) =$ _______

d) $(-13)(-13)(-13)(-13)(-13)(-13)$

_______ factors of _______

The base is _______. There are _______ equal factors, so the exponent is _______.

So, $(-13)(-13)(-13)(-13)(-13)(-13) =$ _______

2. Write each expression as a power.

a) $9 \times 9 \times 9 \times 9 = \_\_\_\_\_^4$

b) $(5)(5)(5)(5)(5)(5) = 5^{\_\_}$

c) $11 \times 11 =$ _____

d) $(-12)(-12)(-12)(-12)(-12) =$ _______

3. Write each power as repeated multiplication.

a) 3^2 = ______________ **b)** 3^4 = ______________

c) 2^7 = ____________________________

d) 10^8 = ____

Identify the base first.

4. State whether the answer will be positive or negative.

a) $(-3)^2$ ______________ **b)** 6^3 ______________

c) $(-10)^3$ ______________ **d)** -4^3 ______________

5. Write each power as repeated multiplication and in standard form.

Predict. Will the answer be positive or negative?

a) $(-3)^2$ = ______________
= _______

b) 6^3 = ________________
= _______

c) $(-10)^3$ = __________________
= __________________

d) -4^3 = __________________
= _______

6. Write each product as a power and in standard form.

a) $(-3)(-3)(-3)$ = ________
= ________

b) $(-8)(-8)$ = ________
= ________

c) $-(8 \times 8 \times 8)$ = ________
= ________

d) $-(-1)(-1)(-1)(-1)(-1)(-1)(-1)$ = ________
= ________

7. Identify any errors and correct them.

a) $4^3 = 12$ ____
____

b) $(-2)^9$ is negative. ____
____

c) $(-9)^2$ is negative. ____
____

d) $3^2 = 2^3$ ____
____

e) $(-10)^2 = 100$ ____
____

2.2 Skill Builder

Patterns and Relationships in Tables

Look at the patterns in this table.

Input		Output
1	×2 →	2
2	×2 →	4
3	×2 →	6
4	×2 →	8
5	×2 →	10

(Input: +1 each row; Output: +2 each row)

The input starts at 1 and increases by 1 each time.

The output starts at 2 and increases by 2 each time.

The input and output are also related.
Double the input to get the output.

Check

1. a) Describe the patterns in the table.
b) What is the input in the last row?
What is the output in the last row?

Input	Output
1	5
2	10
3	15
4	20
_____	_____

(Input: +1; Output: +5)

a) The input starts at _____, and increases by _____ each time.
The output starts at _____, and increases by _____ each time.
You can also multiply the input by _____ to get the output.

b) The input in the last row is 4 + _____ = _____.
The output in the last row is 20 + _____ = _____.

2. a) Describe the patterns in the table.

b) Extend the table 3 more rows.

Input	Output
10	100
9	90
8	80
7	70
6	60

a) The input starts at 10, and decreases by _____ each time.
The output starts at 100, and decreases by _____ each time.
You can also multiply the input by _____ to get the output.

b) To extend the table 3 more rows, continue to decrease the input by _____ each time.
Decrease the output by _____ each time.

Input	Output
5	_____
_____	_____
_____	_____

Writing Numbers in Expanded Form

8000 is 8 thousands, or 8×1000
600 is 6 hundreds, or 6×100
50 is 5 tens, or 5×10

Read it aloud.

Check

1. Write each number in expanded form.

a) 7000 __________

b) 900 __________

c) 400 __________

d) 30 __________

2.2 Powers of Ten and the Zero Exponent

FOCUS **Explore patterns and powers of 10 to develop a meaning for the exponent 0.**

This table shows decreasing powers of 3.

Power	Repeated Multiplication	Standard Form
3^5	$3 \times 3 \times 3 \times 3 \times 3$	243
3^4	$3 \times 3 \times 3 \times 3$	81
3^3	$3 \times 3 \times 3$	27
3^2	3×3	9
3^1	3	3

$\div 3$

$\div 3$

$\div 3$

$\div 3$

Look for patterns in the columns.

The exponent decreases by 1 each time. Divide by 3 each time.

The patterns suggest $3^0 = 1$ because $3 \div 3 = 1$.

We can make a similar table for the powers of any integer base except 0.

The Zero Exponent
A power with exponent 0 is equal to 1.

The base of the power can be any integer except 0.

Example 1 Powers with Exponent Zero

Evaluate each expression.

a) 6^0

b) $(-5)^0$

Solution

A power with exponent 0 is equal to 1.

a) $6^0 = 1$

b) $(-5)^0 = 1$

The zero exponent applies to the number in the brackets.

Check

1. Evaluate each expression.

a) $8^0 =$ _____

b) $-4^0 =$ _____

c) $4^0 =$ _____

d) $(-10)^0 =$ _____

If there are no brackets, the zero exponent applies only to the base.

Example 2 Powers of Ten

Write as a power of 10.

a) 10 000 **b)** 1000 **c)** 100 **d)** 10 **e)** 1

Solution

a) $10\,000 = 10 \times 10 \times 10 \times 10$
$= 10^4$

b) $1000 = 10 \times 10 \times 10$
$= 10^3$

c) $100 = 10 \times 10$
$= 10^2$

Notice that the exponent is equal to the number of zeros.

d) $10 = 10^1$

e) $1 = 10^0$

Check

1. a) $5^1 =$ _____ **b)** $(-7)^1 =$ _____

c) $10^1 =$ _____ **d)** $10^0 =$ _____

Practice

1. a) Complete the table below.

Power	Repeated Multiplication	Standard Form
5^4	$5 \times 5 \times 5 \times 5$	625
5^3	$5 \times 5 \times 5$	_____
5^2	_____________	_____
5^1	_____________	_____

b) What is the value of 5^1? _____

c) Use the table. What is the value of 5^0? _____

2. Evaluate each power.

> *If there are no brackets, the exponent applies only to the base.*

a) $2^0 =$ _____ **b)** $9^0 =$ _____

c) $(-2)^0 =$ _____ **d)** $-2^0 =$ _____

e) $10^1 =$ _____ **f)** $(-8)^1 =$ _____

3. Write each number as a power of 10.

a) $10\ 000 = 10^{\_\_}$ **b)** $1\ 000\ 000 = 10^{\_\_}$

c) Ten million = _____ **d)** One = _____

e) 1 000 000 000 = _____ **f)** 10 = _____

4. Evaluate each power of 10.

a) $-10^6 =$ _______________ **b)** $-10^0 =$ _____

c) $-10^8 =$ _______________ **d)** $-10^1 =$ _____

5. One trillion is written as 1 000 000 000 000.
Write each number as a power of 10.

a) One trillion = 1 000 000 000 000 = ______

b) Ten trillion = 10 × ____________________ = ______

c) One hundred trillion = __________________________ = ______

6. Write each number in standard form.

a) $5 \times 10^4 = 5 \times 10\ 000$
= _______________

b) $(4 \times 10^2) + (3 \times 10^1) + (7 \times 10^0) = (4 \times 100) +$ _______________
= _______________
= _______________

c) $(2 \times 10^3) + (6 \times 10^2) + (4 \times 10^1) + (9 \times 10^0)$
= ______________________________
= ______________________________
= ______________________________

d) $(7 \times 10^3) + (8 \times 10^0) =$ _______________
= _______________
= _______________

2.3 Skill Builder

Adding Integers

To add a positive integer and a negative integer: 7 + (−4)

- Model each integer with tiles.
- Circle zero pairs.

Each pair of 1 ■ tile and 1 ■ tile makes a zero pair. The pair models 0.

There are 4 zero pairs.
There are 3 ■ tiles left.
They model 3.
So, 7 + (−4) = 3

To add 2 negative integers: (−4) + (−2)

- Model each integer with tiles.
- Combine the tiles.

There are 6 ■ tiles.
They model −6.
So, (−4) + (−2) = −6

Check

1. Add.

a) (−3) + (−4) = _____ **b)** 6 + (−2) = _____

c) (−5) + 2 = _____ **d)** (−4) + (−4) = _____

2. a) Kerry borrows $5. Then she borrows another $5.
Add to show what Kerry owes.
(−5) + (−5) = _____ Kerry owes $_____.

When an amount of money is negative, it is owed.

b) The temperature was 8°C. It fell 10°C.
Add to show the new temperature.
8 + (_____) = _____ The new temperature is _____°C.

Subtracting Integers

To subtract 2 integers: $3 - 6$

- Model the first integer.
- Take away the number of tiles equal to the second integer.

Model 3.

Adding zero pairs does not change the value. Zero pairs represent 0.

There are not enough tiles to take away 6.
To take away 6, we need 3 more tiles.
We add zero pairs. Add 3 tiles and 3 tiles.

Now take away the 6 tiles.

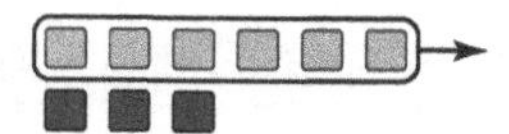

Since 3 tiles remain, we write: $3 - 6 = -3$

When tiles are not available, think of subtraction as the opposite of addition.
To subtract an integer, add its opposite integer.
For example,

$(-3) - (+2) = -5$ — Subtract +2.

$(-3) + (-2) = -5$ — Add −2.

Check

1. Subtract.

a) $(-6) - 2 =$ _____

b) $2 - (-6) =$ _____

c) $(-8) - 9 =$ _____

d) $8 - (-9) =$ _____

Dividing Integers

When dividing 2 integers, look at the sign of each integer:

- When the integers have the same sign, their quotient is positive.
- When the integers have different signs, their quotient is negative.

The same rule applies to the multiplication of integers.

$6 \div (-3)$ These 2 integers have different signs, so their quotient is negative.
$6 \div (-3) = -2$

$(-10) \div (-2)$ These 2 integers have the same sign, so their quotient is positive.
$(-10) \div (-2) = 5$

Check

1. Calculate.

a) $(-4) \div 2$
$=$ _____

b) $(-6) \div (-3)$
$=$ _____

c) $15 \div (-3)$
$=$ _____

2.3 Order of Operations with Powers

FOCUS **Explain and apply the order of operations with exponents.**

We use this order of operations when evaluating an expression with powers:

- Do the operations in brackets first.
- Evaluate the powers.
- Multiply and divide, in order, from left to right.
- Add and subtract, in order, from left to right.

We can use the word BEDMAS to help us remember the order of operations:

B **B**rackets
E **E**xponents
D **D**ivision
M **M**ultiplication
A **A**ddition
S **S**ubtraction

Example 1 Adding and Subtracting with Powers

Evaluate.

a) $2^3 + 1$ **b)** $8 - 3^2$ **c)** $(3 - 1)^3$

Solution

a) $2^3 + 1$ Evaluate the power first: 2^3
$= (2)(2)(2) + 1$ Multiply: $(2)(2)(2)$
$= 8 + 1$ Then add: $8 + 1$
$= 9$

b) $8 - 3^2$ Evaluate the power first: 3^2
$= 8 - (3)(3)$ Multiply: $(3)(3)$
$= 8 - 9$ Then subtract: $8 - 9$
$= -1$

To subtract, add the opposite: $8 + (-9)$

c) $(3 - 1)^3$ Subtract inside the brackets first: $3 - 1$
$= 2^3$ Evaluate the power: 2^3
$= (2)(2)(2)$ Multiply: $(2)(2)(2)$
$= 8$

Check

1. Evaluate.

a) $4^2 + 3 =$ ______ $+ 3$

$=$ ______

b) $5^2 - 2^2 =$ ______ $- (2)(2)$

$=$ ______

c) $(2 + 1)^2 =$ ______

$=$ ______

d) $(5 - 6)^2 =$ ______

$=$ ______

Example 2 Multiplying and Dividing with Powers

Evaluate.

a) $[2 \times (-2)^3]^2$

Curved brackets — Square brackets

b) $(7^2 + 5^0) \div (-5)^1$

When we need curved brackets for integers, we use square brackets to show the order of operations.

Solution

a) $[2 \times (-2)^3]^2$ — Evaluate what is inside the square brackets first: $2 \times (-2)^3$

$= [2 \times (-8)]^2$ — Start with $(-2)^3 = -8$.

$= (-16)^2$

$= 256$

b) $(7^2 + 5^0) \div (-5)^1$ — Evaluate what is inside the brackets first: $7^2 + 5^0$

$= (49 + 1) \div (-5)^1$ — Add inside the brackets: $49 + 1$

$= 50 \div (-5)^1$ — Evaluate the power: $(-5)^1$

$= 50 \div (-5)$

$= -10$

Check

1. Evaluate.

a) $5 \times 3^2 = 5 \times$ _____

$=$ _____

b) $8^2 \div 4 =$ _____ $\div 4$

$=$ _____

c) $(3^2 + 6^0)^2 \div 2^1$

$= ($_____ $+$ _____$)^2 \div 2^1$

$=$ _____ $\div 2^1$

$=$ _____

d) $10^2 + (2 \times 2^2)^2 = 10^2 + (2 \times$ _____$)^2$

$= 10^2 +$ _____

$=$ _____

Example 3 Solving Problems Using Powers

Corin answered the following skill-testing question to win free movie tickets:

$120 + 20^3 \div 10^3 + 12 \times 120$

His answer was 1568.

Did Corin win the movie tickets? Show your work.

Solution

$120 + 20^3 \div 10^3 + 12 \times 120$	Evaluate the powers first: 20^3 and 10^3
$= 120 + 8000 \div 1000 + 12 \times 120$	Divide and multiply.
$= 120 + 8 + 1440$	Add: $120 + 8 + 1440$
$= 1568$	

Corin won the movie tickets.

Check

1. Answer the following skill-testing question to enter a draw for a Caribbean cruise.

$(6 + 4) + 3^2 \times 10 - 10^2 \div 4$

$=$ ____________________

$=$ ____________________

$=$ ____________________

Practice

1. Evaluate.

a) $2^2 + 1 = \_\_\_\_ + 1$
$= \_\_\_\_$

b) $2^2 - 1 = \_\_\_\_ - 1$
$= \_\_\_\_$

c) $(2 + 1)^2 = \_\_\_\_$
$= \_\_\_\_$

d) $(2 - 1)^2 = \_\_\_\_$
$= \_\_\_\_$

2. Evaluate.

a) $4 \times 2^2 = 4 \times \_\_\_\_$
$= \_\_\_\_$

b) $4^2 \times 2 = \_\_\_\_ \times 2$
$= \_\_\_\_$

c) $(4 \times 2)^2 = \_\_\_\_$
$= \_\_\_\_$

d) $(-4)^2 \div 2 = \_\_\_\_ \div 2$
$= \_\_\_\_$

3. Evaluate.

a) $2^3 + (-1)^3 = \_\_\_\_ + (-1)^3$
$= \_\_\_\_ + \_\_\_\_$
$= \_\_\_\_$

b) $(2 - 1)^3 = \_\_\_\_$
$= \_\_\_\_$

c) $2^3 - (-1)^3 = \_\_\_\_ - (-1)^3$
$= \_\_\_\_ - \_\_\_\_$
$= \_\_\_\_$

d) $(2 + 1)^3 = \_\_\_\_$
$= \_\_\_\_$

4. Evaluate.

a) $3^2 \div (-1)^2 = \_\_\_\_ \div (-1)^2$
$= \_\_\_\_ \div \_\_\_\_$
$= \_\_\_\_$

b) $(3 \div 1)^2 = \_\_\_\_$
$= \_\_\_\_$

c) $3^2 \times (-2)^2 = \_\_\_\_ \times (-2)^2$
$= \_\_\_\_ \times \_\_\_\_$
$= \_\_\_\_$

d) $5^2 \div (-5)^1 = \_\_\_\_ \div (-5)^1$
$= \_\_\_\_ \div \_\_\_\_$
$= \_\_\_\_$

5. Evaluate.

a) $(-2)^0 \times (-2) =$ _____ $\times (-2)$

$=$ _____

b) $2^3 \div (-2)^2 =$ ________ $\div (-2)^2$

$=$ _____ $\div$ _____

$=$ _____

c) $(3 + 2)^0 + (3 \times 2)^0 =$ _____ $+$ _____

$=$ _____

d) $(3 \times 5^2)^0 =$ _____

e) $(2)(3) - (4)^2 = (2)(3) -$ ______

$=$ _____ $-$ _____

$=$ _____

f) $3(2 - 1)^2 = 3$_____

$=$ _____

A power with exponent 0 is equal to 1.

g) $(-2)^2 + (3)(4) =$ ________ $+ (3)(4)$

$=$ _____ $+$ _____

$=$ _____

h) $(-2) + 3^0 \times (-2) = (-2) +$ _____ $\times (-2)$

$=$ _____

6. Amaya wants to replace the hardwood floor in her house.
Here is how she calculates the cost, in dollars:
$70 \times 6^2 + 60 \times 6^2$
How much will it cost Amaya to replace the hardwood floor?

$70 \times$ ______ $+ 60 \times$ ______

$=$ ______________

$=$ ______________

Remember the order of operations: BEDMAS

It will cost Amaya $______ to replace the hardwood floor.

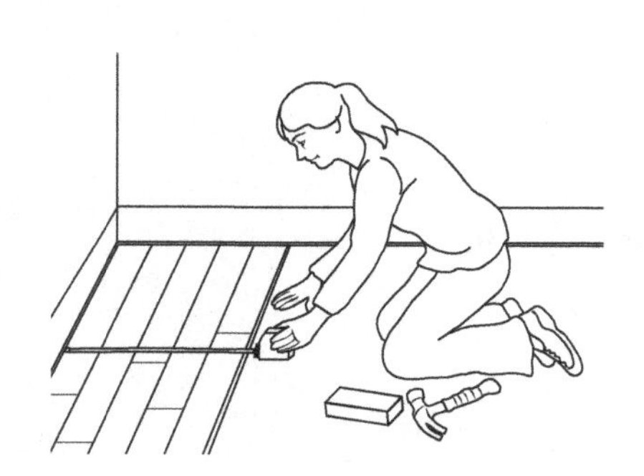

Can you ...

- Use powers to show repeated multiplication?
- Use patterns to evaluate a power with exponent zero, such as 50?
- Use the correct order of operations with powers?

2.1 **1.** Give the base and exponent of each power.

a) 6^2 Base: _____ Exponent: _____
There are _____ factors of _____.

b) 4^5 Base: _____ Exponent: _____
There are _____ factors of _____.

c) $(-3)^8$ Base: _____ Exponent: _____
There are _____ factors of _____.

d) -3^8 Base: _____ Exponent: _____
There are _____ factors of _____.

2. Write as a power.

a) $7 \times 7 \times 7 \times 7 \times 7 \times 7 = 7$___

b) $2 \times 2 \times 2 \times 2 = 2$___

c) $5 =$ _____

d) $(-5)(-5)(-5)(-5)(-5) =$ _____

3. Write each power as repeated multiplication and in standard form.

a) $5^2 = 5 \times$ _____ = _____

b) $2^3 =$ _______________ = _____

c) $3^4 =$ __________________ = _____

2.2 **4. a)** Complete the table.

Power	Repeated Multiplication	Standard Form
7^3	$7 \times 7 \times 7$	343
7^2	7×7	
7^1		

b) What is the value of 7^0? _____

5. Write each number in standard form and as a power of 10.

a) One hundred = 100
= 10——

b) Ten thousand = ____________
= 10——

c) One million = ____________
= 10——

d) One = __
= 10——

6. Evaluate.

a) $6^0 =$ _____

b) $(-8)^0 =$ _____

c) $12^1 =$ _____

d) $-8^0 =$ _____

7. Write each number in standard form.

a) 4×10^3
$= 4 \times$ ________
= ________

b) $(1 \times 10^3) + (3 \times 10^2) + (2 \times 10^1) + (1 \times 10^0)$
$= (1 \times 1000) + (3 \times$ _____$) + ($________$) + ($________$)$
= ________ + _____ + _____ + _____
= ________

c) $(4 \times 10^3) + (2 \times 10^2) + (3 \times 10^1) + (6 \times 10^0)$
$= (4 \times$ ________$) + ($________$) + ($________$) + ($________$)$
= ________ + _____ + _____ + _____
= ________

d) $(8 \times 10^2) + (1 \times 10^1) + (9 \times 10^0)$
= ________ + _______ + ______
= _____________
= _____________

2.3 **8.** Evaluate.

a) $3^2 + 5 =$ ______ $+ 5$

$=$ ______

b) $5^2 - 2^3 =$ ______ $- 2^3$

$=$ ______

$=$ ______

c) $(2 + 3)^3 = (\_)^3$

$=$ ______

d) $2^3 + (-3)^3 =$ ______ $+ (-3)^3$

$=$ ______

$=$ ______

9. Evaluate.

a) $5 \times 3^2 = 5 \times$ ______

$=$ ______

b) $8^2 \div 4 =$ ______ $\div 4$

$=$ ______

c) $(10 + 2) \div 2^2 =$ ______ $\div 2^2$

$=$ ______ $\div$ ______

$=$ ______

d) $(7^2 + 1) \div (2^3 + 2)$

$= ($______ $+ 1) \div ($______ $+ 2)$

$=$ ______ $\div$ ______

$=$ ______

10. Evaluate. State which operation you do first.

a) $3^2 + 4^2$ ______________________

$=$ ______ $+$ ______

$=$ ______

b) $[(-3) - 2]^3$ ______________________

$= ($______$)3$

$=$ ______________

c) $(-2)^3 + (-3)^0$ ______________________

$=$ ______________ $+$ ______

$=$ ______________

d) $[(6 - 3)^3 \times (2 + 2)^2]^0$ ______________________

$=$ ______

2.4 Skill Builder

Simplifying Fractions

To simplify a fraction, divide the numerator and denominator by their common factors.

To simplify $\frac{5 \times 5 \times 5 \times 5}{5 \times 5}$:

This fraction shows repeated multiplication.

Divide the numerator and denominator by their common factors: 5×5.

$\frac{\cancel{5}^1 \times \cancel{5}^1 \times 5 \times 5}{\cancel{5}^1 \times \cancel{5}^1}$

$= \frac{5 \times 5}{1}$

$= 25$

Check

1. Simplify each fraction.

What are the common factors?

a) $\frac{3 \times 3 \times 3}{3}$

$=$

______

b) $\frac{8 \times 8 \times 8 \times 8 \times 8}{8 \times 8 \times 8 \times 8 \times 8}$

$=$

______

c) $\frac{5 \times 5 \times 5 \times 5 \times 5}{5 \times 5 \times 5}$

$=$

______

d) $\frac{2 \times 2 \times 2 \times 2 \times 2 \times 2 \times 2 \times 2}{2 \times 2 \times 2 \times 2 \times 2}$

$=$

______

2.4 Exponent Laws I

FOCUS **Understand and apply the exponent laws for products and quotients of powers.**

Multiply $3^2 \times 3^4$.

$3^2 \times 3^4$ — Write as repeated multiplication.

$= \underbrace{(3 \times 3)}_{\text{2 factors of 3}} \times \underbrace{(3 \times 3 \times 3 \times 3)}_{\text{4 factors of 3}}$

$= \underbrace{3 \times 3 \times 3 \times 3 \times 3 \times 3}_{\text{6 factors of 3}}$

$= 3^6$

Base: 3; Exponent: 6

So, $3^{\mathbf{2}} \times 3^{\mathbf{4}} = 3^{\mathbf{6}}$ — Look at the pattern in the exponents.

2 + 4 = 6

We write: $3^{\mathbf{2}} \times 3^{\mathbf{4}} = 3^{\mathbf{(2+4)}}$

$= 3^{\mathbf{6}}$

This relationship is true when you multiply any 2 powers with the same base.

Exponent Law for a Product of Powers
To multiply powers with the same base, add the exponents.

Example 1 Simplifying Products with the Same Base

Write as a power.

a) $5^3 \times 5^4$ **b)** $(-6)^2 \times (-6)^3$ **c)** $(7^2)(7)$

Solution

a) The powers have the same base: 5
Use the exponent law for products: add the exponents.

$5^3 \times 5^4 = 5^{(3+4)}$

$= 5^7$

To check your work, you can write the powers as repeated multiplication.

b) The powers have the same base: -6

$(-6)^2 \times (-6)^3 = (-6)^{(2+3)}$ Add the exponents.

$= (-6)^5$

c) $(7^2)(7) = 7^2 \times 7^1$ Use the exponent law for products.

$= 7^{(2+1)}$ Add the exponents.

$= 7^3$

7 can be written as 7^1.

Check

1. Write as a power.

a) $2^5 \times 2^4 = 2^{(\_\_ + \_\_)}$
$= 2^{\_\_}$

b) $5^2 \times 5^5 = 5^{\_\_\_}$
$= 5^{\_\_}$

c) $(-3)^2 \times (-3)^3 =$ ________
$=$ ________

d) $10^5 \times 10 =$ ________
$=$ ________

Divide $3^4 \div 3^2$.

We can show division in fraction form.

$3^4 \div 3^2 = \frac{3^4}{3^2}$.

$= \frac{3 \times 3 \times 3 \times 3}{3 \times 3}$ Simplify.

$= \frac{\not{3}^1 \times \not{3}^1 \times 3 \times 3}{\not{3}^1 \times \not{3}^1}$

$= \frac{3 \times 3}{1}$

$= 3 \times 3$

$= 3^2$

So, $3^{\mathbf{4}} \div 3^{\mathbf{2}} = 3^{\mathbf{2}}$ Look at the pattern in the exponents.

$\mathbf{4} - \mathbf{2} = \mathbf{2}$

We write: $3^{\mathbf{4}} \div 3^{\mathbf{2}} = 3^{(\mathbf{4}-\mathbf{2})}$

$= 3^{\mathbf{2}}$

This relationship is true when you divide any 2 powers with the same base.

Exponent Law for a Quotient of Powers
To divide powers with the same base, subtract the exponents.

Example 2 Simplifying Quotients with the Same Base

Write as a power.

a) $4^5 \div 4^3$

b) $(-2)^7 \div (-2)^2$

Solution

Use the exponent law for quotients: subtract the exponents.

a) $4^5 \div 4^3 = 4^{(5-3)}$
$= 4^2$

The powers have the same base: 4

To check your work, you can write the powers as repeated multiplication.

b) $(-2)^7 \div (-2)^2 = (-2)^{(7-2)}$
$= (-2)^5$

The powers have the same base: -2

Check

1. Write as a power.

a) $(-5)^6 \div (-5)^3 = (-5)^{\_\_\_\_}$
$= \_\_\_\_$

b) $\dfrac{(-3)^9}{(-3)^5} = (-3)^{\_\_\_\_}$
$= \_\_\_\_$

$\dfrac{(-3)^9}{(-3)^5}$ is the same as $(-3)^9 \div (-3)^5$

c) $8^4 \div 8^3 = \_\_\_\_$
$= \_\_\_\_$

d) $9^8 \div 9^2 = \_\_\_\_$
$= \_\_\_\_$

Example 3 Evaluating Expressions Using Exponent Laws

Evaluate.

a) $2^2 \times 2^3 \div 2^4$

b) $(-2)^5 \div (-2)^3 \times (-2)$

Solution

a) $2^2 \times 2^3 \div 2^4$
$= 2^{(2+3)} \div 2^4$
$= 2^5 \div 2^4$
$= 2^{(5-4)}$
$= 2^1$
$= 2$

Add the exponents of the 2 powers that are multiplied. Then, subtract the exponent of the power that is divided.

b) $(-2)^5 \div (-2)^3 \times (-2)$
$= (-2)^{(5-3)} \times (-2)$
$= (-2)^2 \times (-2)$
$= (-2)^{(2+1)}$
$= (-2)^{(3)}$
$= (-2)(-2)(-2)$
$= -8$

Subtract the exponents of the 2 powers that are divided.

Multiply: add the exponents.

Check

1. Evaluate.

a) $4 \times 4^3 \div 4^2 = 4^{(\_\_ + \_\_)} \div 4^2$
$= 4^{\_\_} \div 4^2$
$= 4^{(\_\_ - \_\_)}$
$= 4^{\_\_}$
$=$ _____

b) $(-3) \div (-3) \times (-3)$
$= (-3)^{\_\_\_\_} \times (-3)$
$= (-3)^{\_\_} \times (-3)$
$= (-3)^{\_\_\_\_}$
$= (-3)^{\_\_}$
$=$ _____

$(-3) = (-3)^1$

Practice

1. Write each product as a single power.

a) $7^6 \times 7^2 = 7^{(\_\_ + \_\_)}$
$= 7^{\_\_}$

b) $(-4)^5 \times (-4)^3 = (-4)^{\_\_\_}$
$= (-4)^{\_\_}$

To multiply powers with the same base, add the exponents.

c) $(-2) \times (-2)^3 =$ ______
$=$ ______

d) $10^5 \times 10^5 =$ ______
$=$ ______

e) $7^0 \times 7^1 =$ ______
$=$ ______

f) $(-3)^4 \times (-3)^5 =$ ______
$=$ ______

2. Write each quotient as a power.

a) $(-3)^5 \div (-3)^2 = (-3)^{(\_\_ - \_\_)}$
$= (-3)^{\_\_}$

b) $5^6 \div 5^4 = 5^{\_\_\_}$
$= 5^{\_\_}$

To divide powers with the same base, subtract the exponents.

c) $\frac{4^7}{4^4} = 4^{\_\_\_}$
$= 4^{\_\_}$

d) $\frac{5^8}{5^6} =$ ______
$=$ ______

e) $6^4 \div 6^4 =$ ______
$=$ ______

f) $\frac{(-6)^8}{(-6)^7} =$ ______
$=$ ______

3. Write as a single power.

a) $2^3 \times 2^4 \times 2^5 = 2^{(\_\_ + \_\_)} \times 2^5$
$= 2^{\_\_} \times 2^5$
$= 2^{\_\_\_}$
$= 2^{\_\_\_}$

b) $\frac{3^2 \times 3^2}{3^2 \times 3^2} = \frac{3^{\_\_\_}}{3^{\_\_\_}}$
$= \frac{3^{\_\_}}{3^{\_\_}}$
$=$ ______
$=$ ______

Which exponent law should you use?

c) $10^3 \times 10^5 \div 10^2 =$ ______ $\div 10^2$
$=$ ______ $\div 10^2$
$=$ ______
$=$ ______

d) $(-1)^9 \div (-1)^5 \times (-1)^0$
$=$ ______ $\times (-1)^0$
$=$ ______ $\times (-1)^0$
$=$ ______
$=$ ______

4. Simplify, then evaluate.

a) $(-3)^1 \times (-3)^2 \times 2$
$= \_\_\_\_\_\_\_\_ \times 2$
$= \_\_\_\_\_ \times 2$
$= \_\_\_\_\_ \times 2$
$= \_\_\_\_\_$

b) $9^9 \div 9^7 \times 9^0 = \_\_\_\_\_\_\_\_ \times 9^0$
$= \_\_\_\_\_ \times 9^0$
$= \_\_\_\_\_\_$
$= \_\_\_\_\_$
$= \_\_\_\_\_$

See if you can use the exponent laws to simplify.

c) $\frac{5^2}{5^0} = \_\_\_\_\_\_\_\_$
$= \_\_\_\_\_\_\_\_$
$= \_\_\_\_\_\_\_\_$

d) $\frac{5^5}{5^4} \times 5 = 5^{\_\_\_\_} \times 5$
$= 5^{\_\_} \times 5$
$= 5^{\_\_\_\_}$
$= 5^{\_\_}$
$= \_\_\_\_\_$

5. Identify any errors and correct them.

a) $4^3 \times 4^5 = 4^8$

b) $2^5 \times 2^5 = 2^{25}$

c) $(-3)^6 \div (-3)^2 = (-3)^3$

d) $7^0 \times 7^2 = 7^0$

e) $6^2 + 6^2 = 6^4$

f) $10^6 \div 10 = 10^6$

g) $2^3 \times 5^2 = 10^5$

2.5 Skill Builder

Grouping Equal Factors

In multiplication, you can group equal factors.

Order does not matter in multiplication.

For example:

$3 \times 7 \times 7 \times 3 \times 7 \times 7 \times 3$ — Group equal factors.

$= \underbrace{3 \times 3 \times 3} \times \underbrace{7 \times 7 \times 7 \times 7}$ — Write repeated multiplication as powers.

$= \quad 3^3 \quad \times \quad 7^4$

Check

1. Group equal factors and write as powers.

a) $2 \times 10 \times 2 \times 10 \times 2 = 2 \times 2 \times 2 \times$ ________________

$=$ ________________________

b) $2 \times 5 \times 2 \times 5 \times 2 \times 5 \times 2 \times 5 =$ ________________

$=$ ________________

Multiplying Fractions

To multiply fractions, first multiply the numerators, and then multiply the denominators.

$\frac{2}{3} \times \frac{2}{3} \times \frac{2}{3} \times \frac{2}{3} = \frac{2 \times 2 \times 2 \times 2}{3 \times 3 \times 3 \times 3}$ — Write repeated multiplication as powers.

$= \frac{2^4}{3^4}$

There are 4 factors of 2, and 4 factors of 3.

Check

1. Multiply the fractions. Write as powers.

a) $\frac{3}{4} \times \frac{3}{4} \times \frac{3}{4} =$ ________

$=$ ________

b) $\frac{1}{2} \times \frac{1}{2} \times \frac{1}{2} \times \frac{1}{2} \times \frac{1}{2} \times \frac{1}{2}$

$=$ ________________

$=$ ____

2.5 Exponent Laws II

FOCUS **Understand and apply exponent laws for powers of: products; quotients; and powers.**

Multiply $3^2 \times 3^2 \times 3^2$.

$3^2 \times 3^2 \times 3^2 = 3^{2+2+2}$

$= 3^6$

Use the exponent law for the product of powers.
Add the exponents.

We can write repeated multiplication as powers.

So, $\underbrace{3^2 \times 3^2 \times 3^2}_{\text{3 factors of } (3^2)}$

The base is 3^2. ← *This is also a power.*

The exponent is 3.

This is a **power of a power.**

Look at the pattern in the exponents.

$= (3^{\mathbf{2}})^{\mathbf{3}}$

$= 3^{\mathbf{6}}$

2 × 3 = 6

We write: $(3^{\mathbf{2}})^{\mathbf{3}} = 3^{\mathbf{2} \times \mathbf{3}}$

$= 3^{\mathbf{6}}$

Exponent Law for a Power of a Power

To raise a power to a power, multiply the exponents.

For example: $(2^3)^5 = 2^{3 \times 5}$

Example 1 Simplifying a Power of a Power

Write as a power.

a) $(3^2)^4$

b) $[(-5)^3]^2$

c) $-(2^3)^4$

Solution

Use the exponent law for a power of a power: multiply the exponents.

a) $(3^2)^4 = 3^{2 \times 4}$

$= 3^8$

b) $[(-5)^3]^2 = (-5)^{3 \times 2}$

$= (-5)^6$

The base is -5.

c) $-(2^3)^4 = -(2^{3 \times 4})$

$= -2^{12}$

The base is 2.

Check

1. Write as a power.

a) $(9^3)^4 = 9^{\_\_ \times \_\_}$
$= 9^{\_\_}$

b) $[(-2)^5]^3 = (-2)^{\_\_}$
$= (-2)^{\_\_}$

c) $-(5^4)^2 = -(5^{\_\_})$
$= -5^{\_\_}$

Multiply $(3 \times 4)^2$.

The base of the power is a product: $\underbrace{3 \times 4}_{\text{base}}$

Write as repeated multiplication.

$(3 \times 4)^2 = (3 \times 4) \times (3 \times 4)$ Remove the brackets.

$= 3 \times 4 \times 3 \times 4$ Group equal factors.

$= \underbrace{(3 \times 3)}_{\text{2 factors of 3}} \times \underbrace{(4 \times 4)}_{\text{2 factors of 4}}$ Write as powers.

$= 3^2 \times 4^2$

So, $(3 \times 4)^2 = 3^2 \times 4^2$

power product power

Exponent Law for a Power of a Product
The power of a product is the product of powers.
For example: $(2 \times 3)^4 = 2^4 \times 3^4$

Example 2 Evaluating Powers of Products

Evaluate.

a) $(2 \times 5)^2$

b) $[(-3) \times 4]^2$

Solution

Use the exponent law for a power of a product.

a)
$$(2 \times 5)^2 = 2^2 \times 5^2$$
$$= (2)(2) \times (5)(5)$$
$$= 4 \times 25$$
$$= 100$$

b)
$$[(-3) \times 4]^2 = (-3)^2 \times 4^2$$
$$= (-3)(-3) \times (4)(4)$$
$$= 9 \times 16$$
$$= 144$$

Or, use the order of operations and evaluate what is inside the brackets first.

a)
$$(2 \times 5)^2 = 10^2$$
$$= 100$$

b)
$$[(-3) \times 4]^2 = (-12)^2$$
$$= 144$$

Check

1. Write as a product of powers.

   a) $(5 \times 7)^4 =$ _____ $\times$ _____

   b) $(8 \times 2)^2 =$ _____ $\times$ _____

2. Evaluate.

   a) $[(-1) \times 6]^2 =$ $\_\_\_\_\_\_^2$
   $=$ _____

   b) $[(-1) \times (-4)]^3 =$ $\_\_\_\_\_\_^3$
   $=$ _____

Evaluate $\underbrace{\left(\frac{3}{4}\right)}_{\textbf{base}}^2$.

The base of the power is a quotient: $\frac{3}{4}$

Write as repeated multiplication.

$$\left(\frac{3}{4}\right)^2 = \left(\frac{3}{4}\right) \times \left(\frac{3}{4}\right)$$

$$= \frac{3}{4} \times \frac{3}{4}$$

Multiply the fractions.

$$= \frac{3 \times 3}{4 \times 4}$$

Write repeated multiplication as powers.

$$= \frac{3^2}{4^2}$$

So, $\left(\frac{3}{4}\right)^2 = \frac{3^2}{4^2}$

power (→ 3^2)
quotient (→ fraction bar)
power (→ 4^2)

> **Exponent Law for a Power of a Quotient**
> The power of a quotient is the quotient of powers.
> For example: $\left(\frac{2}{3}\right)^4 = \frac{2^4}{3^4}$

Example 3 Evaluating Powers of Quotients

Evaluate.

a) $[30 \div (-5)]^2$

b) $\left(\frac{20}{4}\right)^2$

Solution

Use the exponent law for a power of a quotient.

a) $[30 \div (-5)]^2 = \left(\frac{30}{-5}\right)^2$

$= \frac{30^2}{(-5)^2}$

$= \frac{900}{25}$

$= 36$

b) $\left(\frac{20}{4}\right)^2 = \frac{20^2}{4^2}$

$= \frac{400}{16}$

$= 25$

Or, use the order of operations and evaluate what is inside the brackets first.

a) $[30 \div (-5)]^2 = (-6)^2$

$= 36$

b) $\left(\frac{20}{4}\right)^2 = 5^2$

$= 25$

Check

1. Write as a quotient of powers.

a) $\left(\frac{3}{4}\right)^5 =$ ______

b) $[1 \div (-10)]^3 =$ ______

2. Evaluate.

a) $[(-16) \div (-4)]^2$

$=$ ____^{2} $=$ ____

b) $\left(\frac{36}{6}\right)^3 =$ ____

$=$ ____

You can evaluate what is inside the brackets first.

Practice

1. Write as a product of powers.

a) $(5 \times 2)^4 = 5^{—} \times 2^{—}$

b) $(12 \times 13)^2 =$ ________

c) $[3 \times (-2)]^3 =$ ________

d) $[(-4) \times (-5)]^5 =$ ________

2. Write as a quotient of powers.

a) $(5 \div 8)^0 =$ ______

b) $[(-6) \div 5]^7 =$ ______

c) $\left(\frac{3}{5}\right)^2 =$ ______

d) $\left(\frac{-1}{-2}\right)^3 =$ ______

3. Write as a power.

a) $(5^2)^3 = 5^{\_\_ \times \_\_}$
$= 5^{\_\_}$

b) $[(-2)^3]^5 = (-2)^{\_\_}$
$= \_\_\_\_$

c) $(4^4)^1 = \_\_\_\_$
$= \_\_\_\_$

d) $(8^0)^3 = \_\_\_\_$
$= \_\_\_\_$

4. Evaluate.

a) $[(6 \times (-2)]^2 = \_\_\_\_$
$= \_\_\_\_$

b) $-(3 \times 4)^2 = -(\_\_\_\_)^{\_\_}$
$= \_\_\_\_$

c) $\left(\frac{-8}{-2}\right)^2 = \_\_\_\_$
$= \_\_\_\_$

d) $(10 \times 3)^1 = \_\_\_\_$
$= \_\_\_\_$

e) $[(-2)^1]^2 = \_\_\_\_$
$= \_\_\_\_$
$= \_\_\_\_$

f) $[(-2)^1]^3 = \_\_\_\_$
$= \_\_\_\_$
$= \_\_\_\_$

5. Find any errors and correct them.

a) $(3^2)^3 = 3^5$ ________________

b) $(3 + 2)^2 = 3^2 + 2^2$ ________________

c) $(5^3)^3 = 5^9$ ________________

d) $\left(\frac{2}{3}\right)^8 = \frac{2^8}{3^8}$ ________________

e) $(3 \times 2)^2 = 36$ ________________

f) $\left(\frac{2}{3}\right)^2 = \frac{4}{6}$ ________________

g) $[(-3)^3]^0 = (-3)^3$ ________________

h) $[(-2) \times (-3)]^4 = -6^4$ ________________

Unit 2 Puzzle

Bird's Eye View

This is a view through the eyes of a bird. What does the bird see?

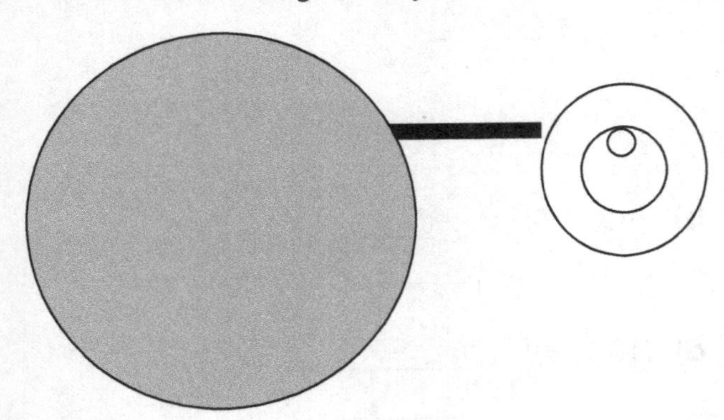

To find out, simplify or evaluate each expression on the left, then find the answer on the right.
Write the corresponding letter beside the question number.
The numbers at the bottom of the page are question numbers.
Write the corresponding letter over each number.

Question		Letter	Answer
1. $5 \times 5 \times 5 \times 5$	______	A	100 000
2. 2^3	______	P	5^6
3. $\frac{3^6}{3^2}$	______	S	0
4. $4 \times 4 \times 4 \times 4 \times 4$	______	E	1
5. $(-2)^3$	______	F	3^4
6. $(-2) + 4 \div 2$	______	G	6
7. $(5^2)^3$	______	I	8
8. $3^2 - 2^3$	______	O	4^6
9. $10^2 \times 10^3$	______	N	4^5
10. $5 + 3^0$	______	R	5^4
11. $4^7 \div 4$	______	Y	-8

__
9

__ __ __ __ __ __
7 8 1 6 11 4

__ __ __ __ __ __
3 1 5 2 4 10

__ __
9 4

__ __ __
8 10 10

Unit 2 Study Guide

Skill	Description	Example
Evaluate a power with an integer base.	Write the power as repeated multiplication, then evaluate.	$(-2)^3 = (-2) \times (-2) \times (-2)$ $= -8$
Evaluate a power with an exponent 0.	A power with an integer base and an exponent 0 is equal to 1.	$8^0 = 1$
Use the order of operations to evaluate expressions containing exponents.	Evaluate what is inside the brackets. Evaluate powers. Multiply and divide, in order, from left to right. Add and subtract, in order, from left to right.	$(3^2 + 2) \times (-5)$ $= (9 + 2) \times (-5)$ $= (11) \times (-5)$ $= -55$
Apply the exponent law for a product of powers.	To multiply powers with the same base, add the exponents.	$4^3 \times 4^6 = 4^{3+6}$ $= 4^9$
Apply the exponent law for a quotient of powers.	To divide powers with the same base, subtract the exponents.	$2^7 \div 2^4 = \frac{2^7}{2^4}$ $= 2^{7-4}$ $= 2^3$
Apply the exponent law for a power of a power.	To raise a power to a power, multiply the exponents.	$(5^3)^2 = 5^{3 \times 2}$ $= 5^6$
Apply the exponent law for a power of a product.	Write the power of a product as a product of powers.	$(6 \times 3)^5 = 6^5 \times 3^5$
Apply the exponent law for a power of a quotient.	Write the power of a quotient as a quotient of powers.	$\left(\frac{3}{4}\right)^2 = \frac{3^2}{4^2}$

Unit 2 Review

2.1 **1.** Give the base and exponent of each power.

a) 6^2 Base _____ Exponent _____

b) $(-3)^8$ Base _____ Exponent _____

2. Write as a power.

a) $4 \times 4 \times 4 = 4$—

b) $(-3)(-3)(-3)(-3)(-3) =$ _______

3. Write each power as repeated multiplication and in standard form.

a) $(-2)^5 =$ ______________________
$=$ ______________________

b) $10^4 =$ ______________________
$=$ ______________________

c) Six squared $=$ __________
$=$ __________
$=$ __________

d) Five cubed $=$ __________
$=$ __________
$=$ __________

2.2 **4.** Evaluate.

a) $10^0 =$ _____

b) $(-4)^0 =$ _____

c) $8^1 =$ _____

d) $-4^0 =$ _____

5. Write each number in standard form.

a) 9×10^3
$= 9 \times$ ________ $\times$ ________ $\times$ ________
$= 9 \times$ ________
$=$ ___________

b) $(1 \times 10^2) + (3 \times 10^1) + (5 \times 10^0)$

$= (1 \times \_\_\_\_) + (3 \times \_\_\_\_) + (5 \times \_\_\_\_)$

$= \_\_\_\_\_\_\_\_\_\_\_\_$

$= \_\_\_\_\_\_\_\_\_\_\_\_$

c) $(2 \times 10^3) + (4 \times 10^2) + (1 \times 10^1) + (9 \times 10^0)$

$= (2 \times \_\_\_\_) + (4 \times \_\_\_\_) + (1 \times \_\_\_\_) + (9 \times \_\_\_\_)$

$= \_\_\_\_\_\_\_\_\_\_\_\_$

$= \_\_\_\_\_\_\_\_\_\_\_\_$

d) $(5 \times 10^4) + (3 \times 10^2) + (7 \times 10^1) + (2 \times 10^0)$

$= \_\_\_\_\_\_\_\_\_\_\_\_\_\_\_\_\_\_\_\_$

$= \_\_\_\_\_\_\_\_\_\_\_\_\_\_\_\_\_\_\_\_$

$= \_\_\_\_\_\_\_\_\_\_\_\_\_\_\_\_\_\_\_\_$

2.3 **6.** Evaluate.

a) $3^2 + 3$

$= \_\_\_\_ + 3$

$= \_\_\_\_$

b) $[(-2) + 4)]^3$

$= \_\_\_\_^3$

$= \_\_\_\_$

c) $(20 + 5) \div 5^2 = \_\_\_\_ \div 5^2$

$= \_\_\_\_ \div \_\_\_\_$

$= \_\_\_\_$

d) $(8^2 - 4) \div (6^2 - 6)$

$= (\_\_\_\_ - 4) \div (\_\_\_\_ - 6)$

$= \_\_\_\_ \div \_\_\_\_$

$= \_\_\_\_$

7. Evaluate.

a) $5 \times 3^2 = 5 \times \_\_\_\_$

$= \_\_\_\_$

b) $10 \times (3^2 + 5^0) = 10 \times \_\_\_\_\_\_$

$= 10 \times \_\_\_\_$

$= \_\_\_\_$

c) $(-2)^3 + (-3)(4) = \_\_\_\_ + \_\_\_\_$

$= \_\_\_\_$

d) $(-3) + 4^0 \times (-3) = (-3) + \_\_\_\_ \times (-3)$

$= (-3) + \_\_\_\_$

$= \_\_\_\_$

2.4 **8.** Write as a power.

a) $6^3 \times 6^7 = 6^{(\_\_ + \_\_)}$
$= 6^{\_\_}$

b) $(-4)^2 \times (-4)^3 = (-4)^{\_\_\_}$
$= (-4)^{\_\_}$

c) $(-2)^5 \times (-2)^4 = (-2)^{\_\_\_}$
$= (-2)^{\_\_}$

d) $10^7 \times 10 =$ __________
$=$ __________

9. Write as a power.

a) $5^7 \div 5^3 = 5^{(\_\_ - \_\_)}$
$= 5^{\_\_}$

b) $\frac{10^5}{10^3} =$ __________
$=$ __________

c) $(-6)^8 \div (-6)^2 =$ __________
$=$ __________

d) $\frac{5^{10}}{5^6} =$ __________
$=$ __________

e) $8^3 \div 8 =$ __________
$=$ __________

f) $\frac{(-3)^4}{(-3)^0} =$ __________
$=$ __________

2.5 **10.** Write as a power.

a) $(5^3)^4 = 5^{\_\_ \times \_\_}$
$= 5^{\_\_}$

b) $[(-3)^2]^6 = (-3)^{\_\_ \times \_\_}$
$= (-3)^{\_\_}$

c) $(8^2)^4 =$ __________
$=$ __________

d) $[(-5)^5]^4 =$ __________
$=$ __________

11. Write as a product or quotient of powers.

a) $(3 \times 5)^2 = 3^{\_\_} \times 5^{\_\_}$

b) $(2 \times 10)^5 =$ ____________

c) $[(-4) \times (-5)]^3 =$ ____________

d) $\left(\frac{4}{3}\right)^5 = \frac{\_\_}{\_\_}$

e) $(12 \div 10)^4 = 12^{\_\_} \div 10^{\_\_}$

f) $[(-7) \div (-9)]^6 =$ ____________

UNIT 3

Rational Numbers

What You'll Learn

How to

- Identify positive and negative decimals and fractions as rational numbers
- Compare and order rational numbers
- Add, subtract, multiply, and divide rational numbers
- Solve problems that involve rational numbers
- Apply the order of operations with rational numbers

Why It's Important

Rational numbers are used by

- building contractors to measure and to estimate costs
- chefs to measure ingredients, plan menus, and estimate costs
- investment professionals to show changes in stock prices

Key Words

fraction
equivalent fraction
numerator
denominator
common denominator
multiple
common multiple
integer
decimal
repeating decimal
terminating decimal
rational number
reciprocal

3.1 Skill Builder

Equivalent Fractions

$\frac{1}{2}$, $\frac{2}{4}$, $\frac{3}{6}$, and $\frac{4}{8}$ are **equivalent fractions.**

They represent the same distance on a number line.

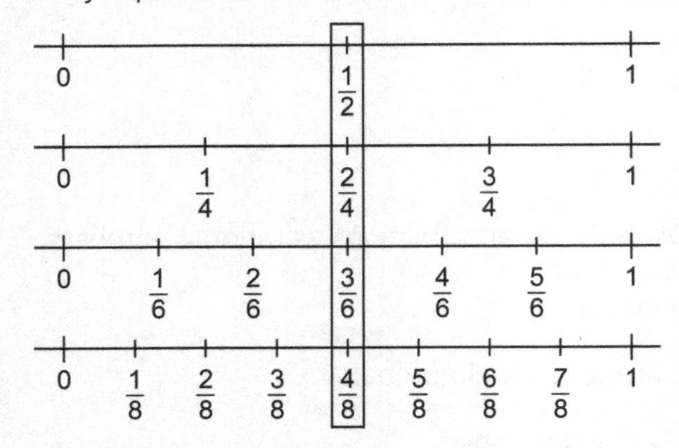

Here is one way to find equivalent fractions. Multiply or divide the numerator and denominator by the same number.

$\frac{1}{2} \overset{\times 3}{=} \frac{3}{6}$ $\frac{4}{8} \overset{\div 2}{=} \frac{2}{4}$

Multiplying or dividing both the numerator and denominator by the same number is like multiplying or dividing by 1. The original quantity is unchanged.

Check

1. Write 2 equivalent fractions.

a) $\frac{7}{10}$ $\frac{7}{10} \overset{\times 2}{=} \frac{\_\_}{\_\_}$ $\frac{7}{10} = \frac{\_\_}{\_\_}$

b) $\frac{12}{15}$ $\frac{12}{15} = \frac{\_\_}{\_\_}$ $\frac{12}{15} = \frac{\_\_}{\_\_}$

2. Write an equivalent fraction with the given denominator.

a) $\frac{3}{5} = \frac{\_\_}{20}$ 5 × 4 = 20, so multiply the numerator and denominator by 4.

b) $\frac{1}{4} = \frac{\_\_}{12}$

c) $\frac{\_\_}{15} = \frac{2}{3}$

d) $\frac{\_\_}{24} = \frac{5}{6}$

Comparing Fractions

Here are 3 ways to compare $\frac{3}{4}$ and $\frac{5}{8}$.

- Using area models:

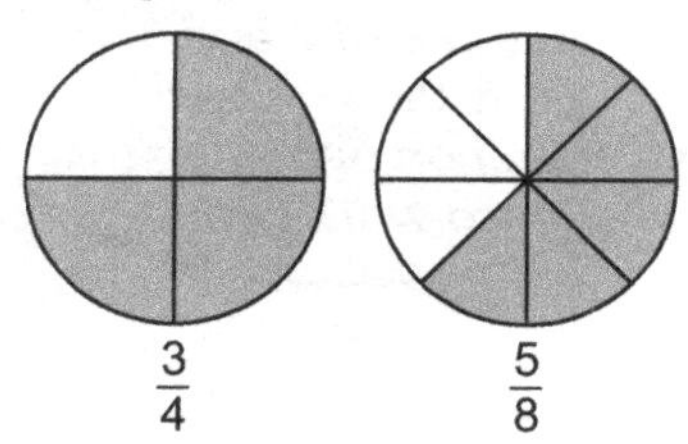

Compare the shaded areas: $\frac{3}{4} > \frac{5}{8}$

- Using number lines:

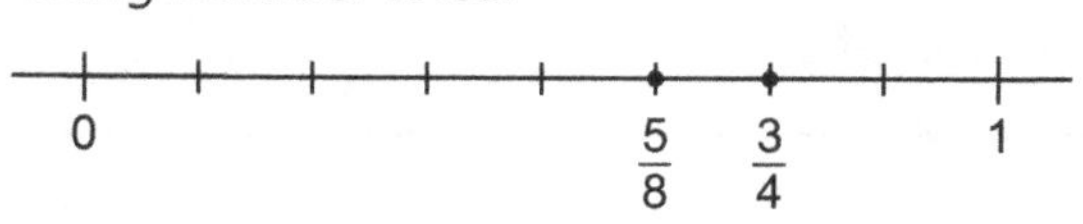

Numbers increase from left to right on a number line.

From the number line: $\frac{5}{8} < \frac{3}{4}$

- Writing equivalent fractions:

$$\frac{3}{4} \overset{\times 2}{=} \frac{6}{8}$$

$\frac{5}{8} < \frac{6}{8}$, so, $\frac{5}{8} < \frac{3}{4}$

Check

Compare the fractions in each pair. Write >, <, or =.

1. a) $\frac{7}{8}$ — $\frac{3}{4}$

b) $\frac{3}{5}$ — $\frac{7}{10}$

c) $\frac{7}{12}$ — $\frac{2}{3}$

d) $\frac{6}{7}$ — $\frac{6}{8}$

2. a) $\frac{2}{5}$ — $\frac{3}{10}$

0 ——— 1

b) $\frac{3}{5}$ — $\frac{9}{10}$

0 ——— 1

Common Denominators

To find a common denominator of $\frac{1}{2}$ and $\frac{2}{3}$:
Look for equivalent fractions with the same denominator.

List the multiples of 2: 2, 4, **6**, 8, 10, 12, 14, …

List the multiples of 3: 3, **6**, 9, 12, 15, …

6 is the least common multiple of 2 and 3. It is the simplest common denominator to work with.

Rewrite $\frac{1}{2}$ and $\frac{2}{3}$ with denominator 6.

$\frac{1}{2} \overset{\times 3}{=} \frac{3}{6}$ $\quad$ $\frac{2}{3} \overset{\times 2}{=} \frac{4}{6}$

Equivalent fractions help us compare, add, or subtract fractions.

Check

1. Write equivalent fraction pairs with a common denominator.

a) $\frac{1}{2}$ and $\frac{3}{8}$

Multiples of 2: 2, 4, 6, 8, 10, …

Multiples of 8: 8, 16, …

A common denominator is ___.

So, $\frac{1}{2} = \frac{\_\_}{\_\_}$ and $\frac{3}{8} = \frac{\_\_}{\_\_}$

b) $\frac{3}{4}$ and $\frac{5}{6}$

Multiples of 4: ______________

Multiples of 6: ______________

____

So, $\frac{3}{4} = \frac{\_\_}{\_\_}$ and $\frac{5}{6} = \frac{\_\_}{\_\_}$

c) $\frac{3}{5}$ and $\frac{2}{3}$

Multiples of ___: ______________

Multiples of ___: ______________

____

So, $\frac{3}{5} = \frac{\_\_}{\_\_}$ and $\frac{2}{3} = \frac{\_\_}{\_\_}$

2. Compare each pair of fractions from question 1.

a) $\frac{1}{2}$ and $\frac{3}{8}$. Since ___ > ___, $\frac{1}{2}$ ___ $\frac{3}{8}$

b) $\frac{3}{4}$ and $\frac{5}{6}$. Since ___ < ___, $\frac{3}{4}$ ___ $\frac{5}{6}$

c) $\frac{3}{5}$ and $\frac{2}{3}$. Since ___ < ___, $\frac{3}{5}$ ___ $\frac{2}{3}$

Converting between Fractions and Decimals

- Fractions to decimals

The fraction bar represents division. For example:

$\frac{1}{6}$ means $1 \div 6$

Use a calculator:

$1 \div 6 = 0.166\ 666\ldots$

$= 0.1\overline{6}$

So, $\frac{1}{6} = 0.1\overline{6}$

The bar over the 6 means that 6 repeats.

$0.1\overline{6}$ is a **repeating decimal.**

$\frac{7}{8}$ means $7 \div 8$

Use a calculator:

$7 \div 8 = 0.875$

So, $\frac{7}{8} = 0.875$

0.875 is a **terminating decimal.**

- Decimals to fractions

Use place value. For example:

0.7 means 7 tenths.

So, $0.7 = \frac{7}{10}$

0.23 means 23 hundredths

So, $0.23 = \frac{23}{100}$

Check

1. Write each fraction as a decimal.

a) $\frac{3}{4} = 3 \div 4$

$=$ ________

b) $\frac{2}{3} =$ ________

$=$ ________

c) $\frac{5}{8} =$ ________

$=$ ________

d) $\frac{5}{9} = 5 \div 9$

$=$ ________

e) $4\frac{1}{5} = 4 + \frac{1}{5}$

$= 4 +$ ________

$= 4 +$ ________

$=$ ________

f) $2\frac{1}{3} = 2 +$ ________

$= 2 +$ ________

$= 2 +$ ________

$=$ ________

2. Which numbers in question 1 are:

a) repeating decimals? ________

b) terminating decimals? ________

3. Write each decimal as a fraction.

a) 0.3 = ________

b) 0.9 = ________

c) 0.11 = ________

d) 0.87 = ________

e) 1.5 = ________

f) 5.7 = ________

3.1 What Is a Rational Number?

FOCUS **Compare and order rational numbers.**

Rational numbers include:

- integers
- positive and negative fractions
- positive and negative mixed numbers
- repeating and terminating decimals

Here is a number line that displays some rational numbers.

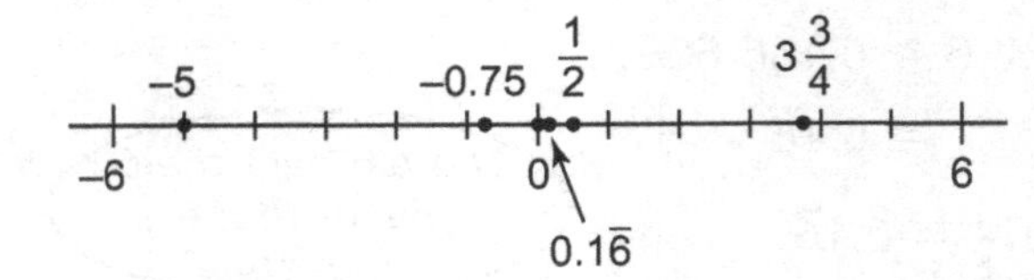

Example 1 Finding a Rational Number between Two Given Numbers

Find 2 rational numbers between $2\frac{1}{3}$ and $3\frac{3}{4}$.

Solution

Label a number line from 2 to 4.

$2\frac{1}{3}$ is one-third of the way from 2 to 3.

$3\frac{3}{4}$ is three-quarters of the way from 3 to 4.

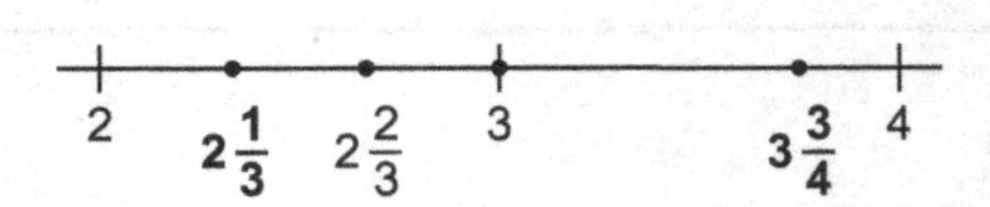

From the number line, 2 rational numbers between $2\frac{1}{3}$ and $3\frac{3}{4}$ are: $2\frac{2}{3}$ and 3

There are many correct solutions. Which ones can you name?

Check

1. Find 2 rational numbers between each pair of numbers.

a) $-2\frac{1}{3}$ and $-1\frac{2}{5}$

Plot points to show $-1\frac{2}{5}$ and $-2\frac{1}{3}$.

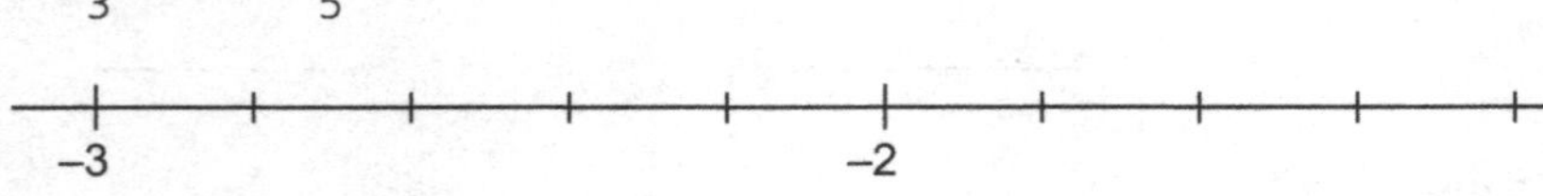

From the number line, 2 values between $-2\frac{1}{3}$ and $-1\frac{2}{5}$ are: _____ and _____

b) −0.3 and 0.6

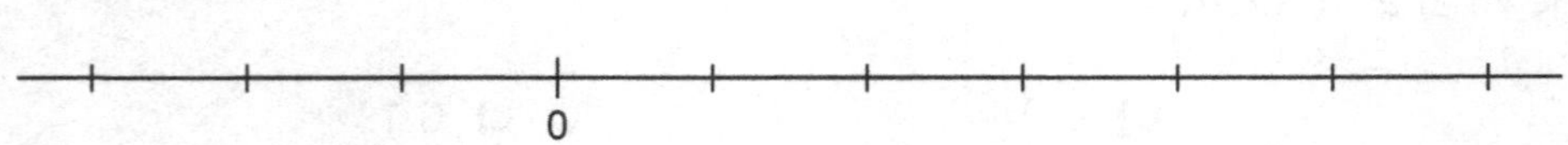

From the number line, 2 values between −0.3 and 0.6 are: _____ and _____

Example 2 Comparing Rational Numbers on a Number Line

Order each set of rational numbers from least to greatest.

a) $0.3, 0.\overline{3}, -1.7, 0.6, -0.6$

b) $3\frac{1}{4}, -\frac{3}{4}, -\frac{4}{8}, 1\frac{3}{4}, -2\frac{3}{8}$

Solution

a) Plot the numbers on a number line.
To plot 0.3 and $0.\overline{3}$, think: $0.\overline{3} = 0.3333...$
So, $0.\overline{3}$ is slightly greater than 0.3.

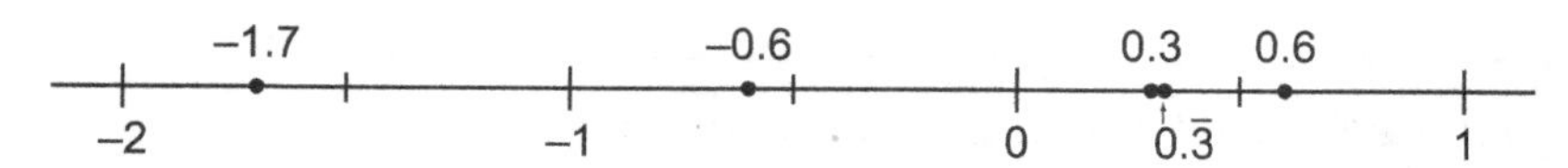

From the number line, the order from least to greatest is: $-1.7, -0.6, 0.3, 0.\overline{3}, 0.6$

b) Plot the numbers on a number line.

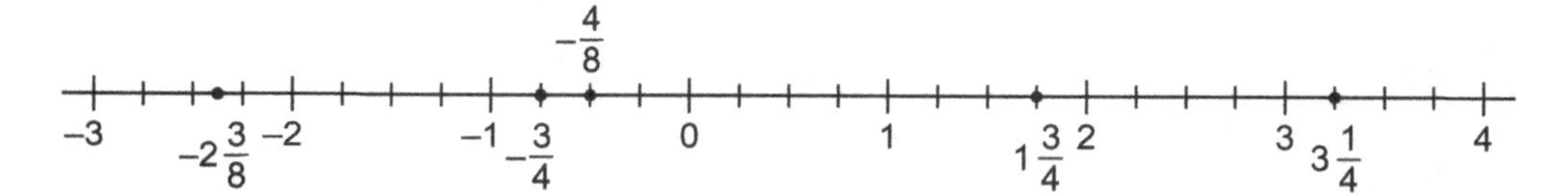

From the number line, the order from least to greatest is: $-2\frac{3}{8}, -\frac{3}{4}, -\frac{4}{8}, 1\frac{3}{4}, 3\frac{1}{4}$

Check

1. Order each set of numbers from least to greatest.

a) $-1.\overline{8}, 0.7, -2, -2.1, -0.3$

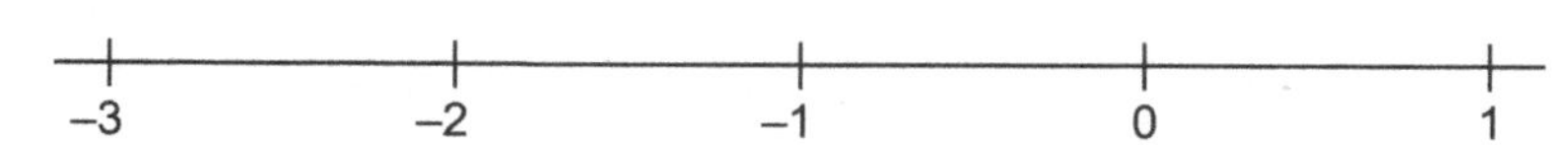

From the number line, the order from least to greatest is: ______________________

b) $-1\frac{9}{10}, -2, -1\frac{4}{5}, \frac{4}{5}, -1\frac{1}{5}$

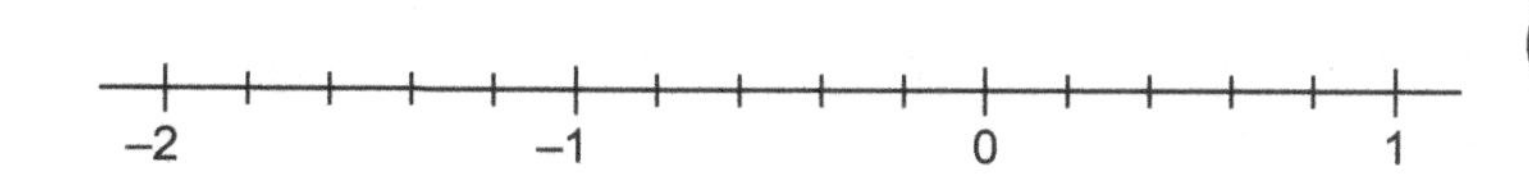

The number line is divided in fifths to help you plot the numbers.

From the number line, the order from least to greatest is: ______________________

Practice

1. Write each rational number as a decimal.

a) $\frac{3}{5} = \_\_ \div \_\_$

$= \_\_\_\_\_\_$

b) $\frac{5}{3} = \_\_\_\_\_\_$

$= \_\_\_\_\_\_$

c) $-\frac{3}{5} = -(\_\_ \div \_\_)$

$= \_\_\_\_\_\_$

d) $\frac{-3}{5} = (\_\_\_) \div \_\_$

$= \_\_\_\_\_\_$

e) $\frac{-5}{3} = (\_\_\_) \div \_\_$

$= \_\_\_\_\_\_$

f) $\frac{3}{-5} = \_\_\_\_\_\_$

$= \_\_\_\_\_\_$

Look for matching answers. What conclusion can you make?

____

2. Plot and compare each pair of rational numbers.

a) $4\frac{2}{5}$ and $4\frac{3}{5}$

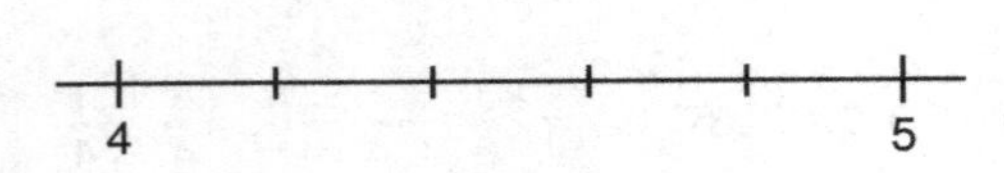

From the number line, $4\frac{2}{5}$ __ $4\frac{3}{5}$

b) $\frac{2}{3}$ and $-\frac{1}{3}$

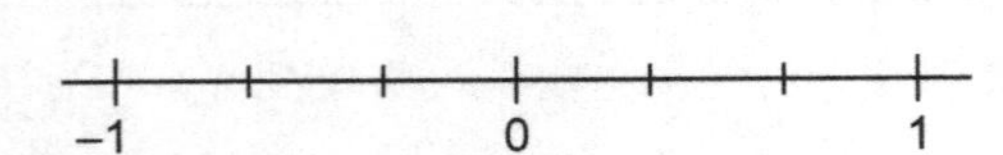

From the number line, ______

c) $-5\frac{5}{6}$ and $-5\frac{1}{6}$

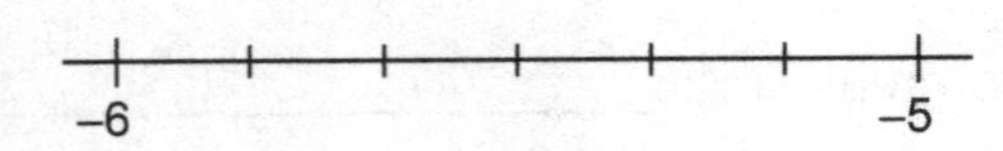

____

3. a) Write a decimal to match each point on the number line.

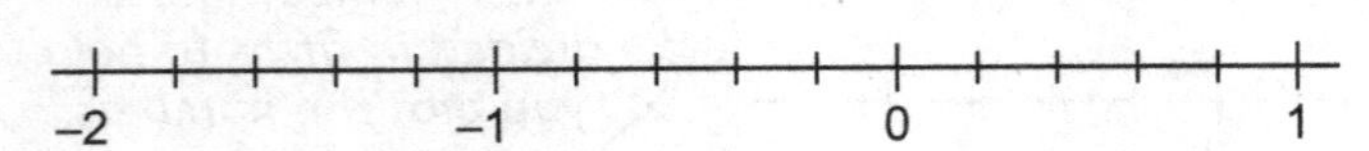

b) Write the numbers in part a from least to greatest.

____

4. Find 2 rational numbers between each pair of numbers.

a) -2.1 and -1.7

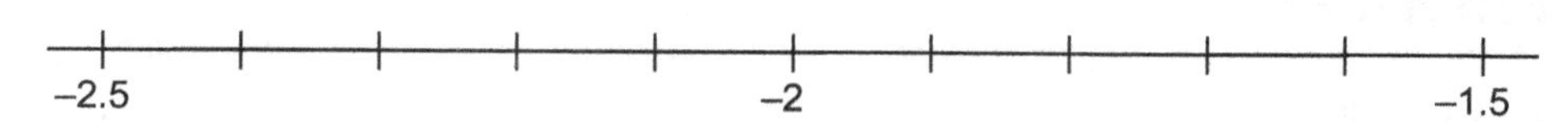

Two possible numbers are: ______________

b) 4.1 and 4.4

Start by plotting the given values on the number line.

Two possible numbers are: ______________

c) $-1\frac{3}{5}$ and $-2\frac{1}{5}$

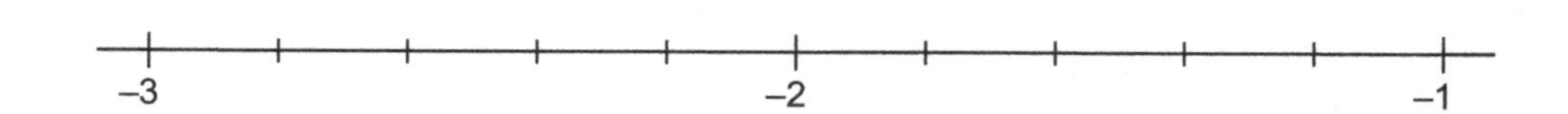

____

5. Order these rational numbers from least to greatest.
$-1\frac{1}{2}, \frac{3}{2}, -1.7, -2, \frac{3}{4}$

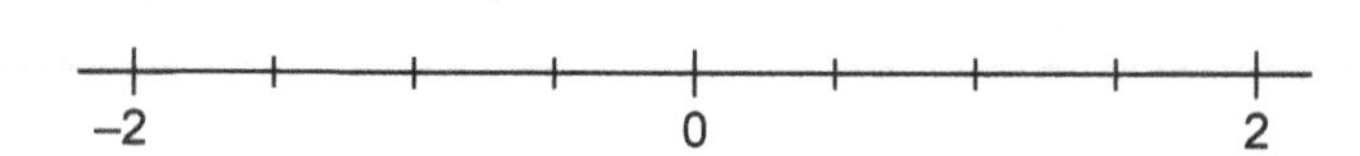

Estimate to place numbers where necessary.

From least to greatest: ____________________

6. Kiki recorded the temperatures at the same time each day over a 5-day period.

-0.8°C, -1.3°C, 2.4°C, -1.5°C, 0.9°C

Order the temperatures from lowest to highest:

____

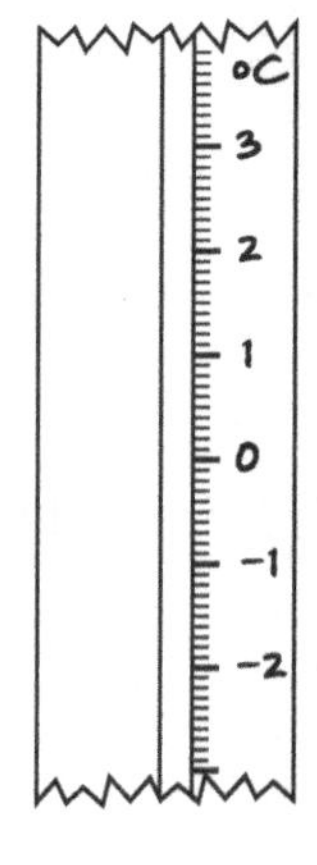

3.2 Skill Builder

Adding Fractions

Here are 2 ways to add $\frac{1}{3}$ and $\frac{1}{6}$.

- Using fraction strips on a number line:
  Place the fraction strips end to end, starting at 0.

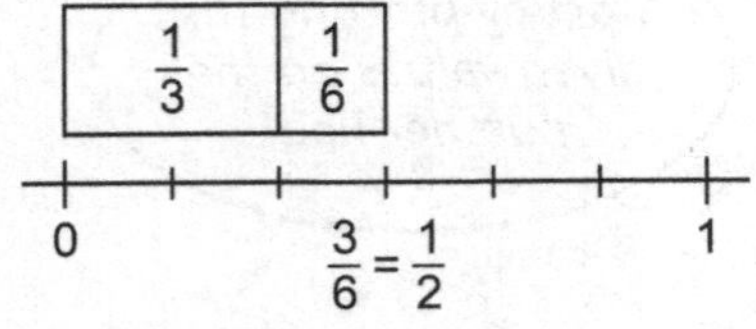

From the number line: $\frac{1}{3} + \frac{1}{6} = \frac{3}{6}$, or $\frac{1}{2}$

- Using common denominators:
  $\frac{1}{3}$ is the same as $\frac{2}{6}$.

  So, $\frac{1}{3} + \frac{1}{6} = \frac{2}{6} + \frac{1}{6}$

  $= \frac{3}{6}$, or $\frac{1}{2}$

Some additions give answers that are greater than 1.

$\frac{2}{3} + \frac{1}{2} = \frac{4}{6} + \frac{3}{6}$

$= \frac{7}{6}$ ⟵ improper fraction

$= 1\frac{1}{6}$ ⟵ mixed number

Rewrite the improper fraction as a mixed number: divide 6 into 7 to see that there is 1 whole, and 1 sixth left over.

Check

1. Find each sum. Use diagrams to show your thinking.

a) $\frac{1}{6} + \frac{4}{6} =$ ____ **b)** $\frac{1}{3} + \frac{1}{2} =$ ____

2. Find each sum. Use the method you like best.

a) $\frac{2}{5} + \frac{4}{5} =$ ____, or ____ **b)** $\frac{2}{4} + \frac{5}{8} =$ ____

$=$ ____, or ____

Adding Mixed Numbers

Mixed numbers combine whole numbers and fractions.

Add: $1\frac{1}{8} + 3\frac{3}{4}$

We can add numbers in any order without changing the answer.

Add the whole numbers and add the fractions.

$1\frac{1}{8} + 3\frac{3}{4} = 1 + 3 + \frac{1}{8} + \frac{3}{4}$ A common denominator is 8.

$= 1 + 3 + \frac{1}{8} + \frac{6}{8}$

$= 4 + \frac{7}{8}$

$= 4\frac{7}{8}$

Check

1. Find each sum. Use diagrams to show your thinking.

a) $1\frac{1}{3} + 1\frac{2}{3} =$ ____________

b) $2\frac{1}{6} + \frac{1}{2} =$ ____________

2. Find each sum.
Use the method you like best.

a) $3\frac{2}{7} + 2\frac{3}{7} =$ ________________
= ________________
= ________________

b) $4\frac{1}{9} + 1\frac{2}{3} =$ ________________
= ________________
= ________________
= ________________

3.2 Adding Rational Numbers

FOCUS **Solve problems by adding rational numbers.**

Integers and fractions are rational numbers.
So, you can use strategies for adding integers, and strategies for adding fractions, to add rational numbers.

Example 1 Adding Rational Numbers on a Number Line

a) $-2.3 + (-1.9)$

b) $-\frac{1}{2} + \left(-\frac{5}{4}\right)$

Solution

a) $-2.3 + (-1.9)$
Use a number line divided in tenths.
Start at -2.3. To add -1.9, move 1.9 to the left.

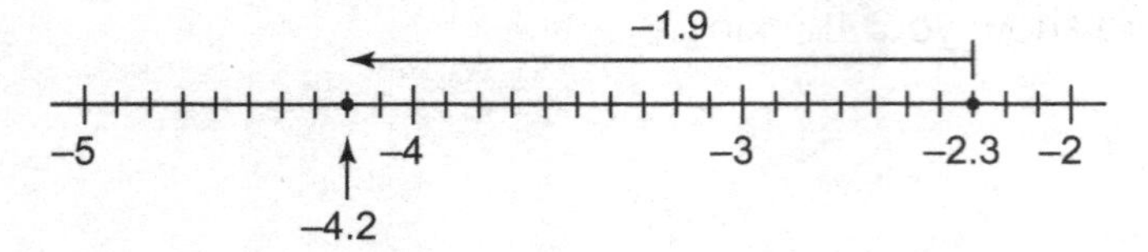

When we add a negative number, we move to the left. When we add a positive number, we move to the right.

So, $-2.3 + (-1.9) = -4.2$.

b) $-\frac{1}{2} + \left(-\frac{5}{4}\right)$
Use a number line divided into fourths.
Start at $-\frac{1}{2}$. To add $-\frac{5}{4}$, move $\frac{5}{4}$ to the left.

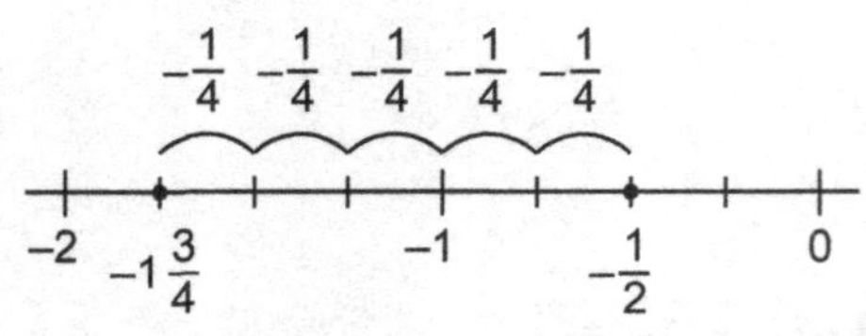

So, $-\frac{1}{2} + \left(-\frac{5}{4}\right) = -1\frac{3}{4}$.

Check

1. Use a number line to add.

a) $-4.5 + 2.3 =$ ________

b) $-\frac{1}{3} + \left(-\frac{7}{3}\right) =$ ________

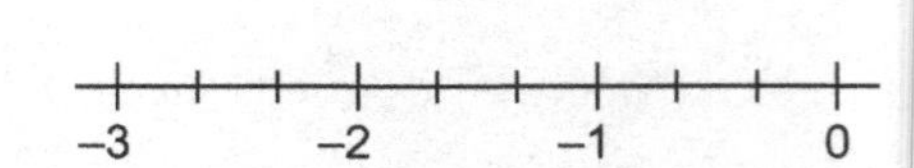

c) $\frac{3}{8} + \left(-\frac{3}{4}\right) =$ ______

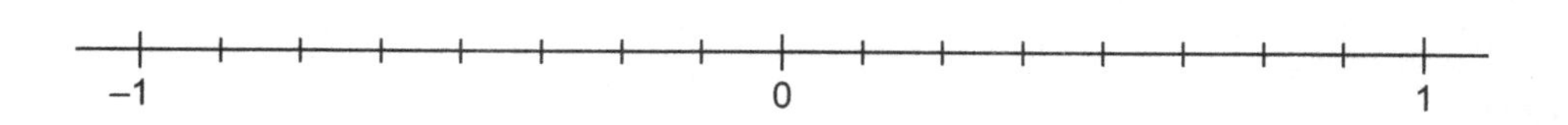

Example 2 Adding Fractions without a Number Line

Add: $-\frac{2}{5} + \left(-\frac{1}{2}\right)$

Solution

To find $-\frac{2}{5} + \left(-\frac{1}{2}\right)$, look for a common denominator.

Use a common denominator of 10.

Multiples of 5: 5, 10, 15, …
Multiples of 2: 2, 4, 6, 8, 10, …

$-\frac{2}{5} \overset{\times 2}{=} -\frac{4}{10}$ and $-\frac{1}{2} \overset{\times 5}{=} -\frac{5}{10}$

So, $-\frac{2}{5} + \left(-\frac{1}{2}\right) = -\frac{4}{10} + \left(-\frac{5}{10}\right)$

$= -\frac{9}{10}$

Think of integer addition: $(-4) + (-5) = -9$

Check

1. Add.

a) $-\frac{7}{12} + \frac{1}{6}$ Use a common denominator of ____.

$\frac{1}{6} = \frac{\_\_}{\_\_}$

$= -\frac{7}{12} + \frac{\_\_}{\_\_}$

$= \frac{\_\_}{\_\_}$

b) $\frac{3}{5} + \left(-\frac{2}{3}\right)$ Use a common denominator of ____.

$\frac{3}{5} = \frac{\_\_}{\_\_}$ and $-\frac{2}{3} = \frac{\_\_}{\_\_}$

$= \frac{\_\_}{\_\_} + \frac{\_\_}{\_\_}$

$= \frac{\_\_}{\_\_}$

Example 3 Adding Mixed Numbers

Calculate: $-2\frac{1}{8} + 3\frac{1}{3}$

Solution

Estimate first to predict the answer:

$-2\frac{1}{8} + 3\frac{1}{3}$ is about $-2 + 3$, or 1.

We expect an answer close to 1.

To calculate, add the whole numbers and add the fractions.
Keep the signs with each part of the mixed number.

$-2\frac{1}{8} + 3\frac{1}{3} = (-2) + 3 + \left(-\frac{1}{8}\right) + \frac{1}{3}$ Use a common denominator of 24.

$-\frac{1}{8} \overset{\times 3}{=} -\frac{3}{24}$ and $\frac{1}{3} \overset{\times 8}{=} \frac{8}{24}$

So, $-2\frac{1}{8} + 3\frac{1}{3} = (-2) + 3 + \left(-\frac{3}{24}\right) + \frac{8}{24}$

$= 1 + \frac{5}{24}$

$= 1\frac{5}{24}$

Check: the answer is reasonably close to the original estimate of 1.

Check

1. Find each sum.

a) $-1\frac{5}{16} + 3\frac{3}{8}$ = ____ + ____ + ____ + ____ Use a common denominator of ____.

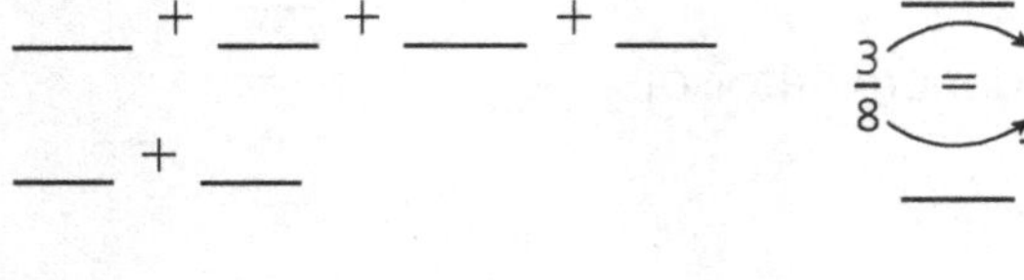

= ____ + ____ + ____ + ____ $\frac{3}{8}$ = ____

= ____ + ____

= ____

Estimate to check if your answer is reasonable.

b) $2\frac{3}{5} + 1\frac{1}{4}$ = ___ + ___ + ___ + ___

= ___ + ___ + ___ + ___

= ___ + ___

= ___

Use a common denominator of ___.

$\frac{3}{5} = \frac{\_\_}{\_\_}$ and $\frac{1}{4} = \frac{\_\_}{\_\_}$

Practice

1. Write the addition statement shown by each number line.

a)

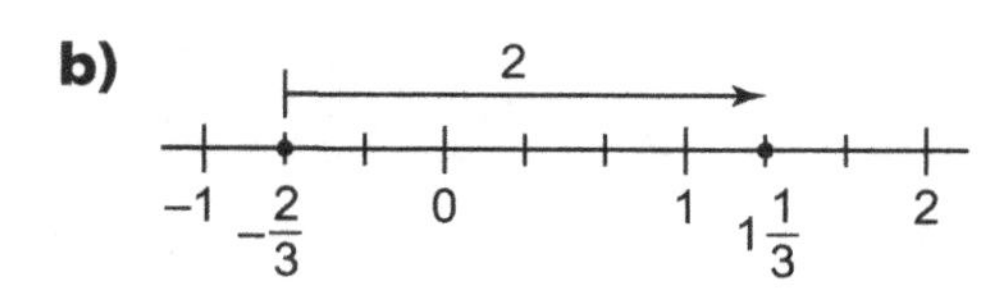

___ + (___) = ___

b)

2
−1 $-\frac{2}{3}$ 0 1 $1\frac{1}{3}$ 2

___

2. Use the number line to add.

a) $-4.5 + (1.2) =$ ___

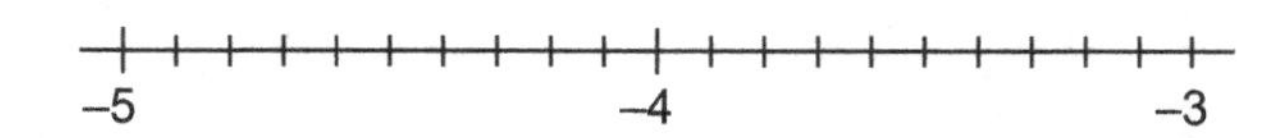

b) $1.7 + (-1.9) =$ ___

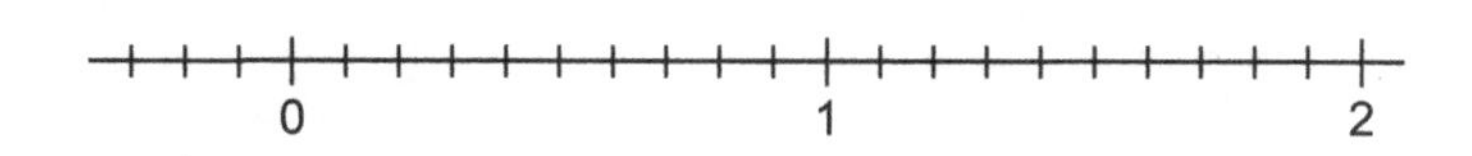

3. Add.

a) i) $4 + 6 =$ ___ **ii)** $4.1 + 6.4 =$ ___ **iii)** $\frac{4}{11} + \frac{6}{11} =$ ___

b) i) $4 + (-6) =$ ___ **ii)** $4.1 + (-6.4) =$ ___ **iii)** $\frac{4}{11} + \left(-\frac{6}{11}\right) =$ ___

c) i) $-4 + 6 =$ ___ **ii)** $-4.1 + 6.4 =$ ___ **iii)** $-\frac{4}{11} + \frac{6}{11} =$ ___

d) i) $-4 + (-6) =$ ___ **ii)** $-4.1 + (-6.4) =$ ___ **iii)** $-\frac{4}{11} + \left(-\frac{6}{11}\right) =$ ___

4. Find each sum.

a) $-4.6 + 5.8 =$ ____ **b)** $2.3 + (-4.6) =$ ______

c) $-0.3 + (-6.2) =$ ______ **d)** $(-26.5) + (-18.1) =$ _______

5. Find each sum.

a) $-\frac{1}{3} + \frac{5}{9}$

$= \underline{\quad\quad} + \frac{5}{9}$

$= \underline{\quad\quad}$

b) $\frac{1}{3} + \left(-\frac{2}{5}\right)$

$= \underline{\quad\quad\quad\quad\quad\quad}$

$= \underline{\quad\quad}$

c) $-\frac{3}{8} + \left(-\frac{1}{3}\right)$

$= \underline{\quad\quad\quad\quad\quad\quad}$

$= \underline{\quad\quad}$

6. Find each sum.

Look for a common denominator first.

a) $-2\frac{2}{5} + 6\frac{1}{2}$

____

____

____

____

b) $-1\frac{1}{6} + \left(-3\frac{1}{4}\right)$

____

____

____

____

c) $\left(-3\frac{1}{3}\right) + \left(-5\frac{1}{7}\right)$

____

____

____

____

3.3 Skill Builder

Converting Mixed Numbers to Improper Fractions

Here are 2 ways to write $2\frac{3}{8}$ as an improper fraction.

- Make a diagram to show $2\frac{3}{8}$.

  Count individual parts.

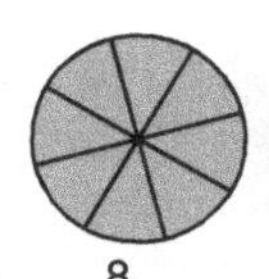
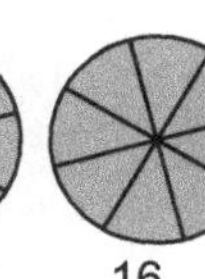
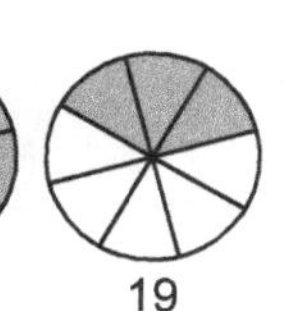

8... 16... 19

- Use calculation.

$$2\frac{3}{8} = \frac{2 \times 8 + 3}{8}$$
$$= \frac{19}{8}$$

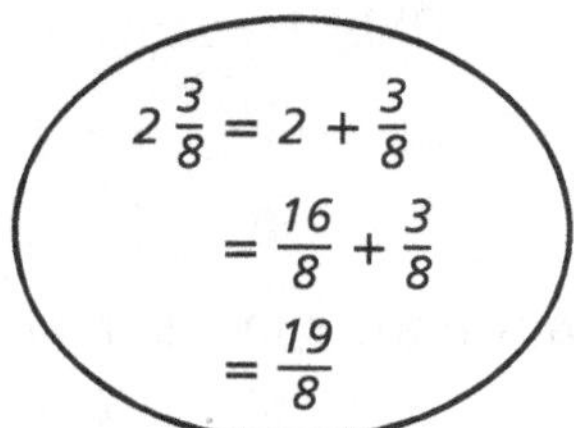

Think of the diagram above:

2 × 8 + 3

2 whole circles shaded	8 pieces in each circle	plus another 3 pieces

Check

1. Write a mixed number and an improper fraction to show each shaded quantity.

a) ___ or ___

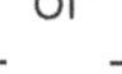
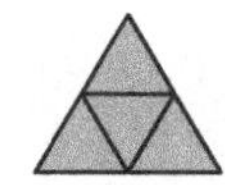
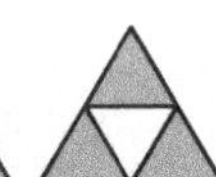
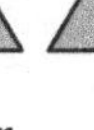

b) ___ or ___

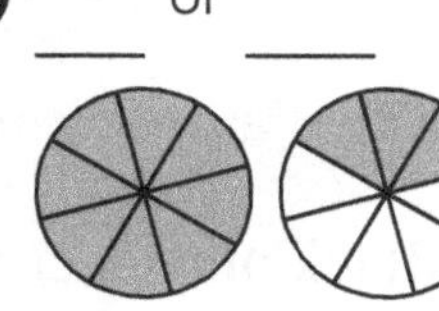

c) ___ or ___

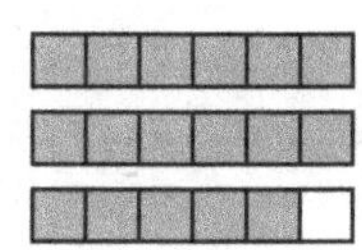

d) ___ or ___

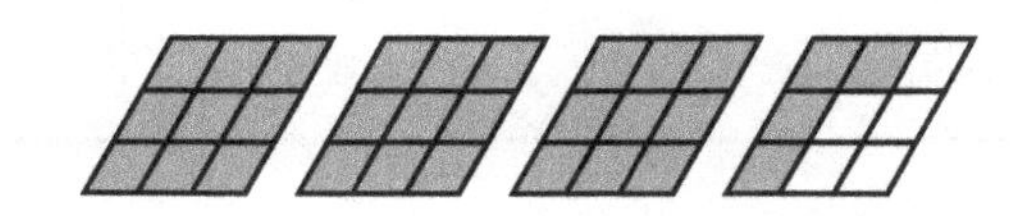

2. Write each mixed number as an improper fraction.

a) $1\frac{2}{5} = \_\_\_ + \frac{2}{5}$

$= \frac{\_\_}{5} + \frac{2}{5}$

$= \frac{\_\_}{5}$

b) $2\frac{2}{3} = \_\_\_ + \_\_\_$

$= \frac{\_\_}{3} + \_\_\_$

$= \_\_\_$

c) $5\frac{3}{4} = \_\_\_ + \_\_\_$

$= \frac{\_\_}{4} + \_\_\_$

$= \_\_\_$

3.3 Subtracting Rational Numbers

FOCUS **Solve problems by subtracting rational numbers.**

To subtract an integer, we add its opposite.

- $-5 - 2$ is the same as $-5 + (-2)$.

  So, $-5 - 2 = -5 + (-2)$
  $= -7$

- $-5 - (-2)$ is the same as $-5 + (+2)$

  So, $-5 - (-2) = -5 + (+2)$
  $= -3$

We can use the same strategy to subtract rational numbers.

Subtracting Rational Numbers
To subtract a rational number, add its opposite.

Example 1 Subtracting Rational Numbers in Fraction Form

Subtract: $\frac{1}{3} - \frac{5}{6}$

Solution

$\frac{1}{3} - \frac{5}{6}$	Add the opposite.
$= \frac{1}{3} + \left(-\frac{5}{6}\right)$	Use 6 as a common denominator.
$= \frac{2}{6} + \left(-\frac{5}{6}\right)$	Think of integer addition: $2 + (-5) = -3$
$= -\frac{3}{6}$	Write the answer in simplest form.
$= -\frac{1}{2}$	

Check

1. Subtract.

a) $-\frac{1}{2} - \frac{7}{8} = -\frac{1}{2} + \left(-\frac{7}{8}\right)$

$= \_\_\_ + \left(-\frac{7}{8}\right)$

$= \_\_\_$

$= \_\_\_$

b) $\frac{4}{5} - \left(-\frac{2}{3}\right) = \_\_\_ + \_\_\_$

$= \_\_\_ + \_\_\_$

$= \_\_\_$

$= \_\_\_$

Example 2 Subtracting Rational Numbers in Mixed Number Form

Subtract: $\frac{3}{4} - 2\frac{5}{8}$

Solution

$\frac{3}{4} - 2\frac{5}{8}$ — Write $2\frac{5}{8}$ as an improper fraction.

$= \frac{3}{4} - \frac{21}{8}$ — Use 8 as a common denominator.

$= \frac{6}{8} - \frac{21}{8}$ — Add the opposite.

$= \frac{6}{8} + \left(-\frac{21}{8}\right)$

$= -\frac{15}{8}$, or $-1\frac{7}{8}$

Check

1. Find the difference.

a) $-\frac{13}{15} - 1\frac{1}{5}$ — Write $1\frac{1}{5}$ as an improper fraction.

$= -\frac{13}{15} - \frac{}{\_\_\_}$ — Use ____ as a common denominator.

$= -\frac{13}{15} - \frac{\_\_}{15}$ — Add the opposite.

$= -\frac{13}{15} + \left(-\frac{\_\_}{15}\right)$

$= \frac{}{\_\_\_}$ — Write the answer as a mixed number.

$= \_\_\_\_$

b) $-2\frac{3}{8} - 3\frac{1}{2}$ — Rewrite $-2\frac{3}{8}$ and $3\frac{1}{2}$ as improper fractions.

$= \frac{}{\_\_\_} - \frac{}{\_\_\_}$ — Use ___ as a common denominator.

$= \frac{}{\_\_\_} - \frac{}{\_\_\_}$ — Add the opposite.

$= \frac{}{\_\_\_} + \_\_\_\_$

$= \frac{}{\_\_\_}$ — Write the answer as a mixed number.

$= \_\_\_\_$

Example 3 Solving a Problem by Subtracting Rational Numbers

In Alberta:

- The lowest temperature ever recorded was −61.1°C at Fort Vermilion in 1911.
- The highest temperature was 43.3°C at Bassano Dams in 1931.

What is the difference between these temperatures?

Solution

Subtract to find the difference between the temperatures.

$43.3 - (-61.1)$ Add the opposite.

$= 43.3 + (61.1)$

$= 104.4$

The difference between the temperatures is 104.4°C.

Use mental math to check.
$40 + 60 = 100$
$3.3 + 1.1 = 4.4$
$100 + 4.4 = 104.4$

Check

1. The lowest temperature ever recorded on Earth was −89.2°C in Antarctica.
The highest temperature ever recorded is 57.8°C in Libya.
What is the difference between these temperatures?

_____ − (_____) = _____ + (_____)

= _____

The difference between the temperatures is _____°C.

Practice

1. Subtract.

a) $1.6 - 3.9 =$ ______ **b)** $1.6 - (-3.9) =$ ______

c) $-2.4 - 4.5 =$ ______ **d)** $2.4 - (-4.5) =$ ______

2. Draw lines to join matching subtraction sentences, addition sentences, and answers.

Subtraction sentence	Addition sentence	Answer
$2.7 - 9.7$	$2.7 + 9.7$	-12.4
$-2.7 - 9.7$	$2.7 + (-9.7)$	-7
$-2.7 - (-9.7)$	$-2.7 + (-9.7)$	7
$2.7 - (-9.7)$	$-2.7 + 9.7$	12.4

3. Find each difference.

a) $7.1 - 4.7 =$ ______ **b)** $-3.2 - 1.9 =$ ______

c) $26.2 - (-8.4) =$ ______ **d)** $(-8.6) - (-7.2) =$ ______

Estimate to check if your answers are reasonable.

4. Subtract.

a) i) $6 - 3 =$ ____ **ii)** $6.3 - 3.1 =$ ____ **iii)** $\frac{6}{7} - \frac{3}{7} =$ ____

b) i) $-6 - 3 =$ ____ **ii)** $-6.3 - 3.1 =$ ______ **iii)** $-\frac{6}{7} - \frac{3}{7} =$ ______

c) i) $6 - (-3) =$ ____ **ii)** $6.3 - (-3.1) =$ ____ **iii)** $\frac{6}{7} - \left(-\frac{3}{7}\right) =$ ______

d) i) $-6 - (-3) =$ ____ **ii)** $-6.3 - (-3.1) =$ ______ **iii)** $-\frac{6}{7} - \left(-\frac{3}{7}\right) =$ ____

5. Determine each difference.

a) $\frac{3}{5} - \left(-\frac{1}{3}\right) = \frac{3}{5} + \frac{1}{3}$

$= \frac{\quad}{\quad} + \frac{\quad}{\quad}$

$= \frac{\quad}{\quad}$

b) $-\frac{17}{20} - \frac{3}{2} = -\frac{17}{20} + \left(-\frac{3}{2}\right)$

$= -\frac{17}{20} + \frac{\quad}{\quad}$

$= \frac{\quad}{\quad}$

c) $\frac{9}{5} - \frac{7}{4} =$ ______

$= \frac{\quad}{\quad} + \frac{\quad}{\quad}$

$= \frac{\quad}{\quad}$

6. Calculate.

a) $2\frac{1}{6} - 1\frac{1}{3} = \frac{\quad}{6} - \frac{\quad}{3}$

$= \frac{\quad}{6} + \left(-\frac{\quad}{3}\right)$

$= \frac{\quad}{\quad} + \frac{\quad}{\quad}$

$= \frac{\quad}{\quad}$

b) $1\frac{1}{2} - \left(-2\frac{1}{3}\right) = \frac{\quad}{2} - \left(-\frac{\quad}{3}\right)$

$= \frac{\quad}{2} + \frac{\quad}{3}$

$= \frac{\quad}{\quad} + \frac{\quad}{\quad}$

$= \frac{\quad}{\quad}$

7. Jenny has a gift card with $24.50 left on it. She makes purchases totaling $42.35. What amount does Jenny still owe the cashier after using the gift card?

Subtraction sentence: ______ − ______ = ______

Jenny still owes the cashier $______.

Can you ...

- Compare and order rational numbers?
- Add and subtract rational numbers?
- Solve problems by adding and subtracting rational numbers?

3.1 **1.** Find 2 rational numbers between each pair of numbers.

a) $-1\frac{1}{3}$ and $\frac{1}{6}$

Plot each number on the number line.

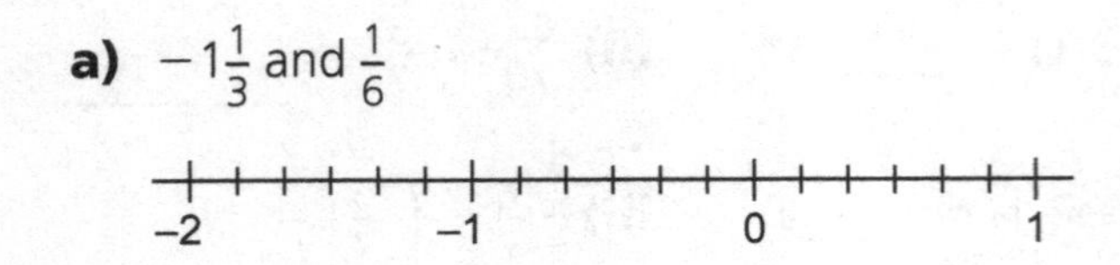

From the number line, 2 values between $-1\frac{1}{3}$ and $\frac{1}{6}$ are: ____ and ____

b) -0.4 and 0.2

–1 0 1

From the number line, 2 values between -0.4 and 0.2 are: ______ and ______

2. Use the number line to order the fractions from least to greatest: $-1\frac{2}{3}, \frac{7}{10}, -\frac{4}{5}$

–2 –1 0 1

For least to greatest, read the points from _____ to ______: ____________

3. a) Write each number as a decimal.

$-\frac{2}{5}$ = ________ $-1\frac{1}{2}$ = ________

$-\frac{5}{3}$ = ________ $-\frac{5}{2}$ = ________

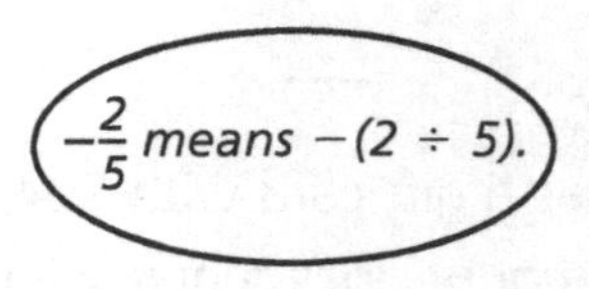

b) Order the decimals in part a from least to greatest.
Use the number line to help you.

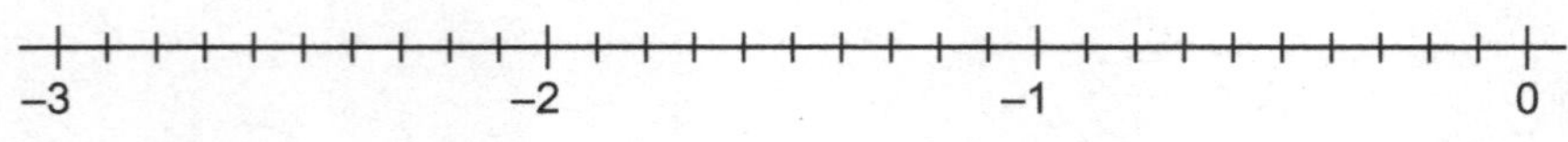

From least to greatest: ______________________________

3.2 **4.** Find each sum.

a) $6.5 + (-4.2) = \_\_\_\_$

b) $-13.6 + (-7.9) = \_\_\_\_$

5. Find each sum. Use equivalent fractions.

a) $-\frac{3}{8} + \frac{1}{4} = -\frac{3}{8} + \frac{\_\_}{}$

$= \_\_\_\_$

b) $\frac{3}{8} + \frac{1}{4} = \frac{\_\_}{} + \frac{\_\_}{}$

$= \_\_\_\_$

c) $-\frac{3}{8} + \left(-\frac{1}{4}\right) = \_\_\_\_$

$= \_\_\_\_$

d) $\frac{3}{8} + \left(-\frac{1}{4}\right) = \_\_\_\_$

$= \_\_\_\_$

6. Add.

a) $\frac{2}{3} + \left(-1\frac{4}{11}\right) = \frac{2}{3} + \left(-\frac{15}{11}\right)$

$= \frac{\_\_}{} + \frac{\_\_}{}$

$= \_\_\_\_$

b) $-1\frac{5}{6} + 3\frac{7}{8} = (\_\_ + \_\_) + \left(\frac{\_\_}{} + \frac{\_\_}{}\right)$

$= \_\_ + \left(\frac{\_\_}{} + \frac{\_\_}{}\right)$

$= \_\_\_\_$

3.3 **7.** Find each difference.

a) $7.6 - 4.2 = \_\_\_\_$

b) $-3.4 - 5.7 = \_\_\_\_$

c) $1.7 - (-9.3) = \_\_\_\_$

d) $-2.3 - (-5.6) = \_\_\_\_$

Estimate to check if your answers are reasonable.

8. Subtract.

a) $-\frac{5}{12} - \frac{1}{6} = -\frac{5}{12} + \frac{\_\_}{}$

$= -\frac{5}{12} + \frac{\_\_}{}$

$= \_\_\_\_$

b) $-2\frac{4}{7} - \left(-3\frac{1}{3}\right) = -2\frac{4}{7} + \_\_\_\_$

$= -\frac{\_\_}{7} + \frac{\_\_}{3}$

$= \frac{\_\_}{} + \frac{\_\_}{}$

$= \_\_\_\_$

9. The table shows Lesley's temperature readings at different times one day.

Time	Temperature (°C)
9:00 A.M.	−5.4
12:00 P.M.	1.3
3:00 P.M.	2.7
9:00 P.M.	−4.2

Find the change in temperature between each pair of given times.
Did the temperature rise or fall each time?

a) 9:00 A.M. and 12:00 P.M.

Change in temperature: $1.3 - (-5.4)$

= ____ + ____

= ____

The temperature ______ by ____°C.

b) 3:00 P.M. and 9:00 P.M.

Change in temperature: _____ − _____

= ______ + ______

= _____

The temperature ______ by ____°C.

c) 9:00 A.M. and 9:00 P.M.

Change in temperature: ______________

____

____

____

3.4 Skill Builder

Writing a Fraction in Simplest Form

A fraction is in simplest form when the only common factor of the numerator and denominator is 1. For example, $\frac{5}{6}$ is in simplest form.

Writing a Fraction in Simplest Form
Look for common factors of the numerator and denominator.
Divide the numerator and denominator by common factors until you cannot go any further.

Write $\frac{24}{30}$ in simplest form.

*Factors of 24: 1, 2, 3, 4, **6**, 8, 12, 24*
*Factors of 30: 1, 2, 3, 5, **6**, 10, 15, 30*

Divide the numerator and the denominator by 6.

$$\frac{24}{30} \overset{\div 6}{\underset{\div 6}{=}} \frac{4}{5}$$

$\frac{4}{5}$ is the simplest form of $\frac{24}{30}$.

Check

1. Write each fraction in simplest form.

a) $\frac{10}{15} \overset{\div 5}{\underset{\div 5}{=}} \frac{\_\_}{\_\_}$ Divide the numerator and the denominator by 5.

b) $\frac{14}{20} = \frac{\_\_}{\_\_}$ Divide the numerator and the denominator by _____.

c) 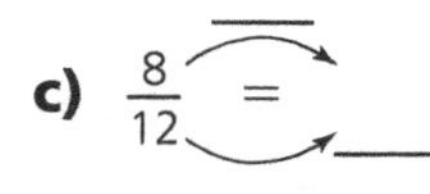 Divide the numerator and the denominator by _____.

d) 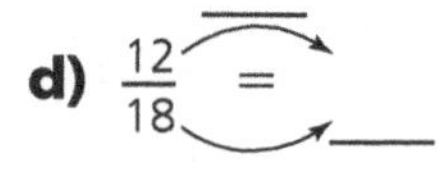 Divide the numerator and the denominator by _____.

Multiplying Proper Fractions

When multiplying fractions, we multiply the numerators, and we multiply the denominators.

$$\frac{2}{5} \times \frac{3}{8} = \frac{2 \times 3}{5 \times 8}$$
$$= \frac{6}{40}, \text{ or } \frac{3}{20}$$

To simplify, look for common factors *before* multiplying.

$$\frac{5}{12} \times \frac{8}{15} = \frac{5 \times 8}{12 \times 15}$$
$$= \frac{\cancel{5}^{1} \times \cancel{8}^{2}}{\cancel{12}^{3} \times \cancel{15}^{3}}$$
$$= \frac{1 \times 2}{3 \times 3}$$
$$= \frac{2}{9}$$

A common factor of 5 and 15 is 5.
A common factor of 8 and 12 is 4.
$5 \div 5 = 1$ $8 \div 4 = 2$
$12 \div 4 = 3$ $15 \div 5 = 3$

Check

1. Find each product.

a) $\frac{3}{4} \times \frac{2}{5}$

Multiply the numerators and multiply the denominators.

$= \frac{3 \times 2}{4 \times 5}$

A common factor of 2 and 4 is ____.

$= \frac{3 \times \cancel{2}}{\cancel{4} \times 5}$

$= \frac{\_\_ \times \_\_}{\_\_ \times \_\_} =$ ________

b) $\frac{9}{14} \times \frac{7}{3}$

Multiply the numerators and multiply the denominators.

A common factor of 9 and 3 is ____.

A common factor of 7 and 14 is ____.

= ________

= ________

2. Multiply.

a) $\frac{6}{7} \times \frac{3}{4} = \frac{\_\_ \times \_\_}{\_\_ \times \_\_}$

= ________

= ________

b) $\frac{4}{5} \times \frac{15}{14} =$ ________

= ________

= ________

c) $\frac{12}{5} \times \frac{5}{18} =$ ________

= ________

= ________

Multiplying Mixed Numbers

Mixed numbers combine whole numbers with fraction parts.
To multiply, write the mixed numbers in fraction form.

Multiply: $2\frac{1}{4} \times \frac{2}{3}$

Rewrite $2\frac{1}{4}$ as an improper fraction: $2\frac{1}{4} = \frac{2 \times 4 + 1}{4}$

$= \frac{9}{4}$

So, $2\frac{1}{4} \times \frac{2}{3} = \frac{9}{4} \times \frac{2}{3}$ — Multiply the numerators and multiply the denominators.

$= \frac{9 \times 2}{4 \times 3}$ — Look for common factors in numerator and denominator.

$= \frac{\cancel{9}^{3} \times \cancel{2}^{1}}{\cancel{4}^{2} \times \cancel{3}^{1}}$

$= \frac{3}{2}$, or $1\frac{1}{2}$

Check

1. Write each mixed number as an improper fraction.

a) $3\frac{4}{5}$

= ______

= ______

b) $3\frac{2}{7}$

= ______

= ______

c) $1\frac{5}{12}$

= ______

= ______

2. Multiply.

a) $3\frac{2}{5} \times \frac{1}{4}$ — Rewrite $3\frac{2}{5}$ as an improper fraction: $3\frac{2}{5} = \frac{17}{5}$

$= \frac{17}{5} \times \frac{1}{4}$ — Multiply the numerators and multiply the denominators.

= ______

b) $1\frac{1}{2} \times 1\frac{1}{3}$ — Rewrite ______ and ______as improper fractions.

= ______ — Multiply the numerators and multiply the denominators.

= ______ — Look for common factors in numerator and denominator.

= ______

3.4 Multiplying Rational Numbers

FOCUS **Multiply rational numbers.**

To predict the sign of the product of two rational numbers,
use the sign rules for multiplying integers:

×	(−)	(+)
(−)	(+)	(−)
(+)	(−)	(+)

- *If the signs are the same, the answer is positive.*
- *If the signs are different, the answer is negative.*

Example 1 Multiplying Rational Numbers in Fraction Form

Multiply: $\left(-\frac{2}{3}\right)\left(-\frac{6}{7}\right)$

Solution

Predict the sign of the product:
Since the fractions have the same sign, their product is positive.

$$\left(-\frac{2}{3}\right)\left(-\frac{6}{7}\right) = \frac{(-2) \times (-\cancel{6})^{-2}}{\cancel{3}^{1} \times 7}$$

$$= \frac{(-2) \times (-2)}{1 \times 7}$$

$$= \frac{4}{7}$$

So, $\left(-\frac{2}{3}\right)\left(-\frac{6}{7}\right) = \frac{4}{7}$

Check

1. Find each product.

a) $\frac{1}{5} \times \left(-\frac{3}{5}\right)$ The fractions have ________________,
so their product is ________________.

$= \frac{\_\_ \times (-3)}{\_\_ \times \_\_}$

$= \_\_\_\_$

b) $\left(-\frac{9}{11}\right)\left(-\frac{7}{12}\right)$

The fractions have ______________,
so their product is ______________.

$= \frac{\_\_\_ \times \_\_\_}{\_\_\_ \times 12}$

A common factor of _____ and 12 is _____.

$= \frac{\_\_\_ \times \_\_\_}{\_\_\_ \times \_\_\_}$

$= \frac{\_\_\_ \times \_\_\_}{\_\_\_ \times \_\_\_}$

$= \_\_\_$

Example 2 Multiplying Rational Numbers in Mixed Number Form

Multiply: $\left(-2\frac{1}{5}\right)\left(-1\frac{3}{4}\right)$

Solution

$\left(-2\frac{1}{5}\right)\left(-1\frac{3}{4}\right)$

Write each mixed number as an improper fraction.

$2\frac{1}{5} = \frac{10}{5} + \frac{1}{5} = \frac{11}{5}$ $\qquad$ $1\frac{3}{4} = \frac{4}{4} + \frac{3}{4} = \frac{7}{4}$

So, $\left(-2\frac{1}{5}\right)\left(-1\frac{3}{4}\right) = \left(-\frac{11}{5}\right)\left(-\frac{7}{4}\right)$

The numbers have the same sign: the product is positive.

$= \frac{(-11) \times (-7)}{5 \times 4}$

$= \frac{77}{20}$, or $3\frac{17}{20}$

$\frac{77}{20} = \frac{60}{20} + \frac{17}{20} = 3\frac{17}{20}$

Check

1. Find each product.

a) $\left(-1\frac{1}{4}\right) \times \frac{6}{7}$

$= \left(-\frac{\_\_\_}{4}\right) \times \frac{6}{7}$

$= \frac{\_\_\_ \times \_\_\_}{\_\_\_ \times \_\_\_}$

$= \_\_\_$, or $\_\_\_$

b) $\left(-2\frac{4}{5}\right)\left(-2\frac{3}{4}\right)$

$= \left(-\frac{\_\_\_}{5}\right)\left(-\frac{\_\_\_}{4}\right)$

$= \_\_\_\_\_\_$

$= \_\_\_\_\_\_$

To multiply rational numbers in decimal form:

- Use the sign rules for integers to find the sign of the product.
- Multiply as you would with whole numbers; estimate to place the decimal point.

Example 3 Multiplying Rational Numbers to Solve a Problem

On March 6, 2009, the price of a share in Bank of Montreal changed by $-\$3.05$.
Joanne owns 50 shares. By how much did the shares change in value that day?

Solution

The change in value is: $50 \times (-3.05)$

Multiply the integers, then estimate to place the decimal point.
$50 \times (-305) = -15\,250$

The product is negative.

Estimate to place the decimal point.
Since -3.05 is close to -3,
$50 \times (-3.05)$ is close to $50 \times (-3)$, or -150.
So, $50 \times (-3.05) = -152.50$

The shares changed in value by $-\$152.50$ that day.

Check

1. On March 13, 2009, the price of a share in Research in Motion changed by $-\$1.13$. Tania owns 80 shares. By how much did those shares change in value that day?

The change in value is: $80 \times (-1.13)$
The product is __________.
To find $80 \times (-1.13)$, multiply: ________ × ________
80 × ________ = __________
Estimate: $80 \times (-1.13)$ is about ________ × ________ = ________
So, $80 \times (-1.13)$ = ___________
The shares changed in value by _________ that day.

Practice

1. Is the product positive or negative?

a) $(-2.5) \times 3.6$ different signs; the product is ________________.

b) $(-4.1) \times (-6.8)$ the same sign; the product is ________________.

c) $\left(-\frac{3}{4}\right)\left(-\frac{7}{9}\right)$ ________________; the product is ________________.

d) $\left(-2\frac{1}{3}\right) \times 6\frac{1}{2}$ ________________; the product is ________________.

2. Which of these expressions have the same product as $\frac{5}{8} \times \left(-\frac{7}{3}\right)$? Why?

a) $\left(-\frac{7}{3}\right) \times \frac{5}{8}$ _____, since ____

____

b) $\left(-\frac{5}{8}\right)\left(-\frac{7}{3}\right)$ _____, since ____

____

c) $\frac{7}{3} \times \frac{5}{8}$ _____, since ____

____

d) $\frac{7}{3} \times \left(-\frac{5}{8}\right)$ _____, since ____

____

3. Find each product.

Think: Is the product positive or negative?

a) $\frac{2}{7} \times \left(-\frac{5}{6}\right)$

$\frac{2}{7} \times \left(-\frac{5}{6}\right) =$ ________________

$=$ ________________

$=$ ________________

$=$ ________________

b) $\left(-\frac{4}{5}\right)\left(-\frac{11}{12}\right)$

$\left(-\frac{4}{5}\right)\left(-\frac{11}{12}\right) =$ ________________

$=$ ________________

$=$ ________________

$=$ ________________

4. Find each product.

a) $\left(-\frac{8}{9}\right) \times 1\frac{1}{2}$

$\left(-\frac{8}{9}\right) \times 1\frac{1}{2} = \left(-\frac{8}{9}\right) \times \frac{\_\_}{2}$

= ____________

= ____________

= ____________

b) $\left(-2\frac{5}{6}\right)\left(-1\frac{1}{5}\right)$

$\left(-2\frac{5}{6}\right)\left(-1\frac{1}{5}\right) = \left(-\frac{\_\_}{6}\right)\left(-\frac{\_\_}{5}\right)$

= ________________

= ________________

= ________________

5. Multiply.

a) $0.4 \times (-3.2)$

To find $0.4 \times (-3.2)$, multiply: $4 \times (-32) =$ _________
$0.4 \times (-3.2)$ is about ______ × ______ = ______
So, $0.4 \times (-3.2) =$ _________.

b) $(-3.03) \times (-0.7)$

To find $(-3.03) \times (-0.7)$, multiply: _________ × _________ = _________
$(-3.03) \times (-0.7)$ is about (___) × (___) = ___
So, $(-3.03) \times (-0.7) =$ _________.

6. On a certain day, the temperature changed by an average of −2.2°C/h. What was the total temperature change in 8 h?

The total change in temperature is: ______ × _______
The product is ____________.
To find ____________, multiply: ______ × ______ = ______
$8 \times (-2.2)$ is about ______ × ______ = ______.
So, $8 \times (-2.2) =$ ______
The temperature____ by ______°C in 8 h.

3.5 Skill Builder

Dividing Fractions

Here are two ways to divide $2 \div \frac{2}{3}$.

- Use a number line.

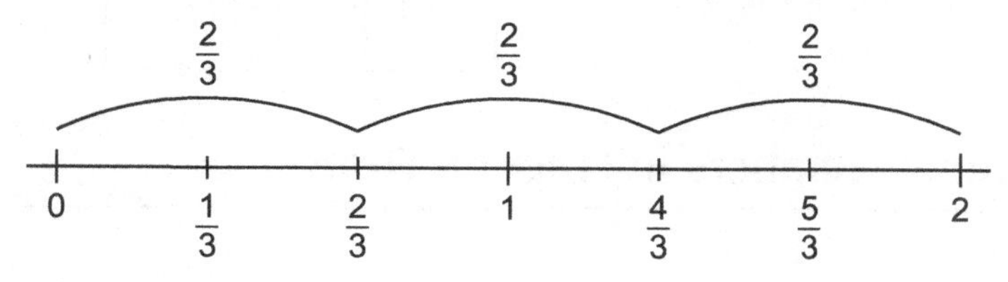

How many groups of two-thirds are there in 2?

There are 3 groups of two-thirds in 2. So, $2 \div \frac{2}{3} = 3$

- Multiply by the reciprocal of $\frac{2}{3}$.

$2 \div \frac{2}{3}$ The reciprocal of $\frac{2}{3}$ is $\frac{3}{2}$.

$= 2 \times \frac{3}{2}$

$= \frac{2}{1} \times \frac{3}{2}$

$= \frac{\cancel{2}^1 \times 3}{1 \times \cancel{2}^1}$ Look for common factors.

$= 3$

Check

1. Find each quotient. Use any method.

a) $2 \div \frac{1}{6} =$ ______

b) $\frac{1}{3} \div 2 =$ ______

c) $\frac{1}{3} \div \frac{5}{3} = \frac{1}{3} \times$ ______

$=$ ______

$=$ ______

d) $4 \div \frac{2}{3} = 4 \times$ ______

$=$ ______

$=$ ______

3.5 Dividing Rational Numbers

FOCUS **Divide rational numbers.**

÷	(−)	(+)
(−)	(+)	(−)
(+)	(−)	(+)

Division is the opposite of multiplication.
So, the sign rules for dividing rational numbers are the same as those for multiplying rational numbers.

Example 1 Dividing Rational Numbers in Fraction Form

Divide: $\frac{3}{4} \div \left(-\frac{9}{8}\right)$

Solution

$\frac{3}{4} \div \left(-\frac{9}{8}\right)$

The fractions have different signs, so the quotient is negative.

$\frac{3}{4} \div \left(-\frac{9}{8}\right) = \frac{3}{4} \times \left(-\frac{8}{9}\right)$ Multiply by the reciprocal.

$= \frac{\cancel{3}^{1} \times (-\cancel{8})^{-2}}{\cancel{4}^{1} \times \cancel{9}^{3}}$ Look for common factors.

$= \frac{1 \times (-2)}{1 \times 3}$

$= -\frac{2}{3}$

So, $\frac{3}{4} \div \left(-\frac{9}{8}\right) = -\frac{2}{3}$

Dividing by $-\frac{9}{8}$ is the same as multiplying by $-\frac{8}{9}$.

Check

1. Divide.

a) $\frac{2}{5} \div \left(-\frac{3}{4}\right)$

$= \frac{2}{5} \times$ ______

$=$ ______

b) $\left(-\frac{2}{9}\right) \div \left(-\frac{4}{7}\right)$

$=$ ______ $\times$ ______

$=$ ______

$=$ ______

Think: Is the quotient positive or negative?

Example 2 Dividing Rational Numbers in Decimal Form

Divide:
$(-5.1) \div 3$

Solution

$(-5.1) \div 3$

Since the signs are different, the quotient is negative.

Divide integers: $(-51) \div 3 = -17$

Estimate to place the decimal point.
-5.1 is close to -6, so $(-5.1) \div 3$ is close to $(-6) \div 3 = -2$
So, $(-5.1) \div 3 = -1.7$

Check

1. Divide: $(-7.5) \div 5$

$(-7.5) \div 5$

Divide integers: ______ ÷ ___ = ______

Estimate to place the decimal point.

$(-7.5) \div 5$ is about _____ ÷ ___ = _____

So, $-7.5 \div 5 =$ ______

Think: Is the quotient positive or negative?

Practice

1. Is the quotient positive or negative?

a) $(-7.5) \div (-3)$ Same sign; the quotient is ______________.

b) $8.42 \div (-2)$ ______________; the quotient is ______________.

c) $\left(-\frac{9}{10}\right) \div \frac{3}{5}$ ______________; the quotient is ______________.

d) $(-16) \div \left(-\frac{4}{5}\right)$ ______________; the quotient is ______________.

2. Which of these expressions have the same answer as $\left(-\frac{3}{10}\right) \div \frac{2}{5}$?

a) $-\frac{3}{10} \times \frac{5}{2}$

____, since ____

____

b) $-\frac{3}{10} \div \left(-\frac{2}{5}\right)$

____, since ____

____

c) $\frac{2}{5} \div \left(-\frac{3}{10}\right)$

____, since ____

____

d) $\frac{3}{10} \div \left(-\frac{2}{5}\right)$

____, since ____

____

3. Find each quotient.

a) $\left(-\frac{2}{3}\right) \div \frac{7}{6}$

$= \left(-\frac{2}{3}\right) \times$ ____

$= \frac{\_\_\_ \times \_\_\_}{\_\_\_ \times \_\_\_}$

$=$ ____

b) $\left(-\frac{15}{16}\right) \div \left(-\frac{5}{8}\right)$

$= \left(-\frac{15}{16}\right) \times$ ____

$= \frac{\_\_\_ \times \_\_\_}{\_\_\_ \times \_\_\_}$

$=$ ____

4. Divide.

a) $\left(-\frac{8}{9}\right) \div \frac{1}{3}$

$= \left(-\frac{8}{9}\right) \times$ ____

$= \frac{\_\_\_\_ \times \_\_\_\_}{\_\_\_\_ \times \_\_\_\_}$

= __________

Think: Is the quotient positive or negative?

b) $\left(-\frac{2}{5}\right) \div \left(-\frac{3}{7}\right)$

= ____ × ____

= ______

5. Use integers to determine each quotient.
Estimate to place the decimal point in the answer.

a) $(-2.94) \div 0.7$

$(-2.94) \div 0.7$
The quotient is __________.
To find $(-2.94) \div 0.7$, divide: ______ ÷ ____ = ______
$(-2.94) \div 0.7$ is about _____ ÷ ____ = ______.
So, $(-2.94) \div 0.7 =$ ______

b) $(-5.52) \div (-0.8)$

$(-5.52) \div (-0.8)$
The quotient is __________.
To find $(-5.52) \div (-0.8)$, divide: ______ ÷ ____ = ____
$(-5.52) \div (-0.8)$ is about _____ ÷ _____ = ___.
So, $(-5.52) \div (-0.8) =$ _____

3.6 Order of Operations with Rational Numbers

The order of operations for rational numbers is the same as for integers and fractions.
Think BEDMAS to remember the correct order of operations.
We use this order of operations to evaluate expressions with more than one operation.

B Do the operations in brackets first.
E Next, evaluate any exponents.
D
M Then, divide and multiply in order from left to right.
A
S Finally, add and subtract in order from left to right.

Example 1 Using the Order of Operations with Decimals

Evaluate.

a) $(-2.4) \div 1.2 - 7 \times 0.2$

b) $(-3.4 + 0.6) + 4^2 \times 0.2$

Solution

a) $\mathbf{(-2.4) \div 1.2} - 7 \times 0.2$ — Divide first.

$= -2 - \mathbf{7 \times 0.2}$ — Then multiply.

$= -2 - 1.4$ — To subtract, add the opposite.

$= -2 + (-1.4)$

$= -3.4$

b) $\mathbf{(-3.4 + 0.6)} + 4^2 \times 0.2$ — Brackets first.

$= -2.8 + \mathbf{4^2} \times 0.2$ — Then evaluate the power.

$= -2.8 + \mathbf{16 \times 0.2}$ — Then multiply.

$= -2.8 + 3.2$ — Add.

$= 0.4$

Check

1. Evaluate.

a) $3.8 + 0.8 \div (-0.2)$
$= 3.8 + (\_\_\_\_)$
$= \_\_\_\_\_$

b) $4.6 - 3^2 + 3.9 \div (-1.3)$
$= 4.6 - \_\_\_\_ + 3.9 \div (-1.3)$
$= 4.6 - \_\_\_\_ + (\_\_\_\_)$
$= -4.4 + (\_\_\_\_)$
$= \_\_\_\_\_$

Example 2 Using the Order of Operations with Fractions

Evaluate:

a) $\left(\frac{3}{4}-\frac{7}{8}\right) \div \left(-\frac{5}{16}\right)$

b) $\left(-\frac{2}{3}\right) \times \frac{1}{6} + \frac{1}{2}$

Solution

a) $\left(\frac{3}{4}-\frac{7}{8}\right) \div \left(-\frac{5}{16}\right)$ — Subtract in the brackets first. Use a common denominator of 8.

$= \left(\frac{\mathbf{6}}{\mathbf{8}}-\frac{\mathbf{7}}{\mathbf{8}}\right) \div \left(-\frac{5}{16}\right)$

$= \left(-\frac{1}{8}\right) \div \left(-\frac{5}{16}\right)$ — To divide, multiply by the reciprocal of $-\frac{5}{16}$.

$= \left(-\frac{1}{8}\right) \times \left(-\frac{16}{5}\right)$

$= \left(-\frac{1}{\cancel{8}^{1}}\right) \times \left(-\frac{\cancel{16}^{2}}{5}\right)$ — Look for common factors.

$= \frac{2}{5}$ — Both factors are negative, so the product is positive.

b) $\left(-\frac{2}{3}\right) \times \frac{1}{6} + \frac{1}{2}$ — Multiply first.

$= \left(-\frac{\mathbf{2}^{1}}{\mathbf{3}}\right) \times \frac{\mathbf{1}}{\cancel{6}^{3}} + \frac{1}{2}$ — Look for common factors.

$= \left(-\frac{1}{9}\right) + \frac{1}{2}$ — Add. Use a common denominator of 18.

$= -\frac{2}{18} + \frac{9}{18} = \frac{7}{18}$

Check

1. Evaluate.

a) $\frac{3}{4} - \left(-\frac{2}{3}\right)\left(-\frac{1}{4}\right)$ — Multiply first.

$= \frac{3}{4} -$ __________ — Look for common factors.

$= \frac{3}{4} -$ __________ — Subtract. Use a common denominator of 12.

$=$ _____

b) $\left(-\frac{1}{6}\right) \div \frac{1}{5} + \left(-\frac{3}{2}\right)$ Divide first. Multiply by the reciprocal of ____.

$= -\frac{1}{6} \times \frac{\quad}{\quad} + \left(-\frac{3}{2}\right)$

$= \frac{\qquad}{\qquad} + \left(-\frac{3}{2}\right)$

$= \frac{\quad}{\quad} + \left(-\frac{3}{2}\right)$ Add. Use a common denominator of ___.

$= \frac{\quad}{\quad} + \frac{\quad}{\quad}$

$= \frac{\quad}{\quad}$

Example 3 Applying the Order of Operations

The formula $C = (F - 32) \div 1.8$ converts temperatures in degrees Fahrenheit, F, to degrees Celsius, C.
What is 28.4°F in degrees Celsius?

Solution

Substitute $F = 28.4$ in the formula $C = (F - 32) \div 1.8$

$C = (28.4 - 32) \div 1.8$ Subtract in the brackets first. Add the opposite.

$= (28.4 + (-32)) \div 1.8$

$= (-3.6) \div 1.8$ Divide.

$= -2$

28.4°F is equivalent to −2°C.

Check

1. The expression $F = 32 + 9 \times C \div 5$ converts temperatures in degrees Celsius, C, to degrees Fahrenheit, F.

What is −12.5°C in degrees Fahrenheit?

$F = 32 + 9 \times (\_\_\_\_\_) \div 5$ Multiply first.

= ____________________ ____________________

= ____________________ ____________________

= ____________________

−12.5°C is equivalent to ________°F.

Practice

1. In each expression, which operation will you do first?

a) $(-8.6) \times 2.4 - (-6 + 2.5)$

____

b) $2.5 - 6.4 \times 2.1 + 3.5$

____

c) $\frac{4}{3} \times \frac{5}{6} + \frac{2}{7} \div \frac{5}{14}$

____

d) $\frac{5}{3} + \frac{2}{7} \div \left(-\frac{1}{4}\right) - \frac{3}{5}$

____

2. Evaluate each expression.

a) $(-3.6) \div 1.8 + (1.2 - 1.5)$

= ______________

= ______________

= ______________

b) $\left(-\frac{1}{4}\right) \div \frac{3}{8} + \left(-\frac{1}{2}\right)^2$

= ______________

= ______________

= ______________

= ______________

= ______________

3. Evaluate each expression.

a) $(5.6 + 4.4) \div (-2.5)$

= ____ $\div (-2.5)$

= ______

b) $(-4.2) + 6 \times (-1.7)$

$= (-4.2) + ($______$)$

= ______________

c) $9.2 \div 4 - 3.6 \times 2$

= ______________

= ______________

= ______________

d) $7.5 \times [-0.7 + (-0.3) \times 3]$

= ______________

= ______________

= ______________

4. Evaluate each expression.

a) $\frac{1}{5} + \left(-\frac{1}{4}\right) \times \frac{8}{15}$

$= \frac{1}{5} +$ ____________

$= \frac{1}{5} +$ ____________

$= \frac{1}{5} +$ ____________

$=$ ____________

$=$ ____________

b) $\left(-\frac{7}{4}\right) \div \frac{2}{3} + \frac{1}{4}$

$=$ ____________

$=$ ____________

$=$ ____________

$=$ ____________

$=$ ____________

c) $\left(\frac{1}{3}\right)^2 \times \frac{3}{2} - \frac{5}{4}$

$=$ ____________

$=$ ____________

$=$ ____________

$=$ ____________

$=$ ____________

5. A mistake was made in each solution.
Identify the line in which the mistake was made, and give the correct solution.

a) $(-3.2 \div 1.6)^2 - (-4.1)$ ____________________

$= (-2)^2 - (-4.1)$ ____________________

$= 4 + (-4.1)$ ____________________

$= -0.1$ ____________________

b) $\frac{1}{3} + \frac{4}{3} \times \left(-\frac{1}{2}\right)$ ____________________

$= \frac{5}{3} \times \left(-\frac{1}{2}\right)$ ____________________

$= \frac{5 \times (-1)}{3 \times 2}$ ____________________

$= -\frac{5}{6}$ ____________________

6. The formula for the area of a trapezoid is $A = h \times (a + b) \div 2$.
In the formula, h is the height and a and b are the lengths of the parallel sides. Find the area of a trapezoid with height 3.5 cm and parallel sides of length 8 cm and 12 cm.

Substitute $h =$ _____, $a =$ ____, and $b =$ _____ in the formula $A = h \times (a + b) \div 2$.

$A =$ ____________________

$=$ ____________________

$=$ ____________________

$=$ ____________________

The trapezoid has area ____ cm^2.

Unit 3 Puzzle

Rational Numbers Bingo

Evaluate each expression and circle the answer on the Bingo cards. Which card is the winning card?

On the winning card, the answers form a horizontal, vertical, or diagonal line.

Questions

Evaluate as a decimal.

1. $(-8.2) - (-2.4) =$ ___________

2. $3.65 \div (-0.5) =$ ___________

3. $(-1.9) \times 2 =$ ___________

4. $(-3.48) + 5.06 =$ ___________

5. $(-0.80) - 0.64 =$ ___________

Evaluate as a fraction.

6. $\left(-\frac{7}{10}\right) + \frac{6}{5} =$ ___________

7. $\left(-\frac{6}{7}\right)\left(-\frac{14}{15}\right) =$ ___________

8. $\left(-\frac{1}{4}\right) \times \frac{1}{3} =$ ___________

9. $\left(-\frac{4}{5}\right) - \left(-\frac{3}{4}\right) =$ ___________

10. $\frac{1}{9} \div \left(-\frac{2}{3}\right) =$ ___________

Card A

$-1\frac{11}{20}$	−0.16	$2\frac{1}{15}$	−5.8	$-2\frac{1}{12}$
7.3	−1	−1.44	$-\frac{4}{5}$	3.99
−3.8	$1\frac{2}{5}$	FREE SPACE	$1\frac{9}{10}$	$-\frac{1}{12}$
3	$-2\frac{1}{15}$	−10.6	$-\frac{1}{6}$	−1.58
1.58	$\frac{4}{5}$	$-\frac{1}{20}$	−7.3	$\frac{1}{2}$

Card B

3	−5.8	−10.6	$1\frac{9}{10}$	$-2\frac{1}{15}$
$2\frac{1}{15}$	$-\frac{4}{5}$	−3.99	$1\frac{2}{5}$	−7.3
−1.44	1.58	FREE SPACE	$-\frac{1}{6}$	−1
7.3	$7\frac{11}{20}$	$\frac{1}{2}$	−1.58	$-2\frac{1}{12}$
$-\frac{1}{20}$	3.99	$\frac{4}{5}$	$-\frac{1}{12}$	−0.16

The winning card is ___________________.

Unit 3 Study Guide

Skill	Description	Example
Compare and order rational numbers.	Numbers increase in value from left to right on a number line.	Number line from −1 to 1 marked: −0.4, $-\frac{1}{3}$, 0, 0.1, $\frac{1}{4}$, 1 From least to greatest: $-0.4, -\frac{1}{3}, 0.1, \frac{1}{4}$
Add rational numbers.	Model on a number line: Start at the first number. Move right to add a positive number; move left to add a negative number.	Number line marked: −1.2, −1, 0, 0.4; arrow −1.6 $0.4 + (-1.6) = -1.2$
	Look for common denominators to add fractions. With decimals, add digits with the same place value.	$-\frac{2}{5} + \frac{1}{2} = -\frac{4}{10} + \frac{5}{10} = \frac{1}{10}$ $(-18.7) + 13.5 = -5.2$
Subtract rational numbers.	Add the opposite.	$3\frac{1}{3} - \left(-1\frac{2}{5}\right) = 3\frac{1}{3} + \left(+1\frac{2}{5}\right)$ $= 3 + 1 + \frac{5}{15} + \frac{6}{15}$ $= 4\frac{11}{15}$ $-18.7 - 13.5 = -18.7 + (-13.5)$ $= -32.2$
Multiply and divide rational numbers.	Use the same rules for signs as with integers. Then determine the numerical value.	$\left(-\frac{2}{3}\right) \times \frac{9}{8} = \frac{(-\cancel{2})^{-1} \times \cancel{9}^{3}}{\cancel{3}^{1} \times \cancel{8}^{4}}$ $= -\frac{3}{4}$ $(-6.3) \times 7 = -44.1$ $\left(-2\frac{1}{5}\right) \div \left(-3\frac{3}{10}\right) = \left(-\frac{11}{5}\right) \div \left(-\frac{33}{10}\right)$ $= \left(-\frac{\cancel{11}^{1}}{\cancel{5}^{1}}\right) \times \left(-\frac{\cancel{10}^{2}}{\cancel{33}^{3}}\right)$ $= \frac{2}{3}$ $(-5.6) \div 0.7 = -8.0$
Use order of operations to evaluate expressions.	**B** Do the operations in brackets first. **E** Next, evaluate any exponents. **D** **M** Then, divide and multiply in order from left to right. **A** **S** Finally, add and subtract in order from left to right.	$(-2.50 + 1.75) \div (0.1 - (-0.4))^2$ $= -0.75 \div (0.1 + (+0.4))^2$ $= -0.75 \div (0.5)^2$ $= -0.75 \div 0.25$ $= -3$

Unit 3 Review

3.1 **1. a)** Write each number as a decimal.

i) $-\frac{16}{9} =$ ________
$=$ ________

ii) $-\frac{7}{3} =$ ________
$=$ ________

iii) $-2\frac{1}{5} = -\frac{\_\_}{\_\_}$
$=$ ________
$=$ ________

b) Find two rational numbers between $-\frac{16}{9}$ and $-\frac{7}{3}$:

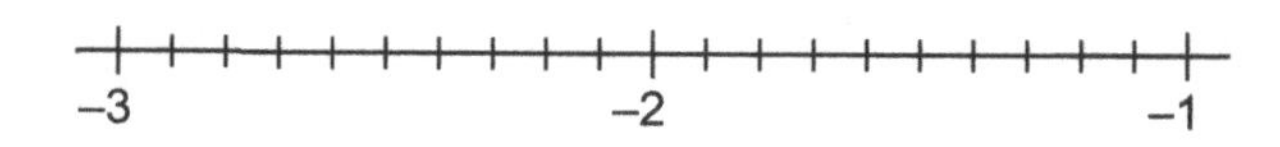

Two rational numbers between $-\frac{16}{9}$ and $-\frac{7}{3}$ are: ____ and _____

2. Order these numbers from least to greatest: -3.9, $-3\frac{4}{5}$, -3.3, $-\frac{7}{2}$

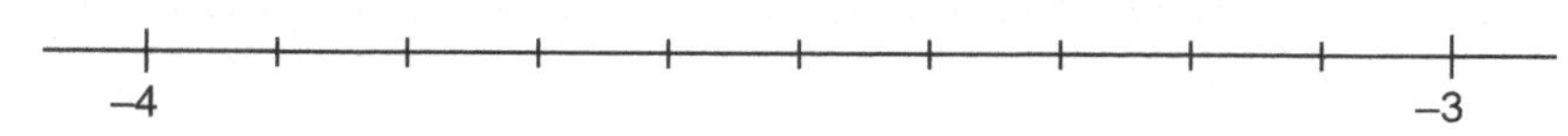

From least to greatest: ____________________

3.2 **3.** Calculate each sum.

a) $(-2.1) + 4.8 =$ ______

b) $25.6 + (-18.9) =$ ______

c) $(-6.4) + (-3.8) =$ ______

4. Add.

a) $-\frac{1}{8} + \left(-\frac{3}{4}\right)$
$= -\frac{1}{8} + \frac{\_\_}{\_\_}$
$= \frac{\_\_}{\_\_}$

b) $-\frac{4}{3} + \frac{11}{12}$
$= \frac{\_\_}{\_\_} + \frac{11}{12}$
$= \frac{\_\_}{\_\_}$

c) $\left(-1\frac{2}{3}\right) + 2\frac{8}{9} = (-1 + 2) + \left(\frac{\_\_}{\_\_} + \frac{\_\_}{\_\_}\right)$
$= (-1 + 2) + \left(\frac{\_\_}{\_\_} + \frac{\_\_}{\_\_}\right)$
$= \frac{\_\_}{\_\_}$
$= \frac{\_\_}{\_\_}$

3.3 **5.** Subtract.

a) $\left(-\frac{7}{12}\right) - \left(-\frac{2}{3}\right) = -\frac{7}{12} + \frac{2}{3}$

$= -\frac{7}{12} + \underline{\quad}$

$= \underline{\quad}$

b) $\frac{3}{5} - 2\frac{1}{7} = \frac{3}{5} + \left(-\frac{\quad}{7}\right)$

$= \underline{\qquad\qquad}$

$= \underline{\qquad\qquad}$

c) $-3\frac{1}{10} - 1\frac{3}{5} = -\frac{\quad}{10} + \left(-\frac{\quad}{5}\right)$

$= \underline{\qquad\qquad}$

$= \underline{\qquad\qquad}$

6. The table shows the elevations of several places on Earth.

Place	Elevation (m)
Mt. Everest	8849.7
Mt. Logan	5959.1
Death Valley	−410.9
Dead Sea	−417.3

Write a subtraction sentence that represents the difference in the elevations of the given locations. Then calculate the difference.

a) Mt. Logan and the Dead Sea

_____ − (_____) = _____ + _____

= _____

The difference in elevations is _____ m.

b) Death Valley and the Dead Sea

_____ − (_____) = _____ + _____

= _____

The difference in elevations is _____ m.

c) Mt. Everest and Mt. Logan

_____ − _____ = _____ + _____

= _____

The difference in elevations is _____ m.

3.4 **7.** What is the sign of each product?

a) $(-3.8) \times (-1.2)$ _____

b) $0.75 \times (-8.6)$ _____

c) $\left(-\frac{1}{3}\right)\left(-\frac{4}{9}\right)$ _____

d) $\left(-1\frac{2}{5}\right) \times \frac{7}{10}$ _____

8. Find each product.

a) $\left(-\frac{2}{5}\right)\left(-\frac{11}{20}\right)$

$= \frac{\_\_\_ \times \_\_\_}{\_\_\_ \times \_\_\_}$

$= \_\_\_\_\_$

b) $\left(-\frac{4}{5}\right) \times \frac{25}{12}$

$= \frac{\_\_\_ \times \_\_\_}{\_\_\_ \times \_\_\_}$

$= \_\_\_\_\_$

c) $-\frac{15}{16} \times 1\frac{1}{3}$

$= -\frac{15}{16} \times \frac{\_\_}{3}$

$= \frac{\_\_\_ \times \_\_\_}{\_\_\_ \times \_\_\_}$

$= \_\_\_\_\_$

d) $-3\frac{2}{3} \times \left(-2\frac{3}{11}\right)$

$= -\frac{\_\_}{3} \times \left(-\frac{\_\_}{11}\right)$

$= \frac{\_\_\_ \times \_\_\_}{\_\_\_ \times \_\_\_}$

$= \_\_\_\_\_$

9. Circle the most reasonable answer.

	Question	Most reasonable answer		
a)	29.5 × 4.8	1.416	14.16	141.6
b)	5.4 × 0.7	0.378	3.78	37.8
c)	305.8 × 3.2	97.856	978.56	9785.6
d)	37.5 × 1.6	0.6	6	60

10. A diver descends at a speed of 0.8 m/min.
How far does the diver descend in 3.5 min?
The distance the diver descends is: _____ × _____

The product is __________. Multiply the whole numbers: _____ × _____ = _____

Estimate: _____ × _____ is about _____ × _____ = _____.

The exact answer is _____ × _____ = _____

The diver descends _____ m in 3.5 min.

3.5 **11.** Divide.

a) $\frac{1}{5} \div \left(-\frac{7}{10}\right)$

$= \frac{1}{5} \times$ ______

= ______

= ______

= ____

b) $\left(-\frac{3}{5}\right) \div \left(-\frac{12}{7}\right)$

= ______

= ______

= ______

= ____

3.6 **12.** Evaluate each expression.

a) $1.1 - 3.1 \times 7$

$= 1.1 -$ ____

$= 1.1 + ($____$)$

= ______

b) $-1.8 \div (-0.3) + [5.1 - (-2.9)]$

$= -1.8 \div (-0.3) + [5.1 +$ ____$]$

$= -1.8 \div (-0.3) +$ __

= __ + __

= __

c) $\left(-\frac{5}{6}\right) \times \frac{1}{4} + \frac{5}{12}$

= ______ $+ \frac{5}{12}$

= ____ $+ \frac{5}{12}$

= ____ + ____

= ______

d) $1\frac{3}{4} + \frac{2}{3} \div \left(-\frac{8}{9}\right)$

$= 1\frac{3}{4} + \frac{2}{3} \times$ ____

$= 1\frac{3}{4} + \frac{\_\_ \times \_\_}{\_\_ \times \_\_}$

$= 1\frac{3}{4} +$ ____

$= \frac{\_\_}{4} +$ ____

= ______

UNIT 4

Linear Relations

What You'll Learn

- Use expressions and equations to write patterns and work with them.
- Use substitution to work with patterns to find more information.
- Graph and analyze linear relations.
- Use interpolation and extrapolation to gather more information from graphs.

Why It's Important

Patterns and linear relations are used by

- book printers, to quote the cost of a job
- managers, to plan for new hiring

Key Words

variable
expression
equation
table of values
relation
linear relation
coordinates
coordinate grid
discrete
origin
vertical
horizontal
oblique
interpolation
extrapolation

4.1 Skill Builder

Algebraic Expressions

We can use a variable, such as n, to represent a number in an expression.

A variable is a letter, and it is always written in italics.

For example:

- 4 more than a number: $4 + n$, or $n + 4$
- 5 times a number: $5n$
- 2 less than a number: $n - 2$
- A number divided by 10: $\frac{n}{10}$

To evaluate an expression, we replace a variable with a number.

Check

1. Find the value of each expression when $x = 3$.

a) $3x + 5$

3(___) + 5 = ____ + 5

= ____

b) $6 + 8x$

6 + 8(___) = 6 + ____

= ____

2. Find the value of each expression when $n = 8$.

a) $8n - 4$

8(___) − ___ = ___ − ___

= ___

b) $20 - 2n$

___ − 2(___) = ___ − ___

= ___

Relationships in Patterns

Here is a pattern made with toothpicks.

Figure 1 Figure 2 Figure 3

A pattern rule for the number of toothpicks is:

Start with □. Add ⊐ each time.

There is a pattern in the numbers as well.

Start with 4. Add 3 each time.

We can also show the pattern using a table of values.

Figure Number	Number of Toothpicks
1	4
2	7
3	10

+3
+3

To extend the pattern, continue to add 3 each time:

Figure 4　　Figure 5

Check

1. a) Draw the next 2 figures to extend each pattern.

i)

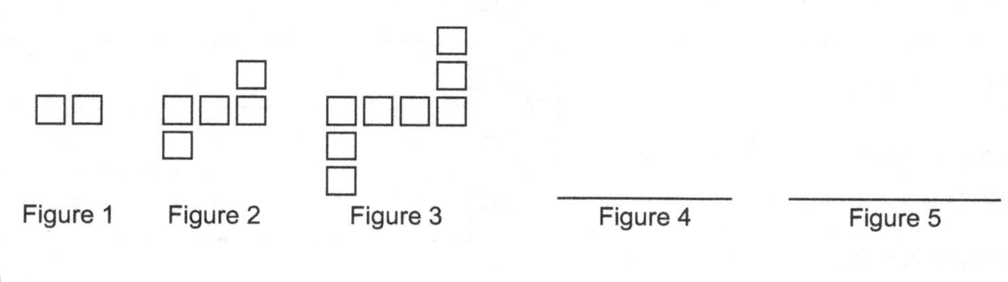

ii)

Figure 1　Figure 2　Figure 3　Figure 4　Figure 5

b) Complete each table of values to show each number pattern above.

i)

Figure Number	Number of Squares
1	2
2	____
3	____
____	____
____	____

ii)

Figure Number	Number of Dots
1	3
2	____
3	____
____	____
____	____

4.1 Writing Equations To Describe Patterns

FOCUS Use equations to describe and solve problems involving patterns.

Example 1 Writing an Equation to Represent a Written Pattern

Here is a pattern made with sticks.

Figure 1

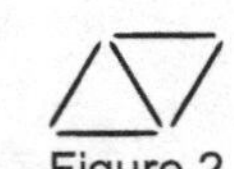
Figure 2

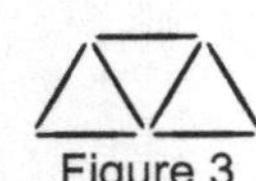
Figure 3

Figure 4

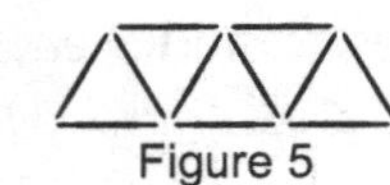
Figure 5

a) Write an equation that relates the number of sticks to the figure number.

b) What is the number of sticks in the 10th figure?

Solution

a) Record the number of sticks in each figure in a table.

As the figure number increases by 1, the number of sticks increases by 2. Repeated addition of 2 is the same as multiplication by 2.

Figure Number, *f*	Number of Sticks, *n*
1	3
2	5
3	7
4	9
5	11

+1 +1 +1 +1

+2 +2 +2 +2

So, the equation $n = 2f$ may represent the relationship.

Check whether the equation is correct.
When $f = 1$, $n = 2(1) = 2$
This is 1 less than 3.
So, add 1.
$2(1) + 1 = 3$

So, an equation is: $n = 2f + 1$

Figure Number, *f*	Number of Sticks, *n*
1	$2(1) + 1 = 3$
2	$2(2) + 1 = 5$
3	$2(3) + 1 = 7$
4	$2(4) + 1 = 9$
5	$2(5) + 1 = 11$

$2f + 1$ represents the number of sticks for any figure number f.

b) To find the number of sticks in the 10th figure, substitute $f = 10$ in the equation:

$$\begin{aligned} n &= 2f + 1 \\ &= 2(10) + 1 \\ &= 20 + 1 \\ &= 21 \end{aligned}$$

There are 21 sticks in the 10th figure.

Check

1. For the table below:

Number of Swings, s	Number of Hits, h
1	5
2	9
3	13
4	17

a) Describe the number of hits in terms of the number of swings.
The number of hits is _____ times the number of swings, plus _____.

b) Write an equation to describe the relationship.
h = _____s + _____

c) Use your equation to find h when $s = 10$.
h = ____________________

____

____

Example 2 Writing an Equation to Represent a Situation

Teagan goes to a carnival. The cost for a ride is shown on a poster at the entrance.

Entrance Fee $5.00
+
$2.00 per ride

a) Write an equation that relates the total cost, C dollars, to the number of rides, r.

b) Teagan goes on 4 rides. What is his total cost?

Solution

a) The cost is \$5.00, plus \$2.00 per ride.
That is, the cost is: $5.00 + 2.00 \times$ (number of rides)
An equation is: $C = 5.00 + 2.00r$

b) Use the equation: $C = 5.00 + 2.00r$
Substitute: $r = 4$
$C = 5.00 + 2.00(4)$
$= 5.00 + 8.00$
$= 13.00$
Teagan's total cost is \$13.00.

Check

1. Marcel takes a summer job at a book packaging plant.
He gets paid \$50 a day, plus \$2 for every box packed.

a) Write an equation that relates the number of boxes packed to Marcel's pay for a day.
Let P represent his pay for one day, and let b represent the number of boxes packed.
$P =$ ____________________

b) Marcel packed 20 boxes one day. How much did he get paid?
$P =$ ____________________
$=$ ____________________
$=$ ____________________
Marcel got paid ______.

Practice

1. In each equation, find the value of T when $n = 6$.

a) $T = 8 + n$
$T = 8 +$ ____
$=$ __________

b) $T = 3n - 2$
$T = 3$____ $- 2$
$=$ __________
$=$ __________

c) $T = 12n + 9$
$T =$ __________
$=$ __________
$=$ __________

d) $T = 7n + 3$
$T =$ __________
$=$ __________
$=$ __________

2. a) This pattern of dots continues. Draw the next 2 figures in the pattern.

Figure 1 Figure 2 Figure 3 ________ Figure 4 ________ Figure 5

b) The pattern is represented in a table of values. Which expression below represents the number of dots in terms of the figure number?

i) $2f$
ii) $3f$
iii) $-3f$
iv) $3f + 1$

____

Figure Number, *f*	Number of Dots, *n*
1	3
2	6
3	9
4	12
5	15

3. a) Look at the pattern of tiles below.
Draw the next 2 figures in the pattern.

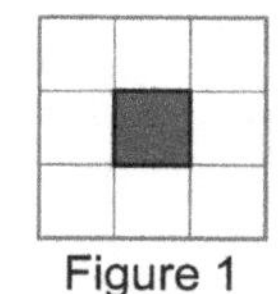
Figure 1

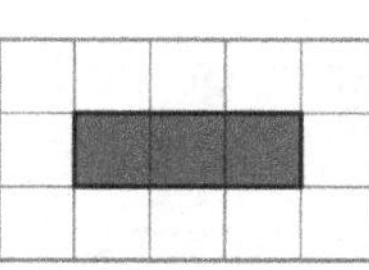
Figure 2

Figure 3 ________ Figure 4 ________ Figure 5

b) Complete the table below.

Number of Shaded Tiles, *s*	Number of White Tiles, *w*
1	8
2	____
____	____
____	____
____	____

c) Write an equation for the number of white tiles, *w*, in terms of the number of shaded tiles, *s*.

$w =$ ____$s +$ ____

d) Use your equation to find *w* when $s = 25$.

$w =$ ____________

$=$ ____________

$=$ ____________

When the number of shaded tiles is 25, there are _____ white tiles.

4. Anabelle is part of the yearbook committee. This year, the set-up cost to print yearbooks is \$400, plus \$3 for each yearbook printed.

a) Write an equation for the total cost in terms of the number of yearbooks printed.

Let *C* represent the total cost, and let *n* represent the number of yearbooks.

$C =$ ______ $+$ ______

b) Anabelle takes 200 orders for yearbooks this year.

What is the total cost to the yearbook committee?

$C =$ ________________

$=$ ________________

$=$ ________________

The total cost is \$ ______.

4.2 Skill Builder

The Coordinate Grid

Plot A(−2, 3) on a coordinate grid.
The first number in the ordered pair tells how far you move left or right on the horizontal axis.
The second number tells how far you move up or down on the vertical axis.

So, to plot A(−2, 3):
Move 2 squares left of the origin, then move 3 squares up.

−2 is the x-coordinate. 3 is the y-coordinate.

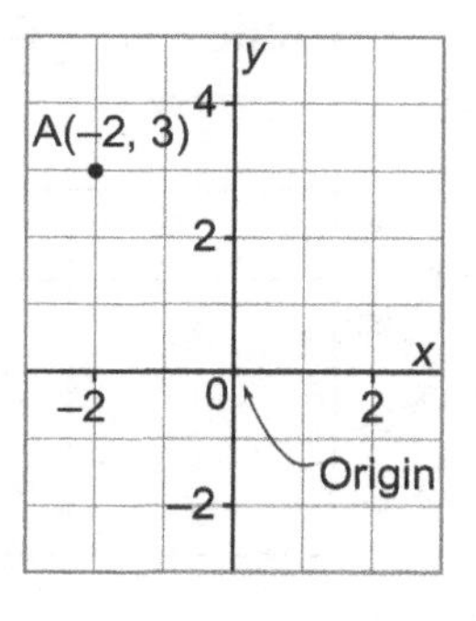

Check

1. What are the coordinates of points A, B, and C?

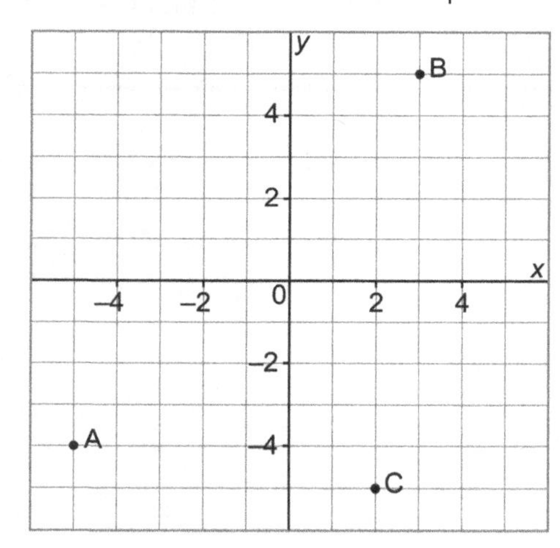

A(__________)
B(__________)
C(__________)

2. Graph these points on the coordinate grid.
A(−3, 0)   B(2, 4)   C(0, −3)

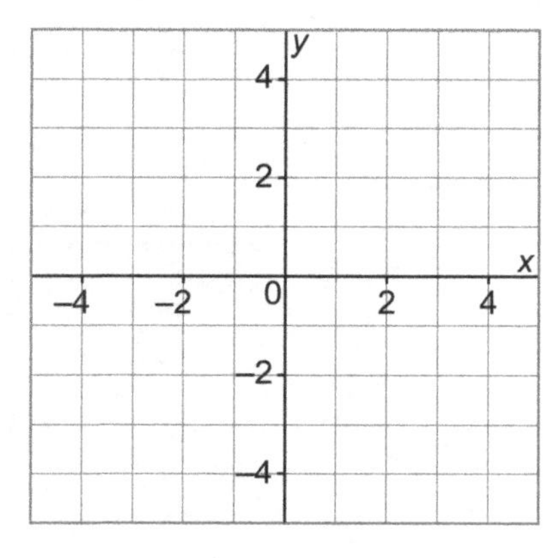

Graphing Relations

This table of values shows how $2n + 1$ relates to n.

Input, n	Output, $2n + 1$
1	3
2	5
3	7
4	9
5	11

+1 +1 +1 +1   +2 +2 +2 +2

The change in input is constant. The change in output is constant.

Graph of $2n + 1$ against n

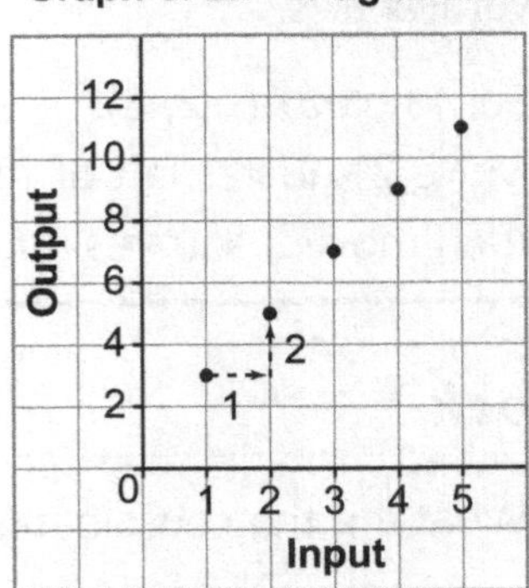

Use the data in the table to graph the relation.

The points lie on a straight line. This is a **linear relation.**
The graph also shows how $2n + 1$ relates to n.
On the graph, we see that each time the input increases by 1, the output increases by 2.

Check

1. a) Graph the data in this table of values.

Input	Output
0	2
1	5
2	8
3	11

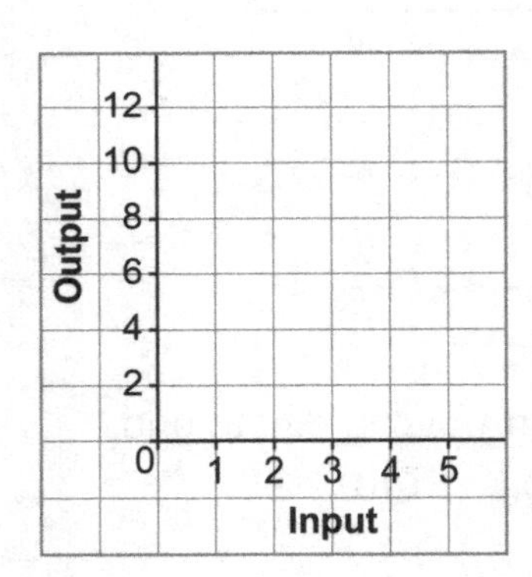

b) Is this a linear relation? Explain.

____

____

4.2 Linear Relations

FOCUS Analyze the graph of a linear relation.

A photographer charges $40 for a sitting fee, plus $20 per sheet of prints.
The charges are shown in the table of values and in the graph.

Number of Sheets, n	Cost, C ($)
0	40
1	60
2	80
3	100

Cost for Number of Sheets of Photos

We cannot order part of a sheet of prints.
So, the points in the graph are not joined with a line.
We say that the data are **discrete.**

For different values of n, we get different values of C.
So the variable C *depends* on the value of the variable n.
When two variables are related in this way, they form a **relation.**

Linear Relation
When the graph of a relation is a straight line, it is called a **linear relation.**

Example 1 Graphing a Linear Relation from a Table of Values

A popular DVD club allows members to purchase DVDs at a reduced price according to the table of values.

a) Graph the data.
b) Should the points be joined? Why or why not?
c) Is the relation linear? Explain.
d) Describe the patterns in the table. How are these patterns shown on the graph?

Number of DVDs Purchased, n	Cost, C ($)
1	20
2	25
3	30
4	35

Solution

a) Plot the points on a grid.

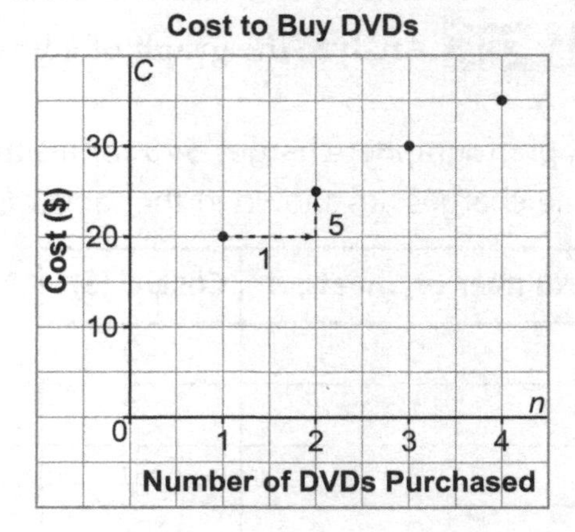

b) The points should not be joined because you cannot buy part of a DVD.

c) The points on the graph lie on a straight line, so this is a linear relation.

d) The table of values shows that:
The number of DVDs purchased increases by 1 each time.
The cost increases by $5 each time.

To get from one point to the next in the graph, move 1 unit right and 5 units up.

Check

1. a) Graph the data from the table of values.

Number of Floors, f	Building Height, h (m)
1	5
2	8
3	11
4	14

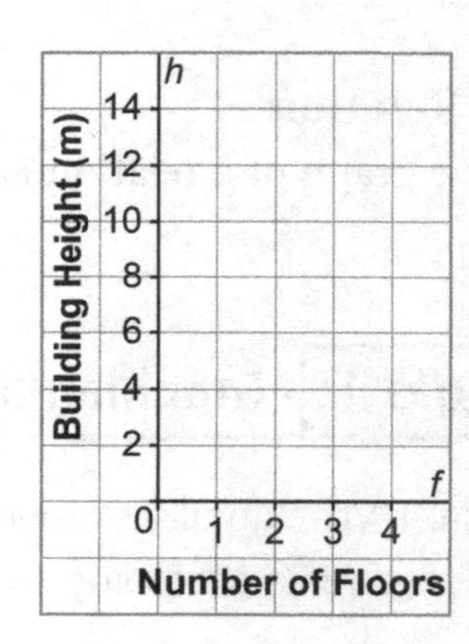

b) Is the relation linear? Explain.

____

c) Should the points on the graph be joined with a line? Explain.

____

Example 2 Graphing a Linear Relation from an Equation

a) Complete the table of values.

b) Graph the relation represented by the data in the table of values.

c) Describe the patterns in the graph and in the table.

d) Is the relation linear? Explain.

x	$y = 5 - x$
-2	
-1	
0	
1	
2	

Solution

a)

x	$y = 5 - x$
-2	$5 - (-2) = 7$
-1	$5 - (-1) = 6$
0	$5 - 0 = 5$
1	$5 - 1 = 4$
2	$5 - 2 = 3$

b)

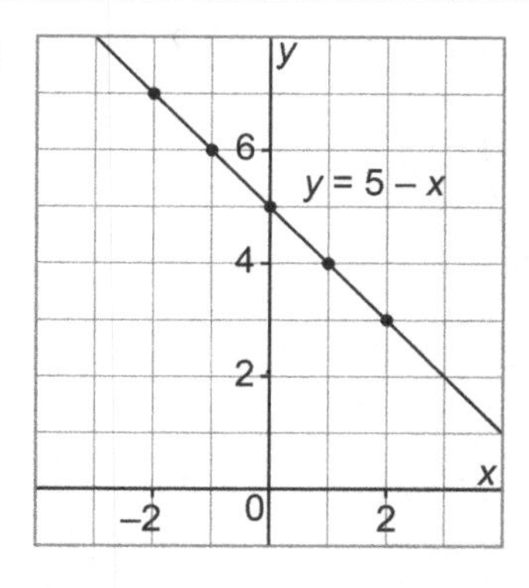

Since we can substitute any value for x, the points can be joined with a line.

c) Each point on the graph is 1 unit right and 1 unit down from the previous point. In the table, when x increases by 1, y decreases by 1.

d) This is a linear relation because its graph is a straight line.

Check

1. Complete the table of values.
Then, graph the relation.

x	$y = 4x - 2$
-1	-6
0	-2
1	______
2	______

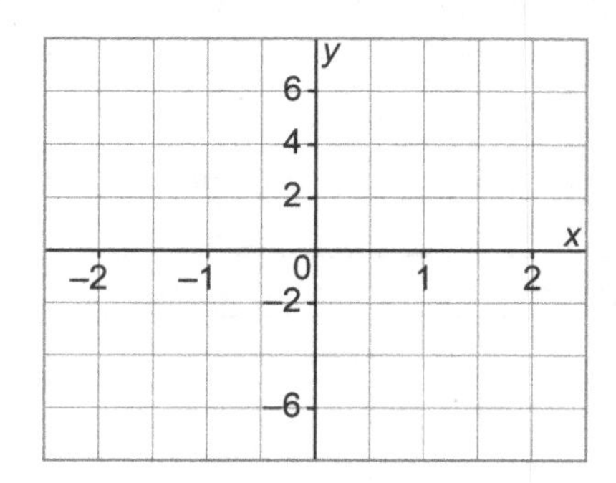

Practice

1. Which graphs represent a linear relation?

a)

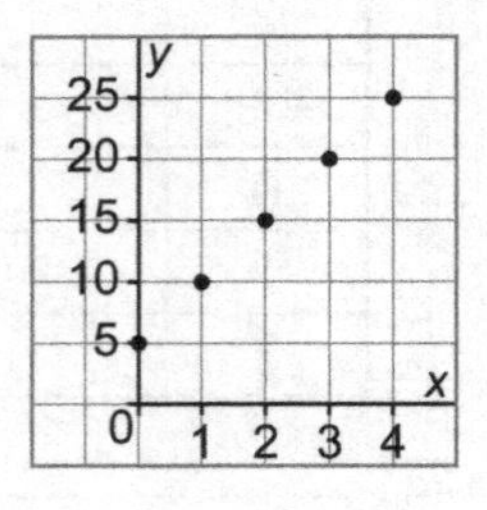

____

b)

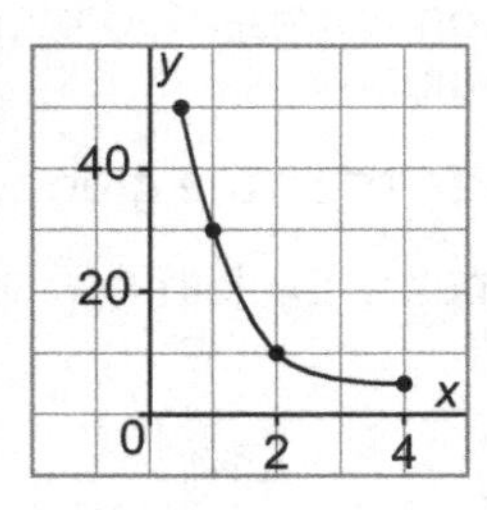

____

c)

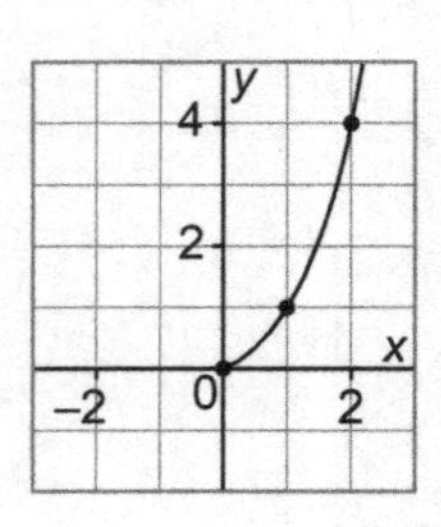

____

d)

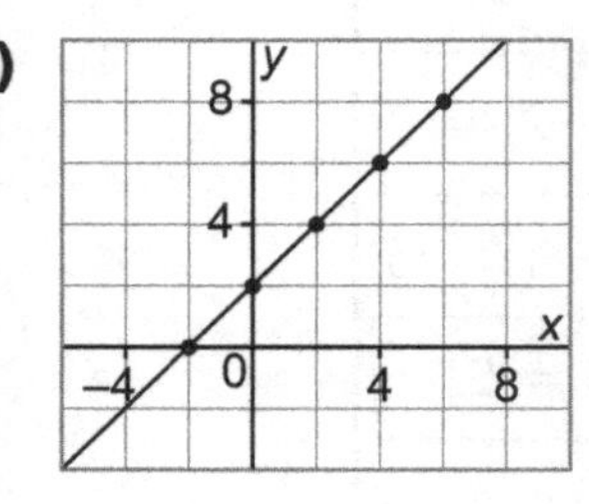

____

2. Describe the patterns in each table of values.
Does each table of values represent a linear relation?

a)

x	y
−3	6
−2	5
−1	4
0	3

x increases by _______ each time.
y decreases by _______ each time.
The relation is _______, because a constant change in x produces a constant change in y.

b)

x	y
0	1
2	4
4	7
6	10

x increases by _______ each time.
y increases by _______ each time.
The relation is _______, because a constant change in x produces a constant change in y.

c)

x	y
1	1
2	3
3	7
4	13

x increases by _______ each time.
y ____
The relation ____

3. Each graph and table of values represents a linear relation.

a) Complete each table of values.

i)

x	y
2	4
3	6
4	___
5	___

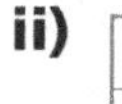

ii)

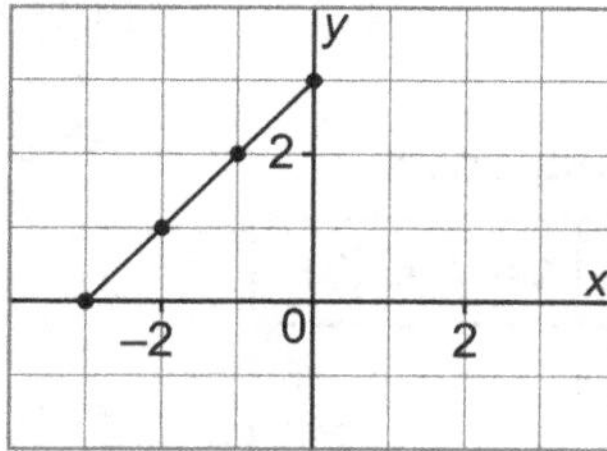

x	y
−3	0
−2	___
−1	2
0	___

b) Describe the patterns in the table.

i) When x increases by _______, y increases by _______.

ii) When _______ increases by _______, _______ increases by _______.

c) Describe the patterns in the graph.

i) To get from one point to the next, move 1 unit right and _________ up.

ii) To get from one point to the next, move _________ right and _________ up.

4. Complete the table of values for each linear relation, then graph it.

a) $y = 4x$

x	y
−1	___
0	___
1	___
2	___

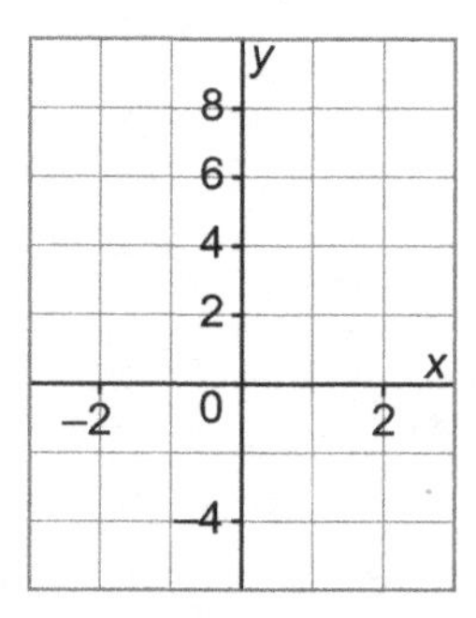

b) $y = -3x$

x	y
−1	___
0	___
___	___
___	___

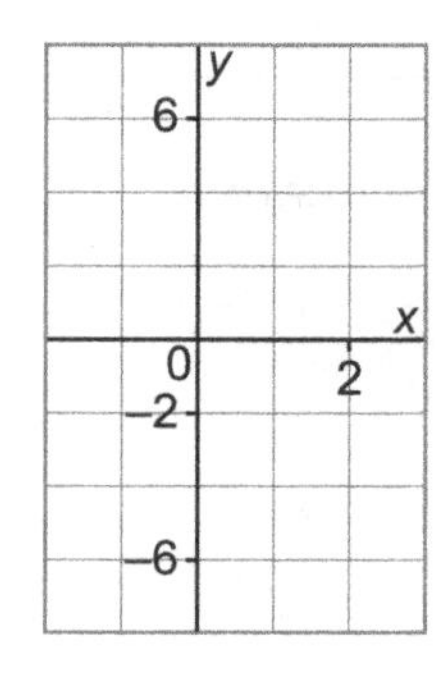

c) $y = 1 - x$

x	y
0	___
___	___
___	___
___	___

5. Complete the table of values.
Graph the data.

x	$y = 2x - 4$
−1	−6
0	−4
1	______
2	______

6. For special events, a bowling alley charges a set fee plus a fee for each hour bowled.

Bowling Costs

Hours, h	Cost, C ($)
1	40
2	50
3	60
4	70

a) Graph the data.

Bowling Costs

Does it make sense to join the points on the graph? Explain.

____

____

b) Is this a linear relation? Why?

____

____

c) Describe the pattern in words and using an equation.
When h increases by ______, C increases by ______.
$C =$ ______$h + 30$

4.3 Skill Builder

Solving Equations

Solving an equation means finding a value of the variable that makes the equation true. To solve the equation $3x - 2 = 7$, find the value of x so that the left and the right side of the equation are balanced.

$3x - 2 = 7$ Add 2 to both sides to isolate x.

$3x - 2 + 2 = 7 + 2$

$3x = 9$ Divide each side by 3.

$\frac{3x}{3} = \frac{9}{3}$

$x = 3$

To verify the equation, substitute $x = 3$ in the original equation.

Left side: $3(3) - 2$ Right side $= 7$

$= 9 - 2$

$= 7$

Since the left side equals the right side, the solution is correct.

Check

1. Solve each equation.

a) $2x + 3 = 11$

$2x + 3 -$ ____ $= 11 -$ ____

____

____

____

Check your solution.

b) $3 - 2x = -9$

____

____

____

____

4.3 Another Form of the Equation for a Linear Relation

FOCUS **Recognize the equations of horizontal, vertical, and oblique lines, and graph them.**

Example 1 Graphing and Describing Vertical Lines

For the equation $x = 4$:

a) Draw the graph.
b) Describe the graph.

Solution

$x = 4$

a) For any value of y, x is always 4.

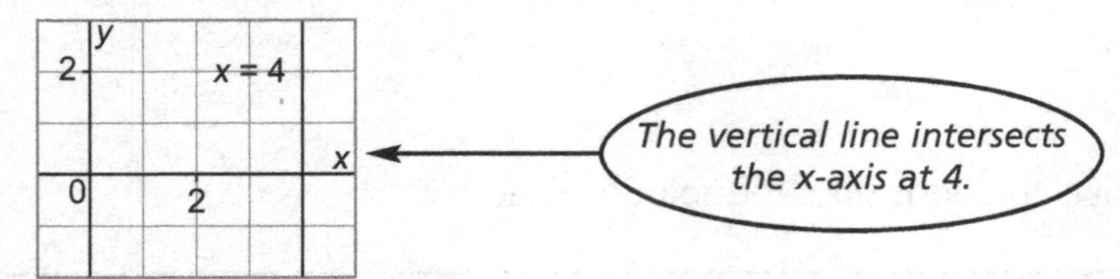

b) The graph is a vertical line.
Every point on the graph has x-coordinate 4.

When the equation is $x =$ a constant, the graph is a **vertical line.**

Check

1. For the equation $x = 1$:
Draw the graph.
Then describe the graph.

The graph is a ______________ line.
Every point on the graph has _____-coordinate _____.

Example 2 Graphing and Describing Horizontal Lines

For the equation $y + 1 = 0$:

a) Draw the graph.
b) Describe the graph.

Solution

$y + 1 = 0$ Solve for y. Subtract 1 from each side.

$y + 1 - 1 = 0 - 1$

$y = -1$

a)

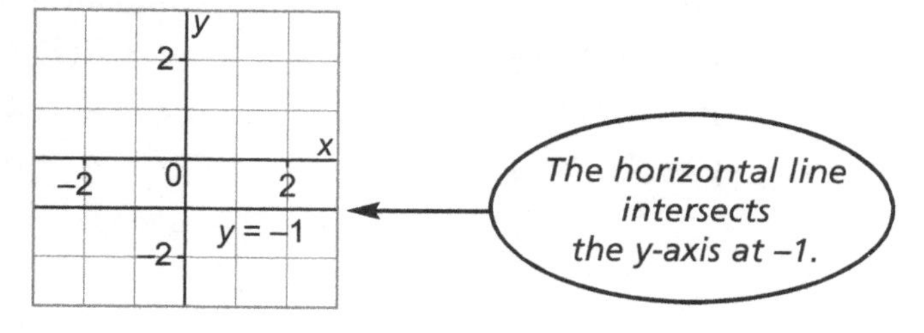

b) The graph is a horizontal line.
Every point on the graph has y-coordinate -1.

When the equation is $y =$ a constant, the graph is a **horizontal line.**

Check

1. For the equation $y = 3$:

a) Draw the graph.
b) Describe the graph.

a)

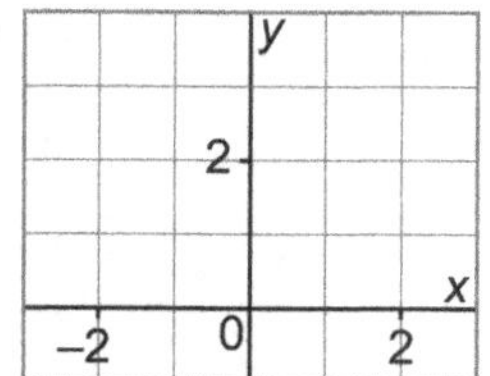

b) The graph is a ________________.
Every point on the graph has _____-coordinate _____.

Example 3 Graphing an Equation

For the equation $y + 2x = 4$:

a) Make a table of values for $x = -2$, 0, and 2.

b) Graph the equation.

Solution

a) $y + 2x = 4$
Substitute each value of x, then solve for y.

Substitute: $x = -2$
$y + 2(-2) = 4$
$y - 4 = 4$
$y - 4 + 4 = 4 + 4$
$y = 8$

Substitute: $x = 0$
$y + 2(0) = 4$
$y + 0 = 4$
$y = 4$

Substitute: $x = 2$
$y + 2(2) = 4$
$y + 4 = 4$
$y + 4 - 4 = 4 - 4$
$y = 0$

x	y
−2	8
0	4
2	0

b) Plot the points on a grid, and join the points.
The graph is an **oblique** line.

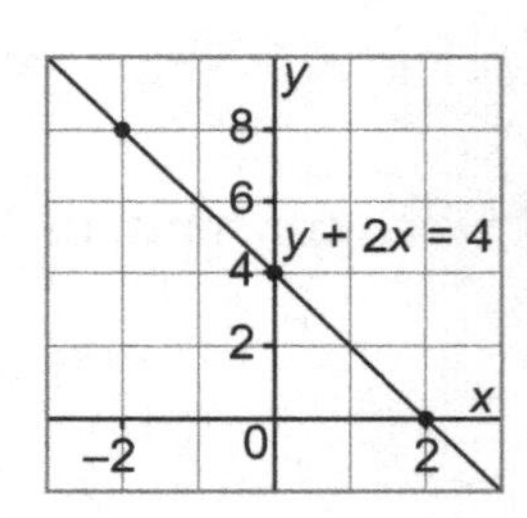

Oblique means slanted.

Check

1. Do not graph the equations.
Does each equation describe a horizontal line, a vertical line, or an oblique line?

a) $y = 4$ ______________

b) $y = 3x - 2$ ______________

c) $x = -1$ ______________

d) $2x + y = -6$ ______________

Practice

1. Which equation describes each graph?

i) $y = 4$ **ii)** $x = 2$ **iii)** $y = 2$ **iv)** $x = 4$

a)

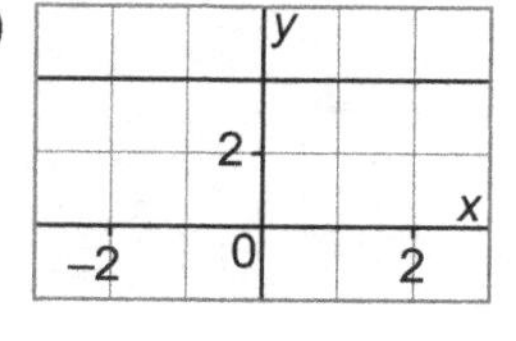

____

b)

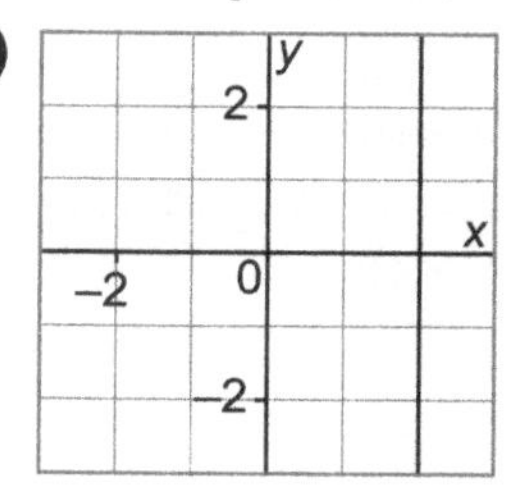

____

2. Write an equation to describe each line.

a)

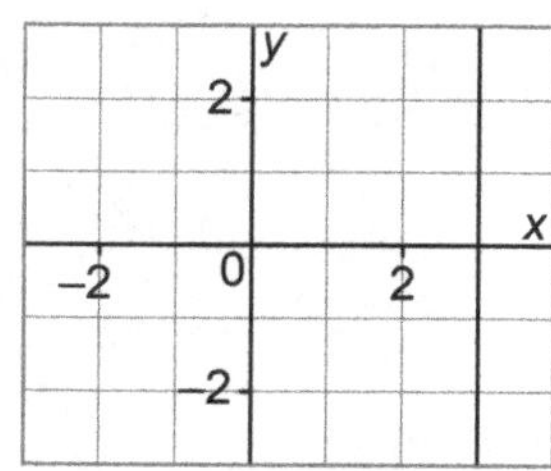

$x =$ ______________

b)

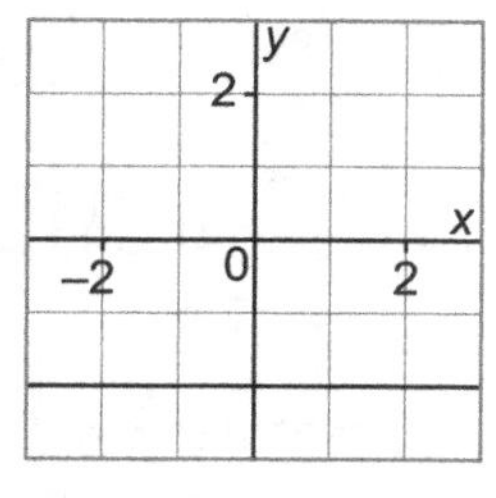

$y =$ ______________

c)

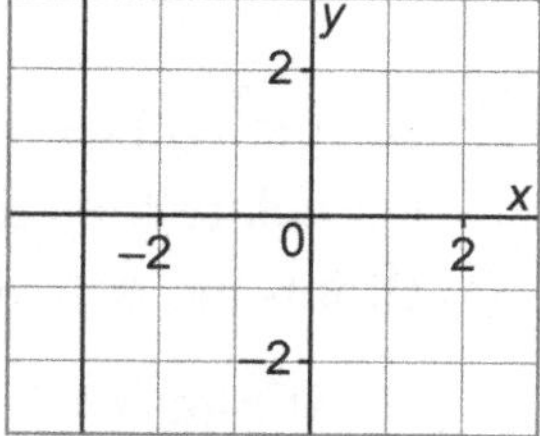

____

3. a) Is each line vertical or horizontal?

i) $x = -1$

____

ii) $y = -4$

____

b) Graph each line. Describe the graph.

i)

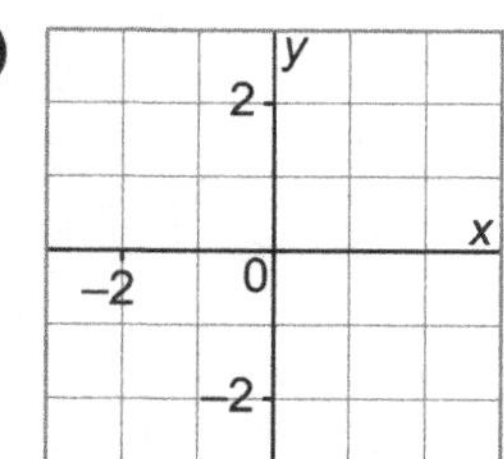

Every point on the graph has x-coordinate ______.

ii)

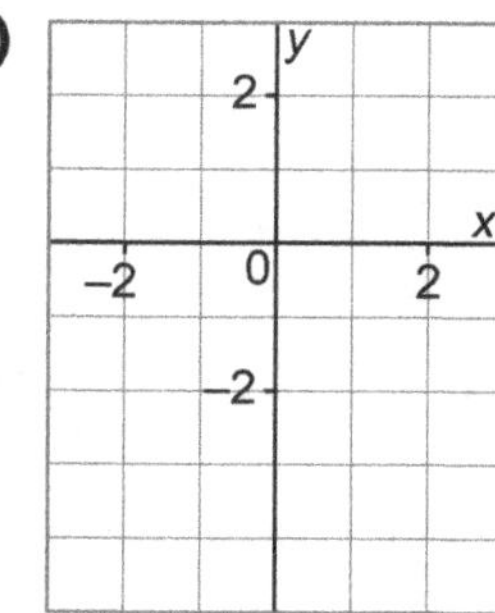

Every point on the graph has ________-coordinate ________.

4. a) Does each equation describe a vertical line, a horizontal line, or an oblique line?

i) $x + 3 = -1$ **ii)** $1 + y = 0$ **iii)** $x + 2y = 8$

_____________   _____________   _____________

b) Graph the first 2 equations in part a.

i) $x + 3 = -1$

____

____

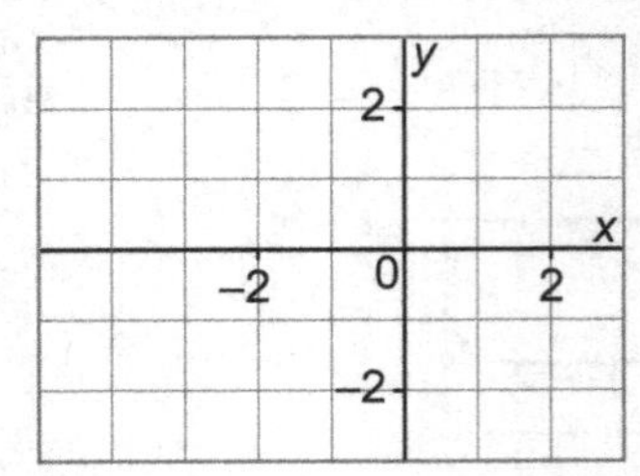

ii) $1 + y = 0$

____

____

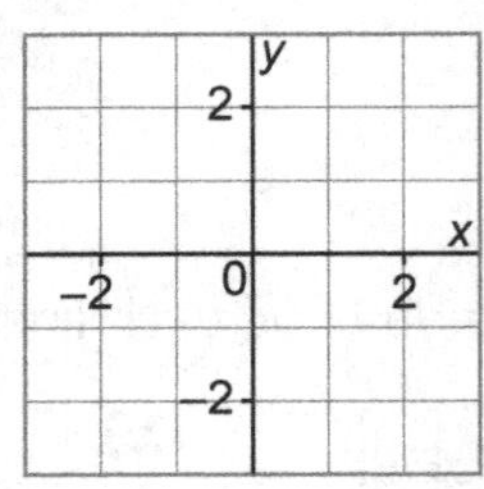

c) For the equation $x + 2y = 8$:
Complete the table of values for $x = -2$, 0, and 2.
Graph the equation.

Substitute: $x = -2$
_____ $+ 2y = 8$

____

____

____

____

Substitute: $x = 0$
_____ $+ 2y = 8$

____

____

____

____

Substitute: $x = 2$
_____ $+ 2y = 8$

____

____

____

____

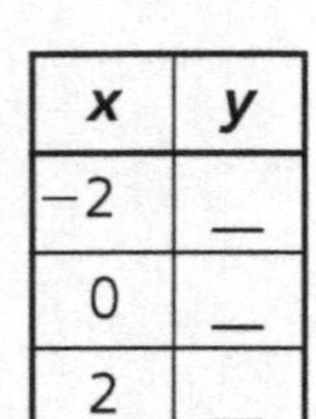

x	y
–2	__
0	__
2	__

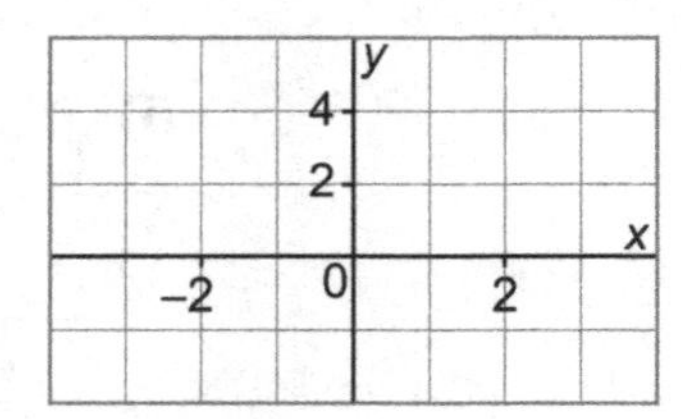

5. a) Explain why this equation describes the graph below.
$y + 3 = 0$

____

____

This is a ____________ line, with _____-coordinate _____,
which matches the graph.

Can you ...

- Use equations to describe and solve problems involving patterns?
- Graph a linear relation?
- Recognize the equations of horizontal, vertical, and oblique lines, and graph them?

4.1 **1.** This pattern of squares continues. Draw the next 2 figures in the pattern.

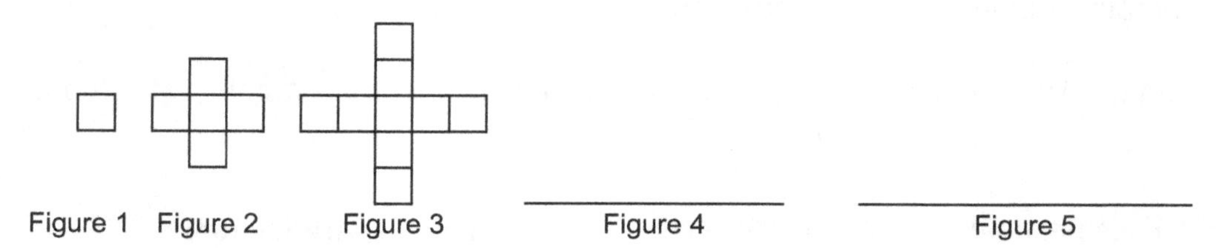

a) Complete the table of values below.

Figure Number, *n*	Number of Squares, *s*
1	1
2	5
3	___
4	___
5	___

(Figure Number: +1, +1, +1, +1; Number of Squares: +4, +4, +4, +4)

b) What patterns do you see?

The figure number increases by _____ each time.
The number of squares increases by _____ each time.

c) Describe how the number of squares relates to the figure number.

The number of squares is _____ times the figure number, less _____.

d) Write an equation for this pattern.
$s =$ _____$n -$ _____

2. The pattern in the table of values continues. Complete the table.

Number of Red Buttons, r	Number of Blue Buttons, b
2	10
3	13
4	16
5	19
____	____
____	____

a) What patterns do you see? The number of red buttons increases by ____ each time. The number of blue buttons increases by ____ each time.

b) Write an equation that relates the number of blue buttons to the number of red buttons.
$b =$ ____$r +$ ____

4.2 **3.** A home service provider charges for the service according to the table of values.

Home Service Charges

Hours, h	Charges, C (\$)
0	60
1	150
2	240
3	330

a) Graph the data.

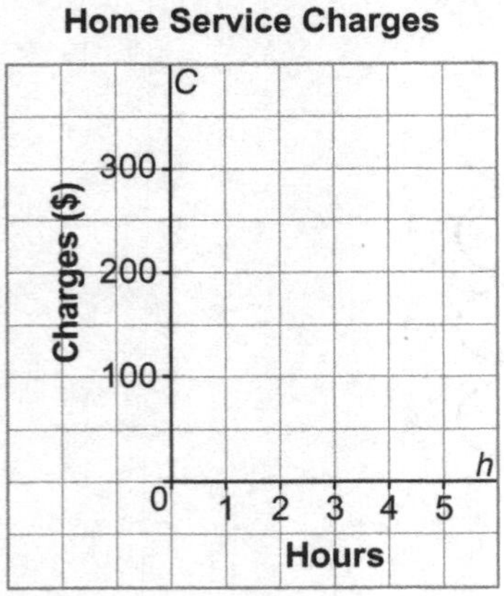

Home Service Charges

b) Is this an example of a linear relation? Why?

____

c) Describe the patterns in the table.
As h increases by ____, C increases by ____.

d) How is the pattern shown in the graph?
On the graph, to get from one point to the next, move 1 unit right, and __________ up.

e) Write an equation for this pattern.
$C =$ ____$h +$ ____

4. a) This table of values represents a linear relation.
Graph the data.

x	y
−2	−2
−1	0
0	2
1	4

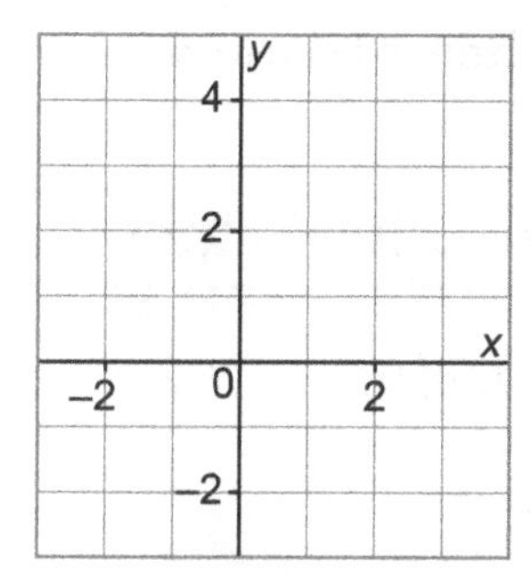

b) How do the patterns in the graph relate to the patterns in the table?
In the table, as x increases by _____, y increases by _____.
On the graph, to get from one point to the next,
move 1 unit right and _________ up.

c) Write an equation for this pattern.
$y =$ _____$x +$ _____

4.3 **5.** Does each equation describe a horizontal line, a vertical line, or an oblique line?

a) $x + 4 = 0$ ______________________

b) $y = -6$ ______________________

c) $x + y = 2$ ______________________

d) $2y = 4$ ______________________

6. Write an equation to describe each line.

a)

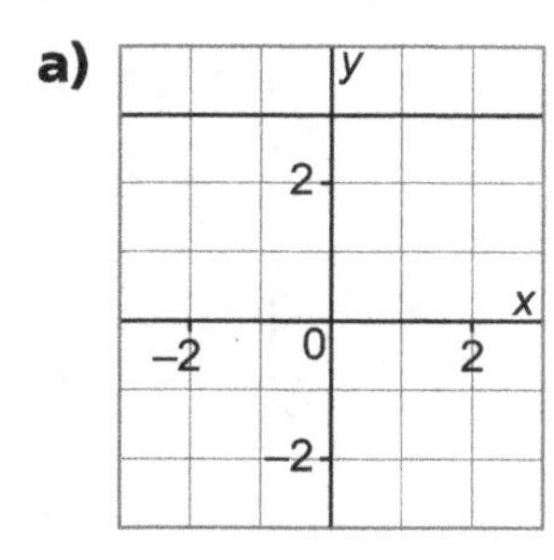

____

b)

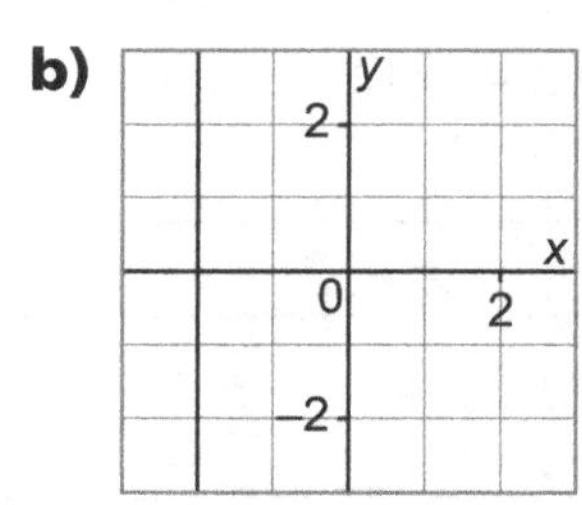

____

c)

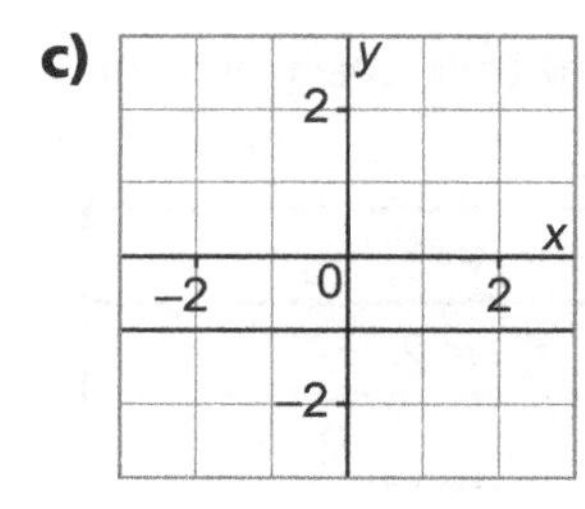

____

4.4 Matching Equations and Graphs

FOCUS **Match equations and graphs of linear relations.**

Example 1 Matching Equations with Graphs

Match each graph on the grid with its equation.

$y = x$

$y = -x$

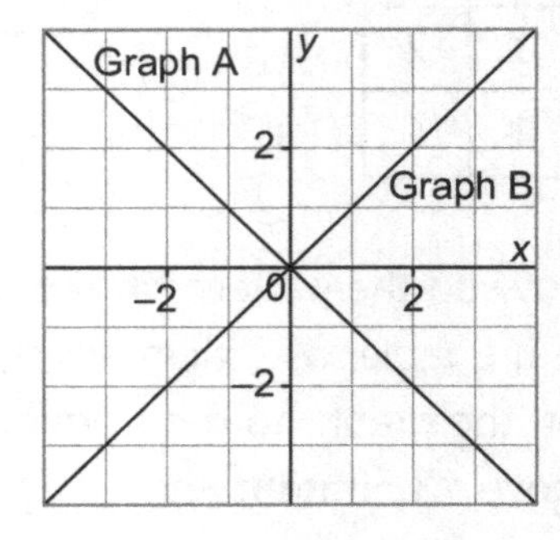

Solution

Substitute $x = -1$, $x = 0$, and $x = 1$ in each equation.

$y = x$

x	y
−1	−1
0	0
1	1

We chose to use x-values of −1, 0, and 1 because they're often easy to substitute.

Points (−1, −1), (0, 0), and (1, 1) lie on Graph B.
So, $y = x$ matches Graph B.

$y = -x$

x	y
−1	1
0	0
1	−1

Points (−1, 1), (0, 0), and (1, −1) lie on Graph A.
So, $y = -x$ matches Graph A.

Check

1. Which equation describes the graph at the right?

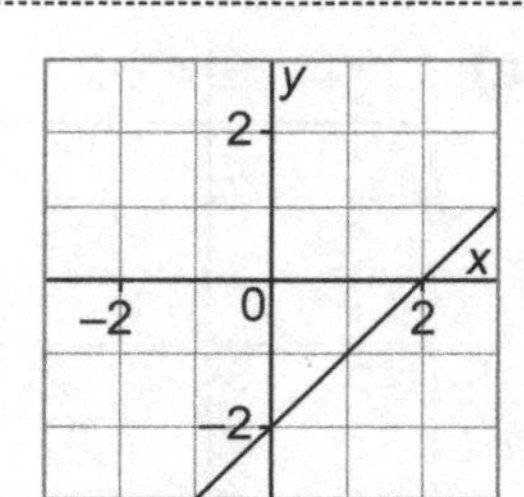

$y = x + 2$

x	$y = x + 2$
0	$y = 0 + 2 =$ ____
1	$y =$ ____ $+ 2 =$ ____
2	$y =$ ____ $+ 2 =$ ____

$y = x - 2$

x	$y = x - 2$
0	$y =$ ____ $- 2 =$ ____
1	$y =$ ________ $=$ ____
2	$y =$ ________ $=$ ____

Points (______), (______), and (______) do not lie on the graph.
Points (______), (______), and (______) lie on the graph.

So, the equation $y =$ ______ describes the graph.

Example 2 Identifying a Graph Given Its Equation

Which graph on this grid has the equation $y = x - 1$?

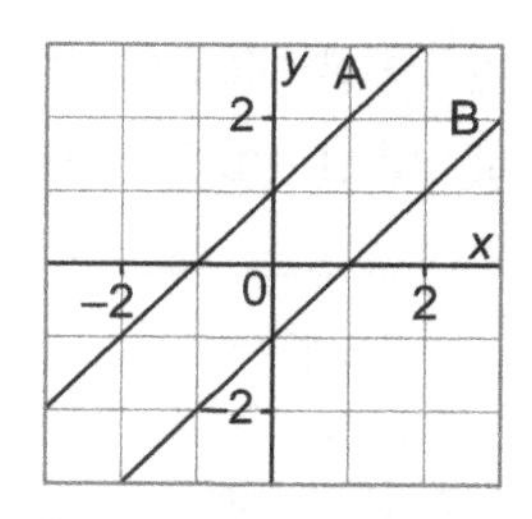

Solution

Pick 2 points on each graph and check if their coordinates satisfy the equation.

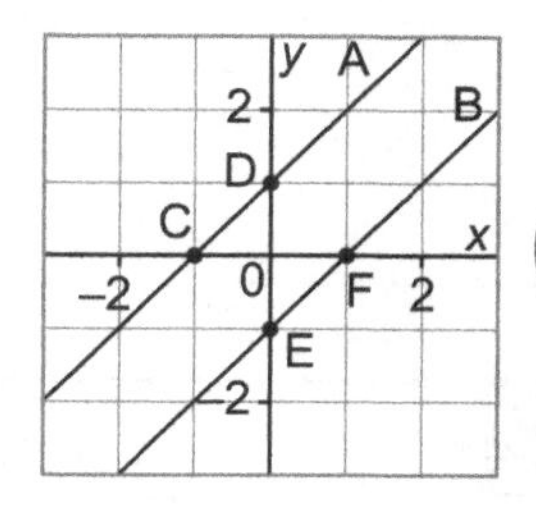

Since C does not work, we do not have to check for D.

For Graph A, use: C(−1, 0) and D(0, 1)
Check if C(−1, 0) satisfies the equation $y = x - 1$.
Substitute $x = -1$ and $y = 0$ in $y = x - 1$
Left side: $y = 0$ Right side: $x - 1 = (-1) - 1$
$= -2$

The left side does not equal the right side.
So, Graph A does not have equation $y = x - 1$.

Verify that the other graph does match the equation.
For Graph B, use: E(0, −1) and F(1, 0)
Check if E(0, −1) satisfies the equation $y = x - 1$.
Substitute $x = 0$ and $y = -1$ in $y = x - 1$
Left side: $y = -1$ Right side: $x - 1 = 0 - 1$
$= -1$

The left side does equal the right side.
So, E(0, −1) lies on the line represented by $y = x - 1$.

Check if F(1, 0) satisfies the equation $y = x - 1$.
Substitute $x = 1$ and $y = 0$ in $y = x - 1$
Left side: $y = 0$ Right side: $x - 1 = 1 - 1$
$= 0$

The left side does equal the right side.
So, F(1, 0) lies on the line represented by $y = x - 1$.
So, Graph B has equation $y = x - 1$.

Check

1. Show that this graph has equation $y = 2x + 1$.

Use the points labelled on the graph.

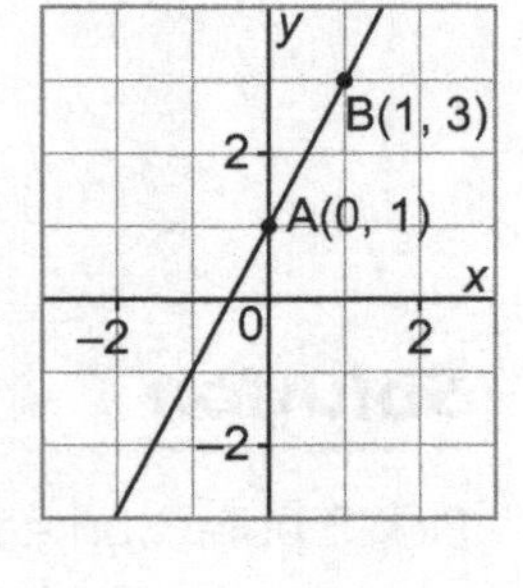

For A(0, 1): Substitute $x = 0$ and $y = 1$ in $y = 2x + 1$.

Left side: $y =$ __________ Right side: $2x + 1 =$ __________

$=$ __________

$=$ __________

For B(1, 3): Substitute $x =$ ______ and $y =$ ______ in $y = 2x + 1$.

Left side: $y =$ ______ Right side: $2x + 1 =$ __________

$=$ __________

$=$ __________

____

Practice

1. Show that the equation $y = x + 2$ matches the graph.

Fill in the table of values.

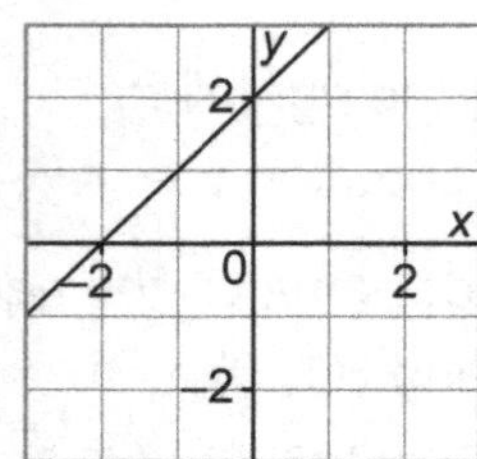

x	$y = x + 2$
–2	$y = -2 + 2 =$ ___
–1	$y =$ ___ $+ 2 =$ ___
0	$y =$ ____ $=$ ___

From the table:

Points (______), (______), and (______) lie on the graph.

So, $y = x + 2$ matches the graph.

2. Match each equation with a graph.

$y = 3x$ $y = -3x$

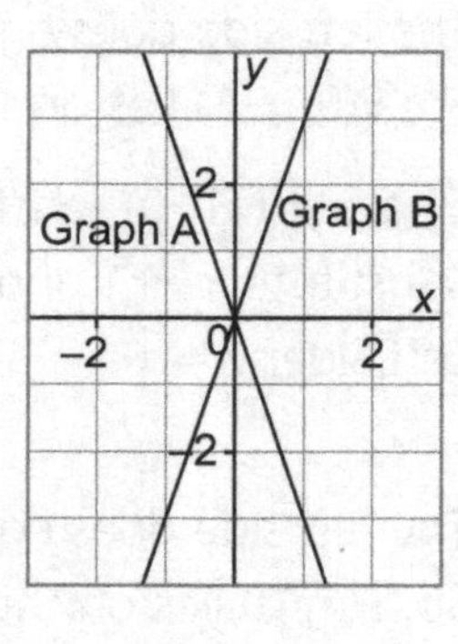

Fill in the tables of values.

x	$y = 3x$
–1	$y = 3($___$) =$ ___
0	$y = 3($___$) =$ ___
1	$y = 3($___$) =$ ___

x	$y = -3x$
–1	$y = -3($___$) =$ ___
0	$y =$ ___$($___$) =$ ___
1	$y =$ ______ $=$ ___

From the tables:

$y = 3x$ has points (________), (________), and (________).

These points lie on Graph ________.

So, $y = 3x$ matches Graph ________.

$y = -3x$ has points (________), (________), and (________).

These points lie on Graph ________.

So, $y = -3x$ matches Graph ________.

3. Match each equation with a graph.

$y = 1 - x$

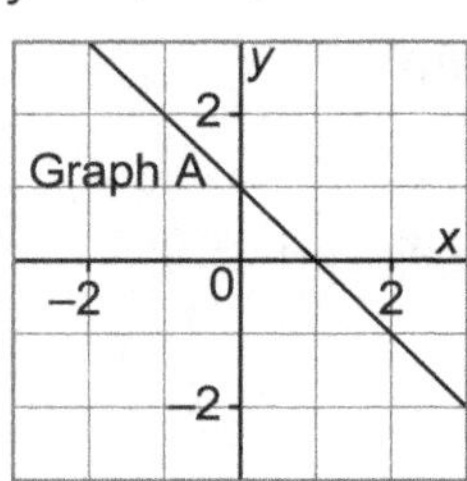

$y = x - 1$

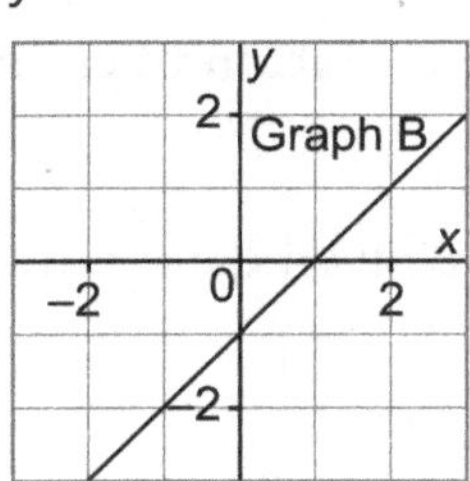

Fill in the tables of values.

x	$y = 1 - x$
-1	$y = 1 - (\_\_\_) = \_\_\_\_$
0	$y = 1 - \_\_\_\_ = \_\_\_\_$
1	$y = 1 - \_\_\_\_ = \_\_\_\_$

x	$y = x - 1$
-1	$y = \_\_\_\_\_\_\_\_ = \_\_\_\_$
0	$y = \_\_\_\_\_\_\_\_ = \_\_\_\_$
1	$y = \_\_\_\_\_\_\_\_ = \_\_\_\_$

From the tables:

$y = 1 - x$ has points (______), (______), and (______).

These points lie on Graph ______.

So, $y = 1 - x$ matches Graph ______.

$y = x - 1$ has points (______), (______), and (______).

These points lie on Graph ______.

So, $y = x - 1$ matches Graph ______.

4. Which graph has equation $y = x - 3$?

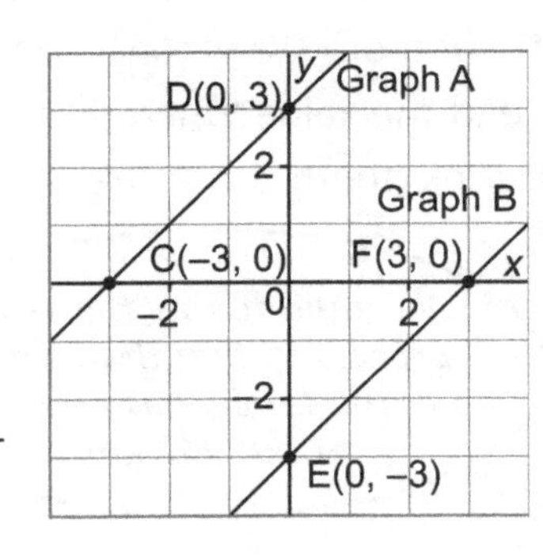

For C(−3, 0):

Left side: $y =$ ______ Right side: $x - 3 =$ ______

$=$ ______

The left side _________ equal the right side.

____

For E(0, −3):

Left side: $y =$ ______ Right side: $x - 3 =$ ______

$=$ ______

The left side ________ the right side.

For F(3, 0):

Left side: $y =$ ______ Right side: $x - 3 =$ ______

$=$ ______

____

So, Graph ____ has equation $y = x - 3$.

4.5 Using Graphs to Estimate Values

FOCUS **Use interpolation and extrapolation to estimate values on a graph.**

When we estimate values between 2 given data points on a graph of a linear relation, we use **interpolation.**

Example 1 Using Interpolation to Solve Problems

This graph shows the distance travelled by Bobbie's family on a trip from Calgary to Moose Jaw. How long did it take his family to travel 320 km?

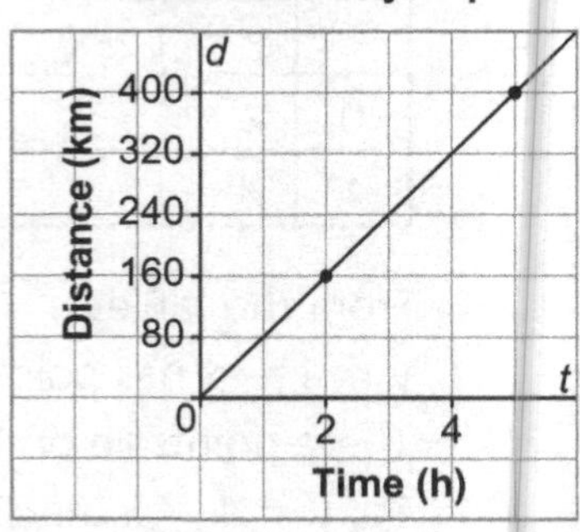

Solution

To find how long it took to travel 320 km:

- Locate the point on the vertical axis that represents 320 km.
- Draw a horizontal line to the graph.
- Then draw a vertical line from the graph to the horizontal axis.

Read the value where the vertical line meets the horizontal axis.
It took about 4 h to travel 320 km.

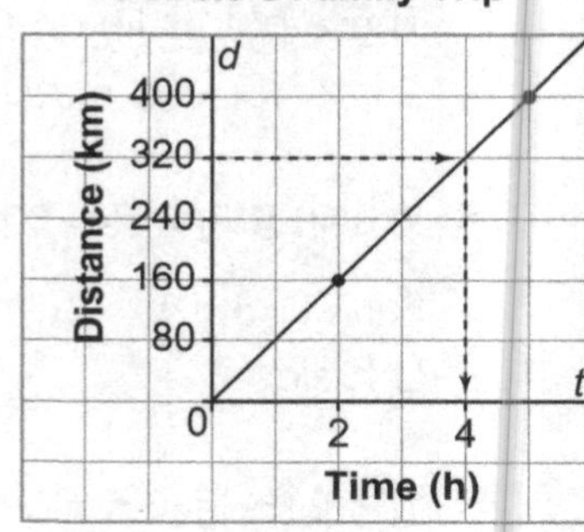

We could follow the same process to find that, after 3 h, the family has travelled about 240 km.

Check

1. Use the graph to find the following values.

a) The cost of 15 L of fuel.
About $______.

b) The quantity of fuel that can be purchased for $10.
About ______ L.

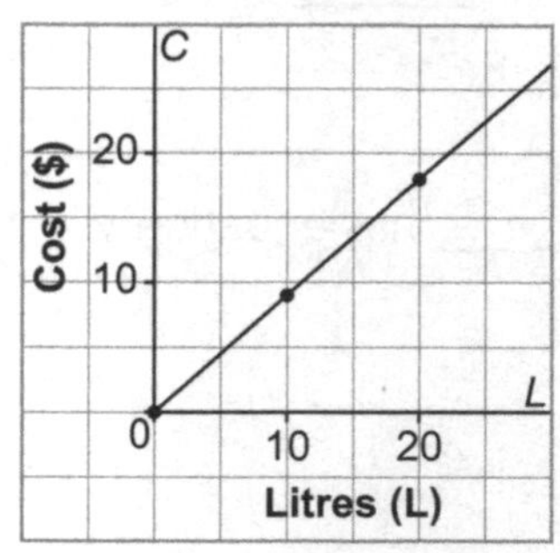

When we extend a graph of a linear relation to estimate values that lie beyond the graph, we use **extrapolation.**

Example 2 Using Extrapolation to Solve Problems

On his family trip from Calgary to Moose Jaw, Bobbie wants to predict how long it will take to travel 640 km.

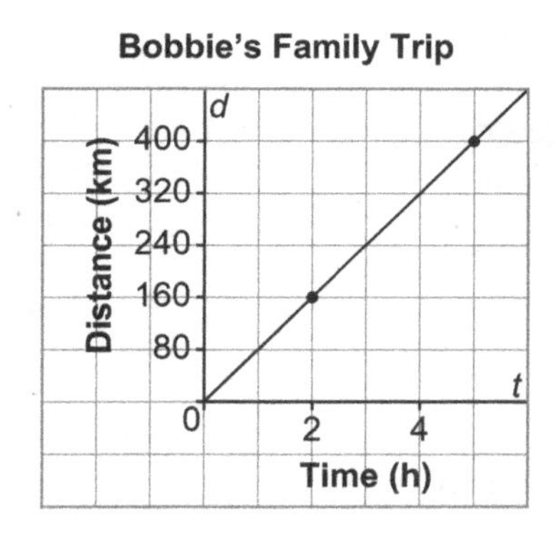

We assume that Bobbie's family will continue to travel at the same average speed.

Solution

Since the relation appears to be linear, we can extend the graph.

- Locate the point on the vertical axis that represents 640 km.
- Draw a horizontal line to the graph.
- Then draw a vertical line from the graph to the horizontal axis.

Read the value where the vertical line meets the horizontal axis. It will take about 8 h to travel 640 km.

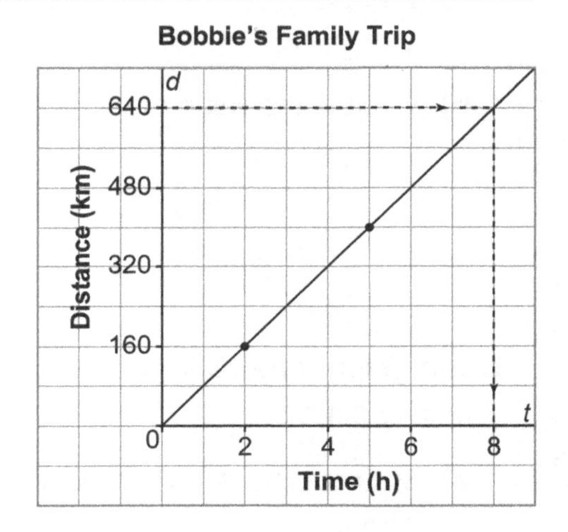

Check

1. Use the graph to find the cost of 30 L of fuel.

____

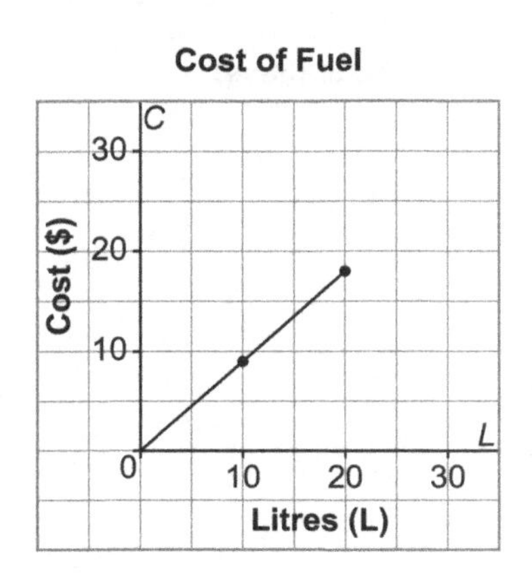

Practice

1. Use this graph of a linear relation.

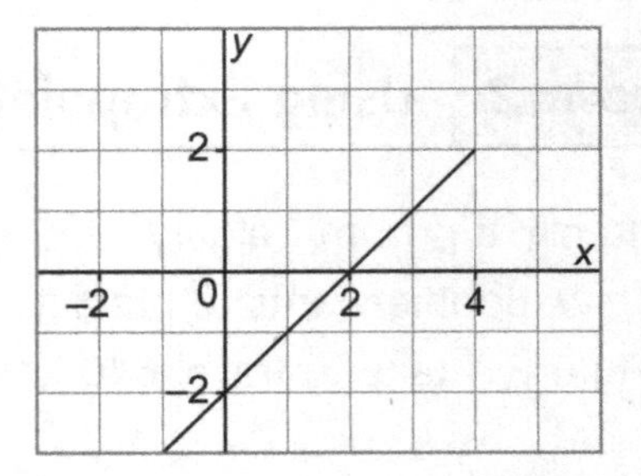

a) What is the value of x when $y = 3$?

$x =$ ________

b) What is the value of y when $x = 1$?

$y =$ ________

2. This graph shows a linear relation.

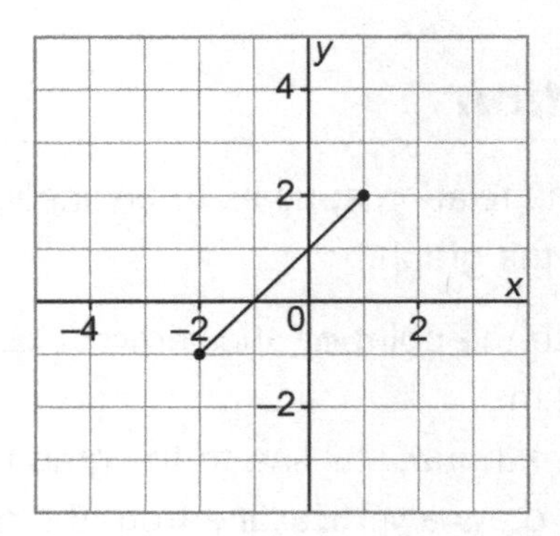

a) What is the value of x when $y = 4$?

$x =$ ________

b) What is the value of y when $x = -4$?

$y =$ ________

3. This graph shows a linear relation for different drilling depths.

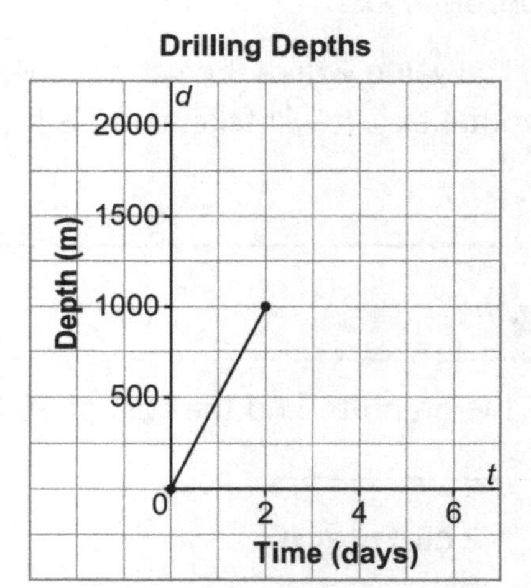

a) Estimate the depth drilled in 1 day.

About ________ m

b) Estimate the time taken to drill to a depth of 750 m.

About ________ days

c) Estimate the depth that will be drilled in 3 days.

About ________ m

d) Estimate the time it will take to drill 2000 m.

About ________ days

Unit 4 Puzzle

A Graphing Perspective

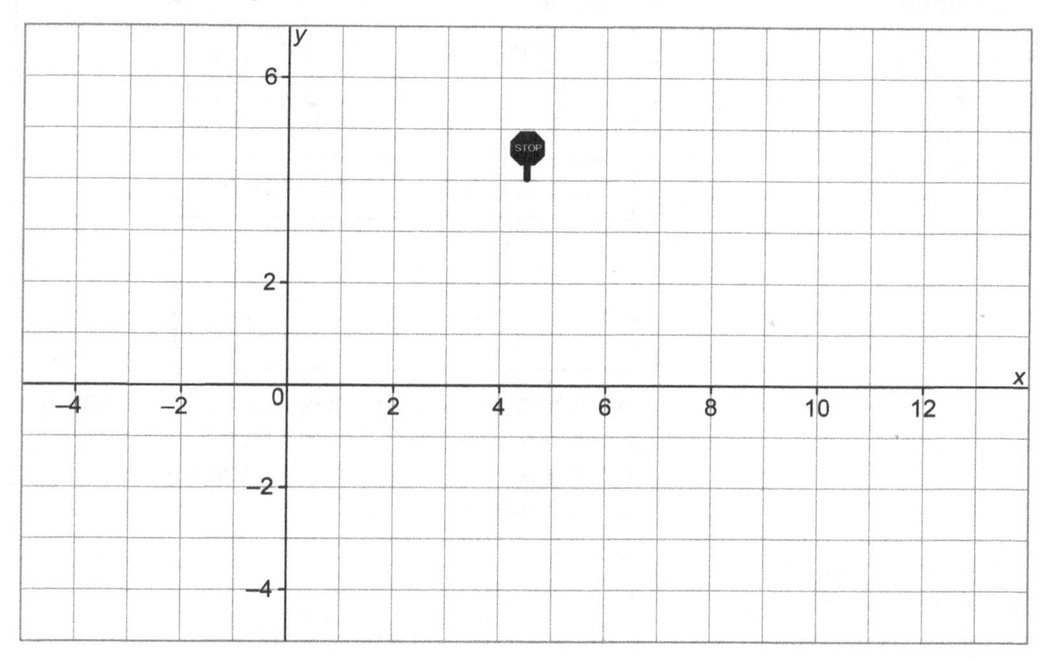

1. On the grid, plot the lines represented by:

a) $y = 4$

b) $x = y$

x	y
−4	___
0	___
4	___

c)

x	y
5	4
9	0
13	−4

2. Plot (4.5, 4) and (4.5, 3), and join the points with a line.

3. Plot (4.5, 2) and (4.5, 0), and join the points with a line.

4. Plot (4.5, −1) and (4.5, −4), and join the points with a line.

What do you see?

____

Unit 4 Study Guide

<table>
<tr><th>Skill</th><th>Description</th><th>Example</th></tr>
<tr><td>Generalize a pattern</td><td>Recognize and extend a pattern using a drawing and a table of values. Describe the pattern. Write an equation for the pattern.</td><td>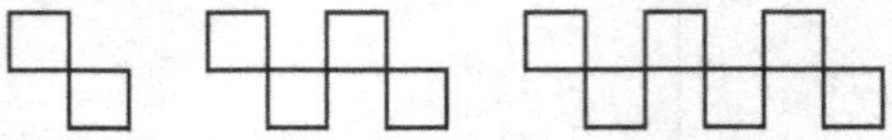
Figure 1 Figure 2 Figure 3
<table>
<tr><th>Figure Number, n</th><th>Figure Value, v</th></tr>
<tr><td>1</td><td>2</td></tr>
<tr><td>2</td><td>4</td></tr>
<tr><td>3</td><td>6</td></tr>
</table>
As the figure number increases by 1, the figure value increases by 2. The pattern is: multiply the figure number by 2 to get the figure value. An equation is: $v = 2n$</td></tr>
<tr><td>Linear relations</td><td>The points on the graph of a linear relation lie on a straight line. To graph a linear relation, create a table of values first. In a linear relation, a constant change in x produces a constant change in y.</td><td>
<table>
<tr><th>x</th><th>y</th></tr>
<tr><td>−2</td><td>0</td></tr>
<tr><td>−1</td><td>1</td></tr>
<tr><td>0</td><td>2</td></tr>
</table>
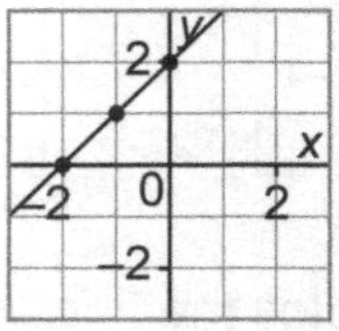

As x increases by 1, y increases by 1.</td></tr>
<tr><td>Horizontal and vertical lines</td><td>A vertical line has equation $x = a$
A horizontal line has equation $y = b$</td><td>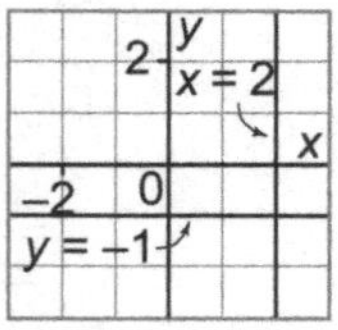

The graph of $x = 2$ is a vertical line. Every point on the line has x-coordinate 2.
The graph of $y = -1$ is a horizontal line. Every point on the line has y-coordinate -1.</td></tr>
<tr><td>Interpolation and extrapolation</td><td>When we estimate values between 2 given points on a graph, we use interpolation. When we estimate values beyond given points on a graph, we use extrapolation.</td><td>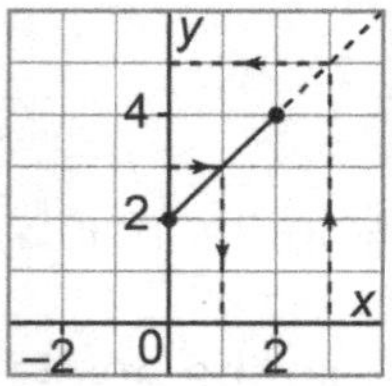

When $y = 3$, $x = 1$
Extend the graph to find that, when $x = 3$, $y = 5$</td></tr>
</table>

Unit 4 Review

4.1 **1.** This pattern continues.

a) Draw the next 2 figures in the pattern.

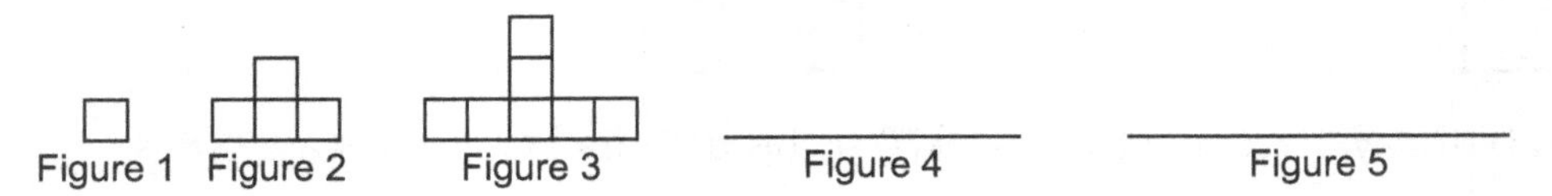

b) Complete the table of values.

Figure Number, n	Number of Squares, s
1	1
2	4
3	7
____	____
____	____

c) Describe the patterns in the table.
The figure number increases by _____ each time.
The number of squares increases by _____ each time.

d) Write an equation that relates the number of squares to the figure number.
$s =$ _____$n -$ _____

e) What is the number of squares in figure 10?
When $n = 10$:
$s =$ ______ $- 2 =$ ______ $- 2 =$ ______
There are ______ squares in figure 10.

2. The pattern in this table of values continues.

a) Complete the table.

b) Which expression below represents the number of squares in terms of the figure number? ________

i) $5n$

ii) $5n - 4$

iii) $n + 4$

iv) $n - 4$

Figure Number, n	Number of Squares, s
1	5
2	6
3	7
4	____
5	____

4.2 **3.** Complete each table of values.

a) $y = x + 1$

x	y
1	___
2	___
3	___
4	___

b) $y = x - 1$

x	y
2	___
4	___
6	___
8	___

4. On his first birthday, Hayden was given $20 by his grandfather. Each year's gift is $10 more than the year before. The data is given in the table below.

Grandfather's Gifts

Birthday, n	Gift, g ($)
1	20
2	30
3	40
4	50

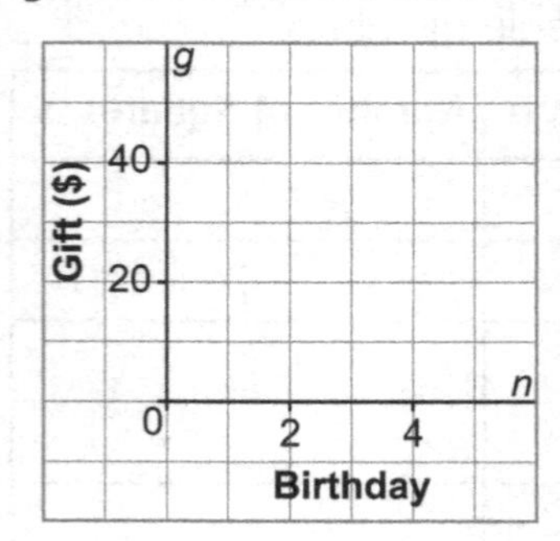

a) Graph the data.

b) Is the graph linear? Explain your thinking.
The points ______________________, so the graph is ________.

c) Should the points be joined? Explain why or why not.

____

____

d) How are the patterns in the table shown in the graph? In the table, as the birthday increases by ____, the gift value increases by ____. Each point on the graph is ______________ and ______________ from the previous point.

4.3 **5.** Write an equation to describe each line.

a)

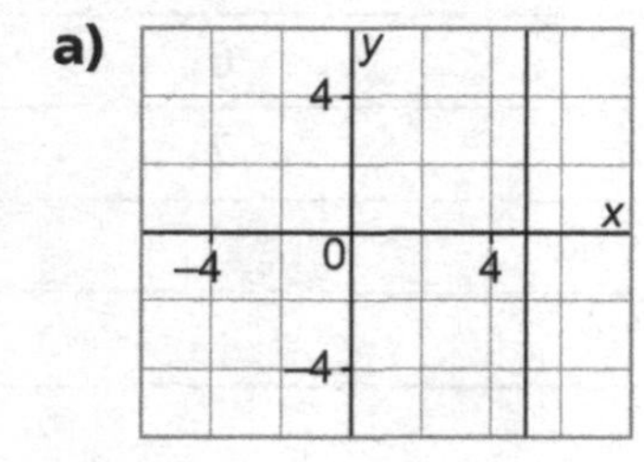

____

b)

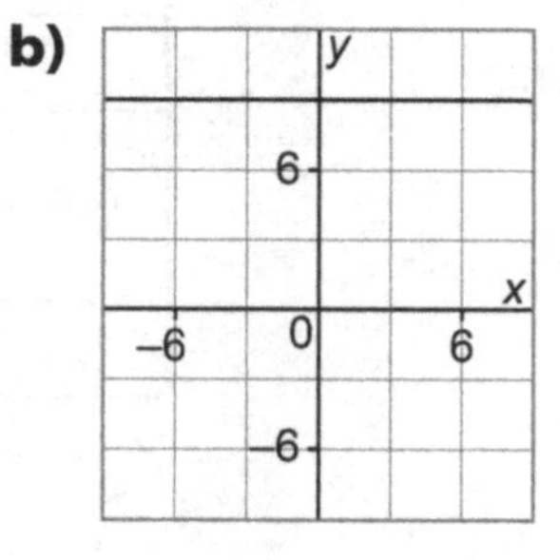

____

c)

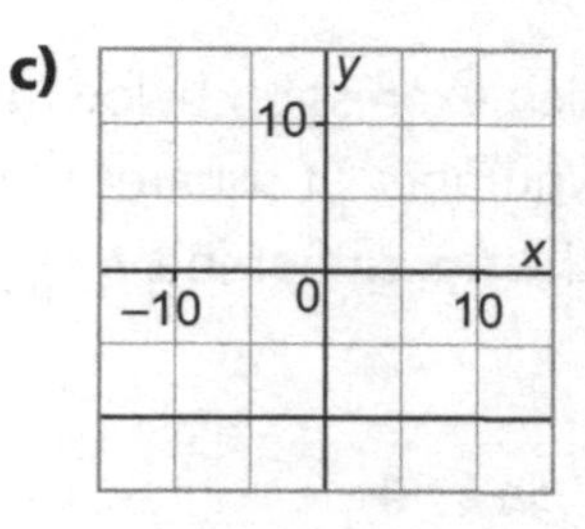

____

6. Does each equation represent a horizontal line, a vertical line, or an oblique line?

a) $x = 2$ ____________ **b)** $y = 2x + 2$ ____________ **c)** $y = 3$ ____________ **d)** $x = -1$ ____________

Draw a graph for each equation above.

a)

b)

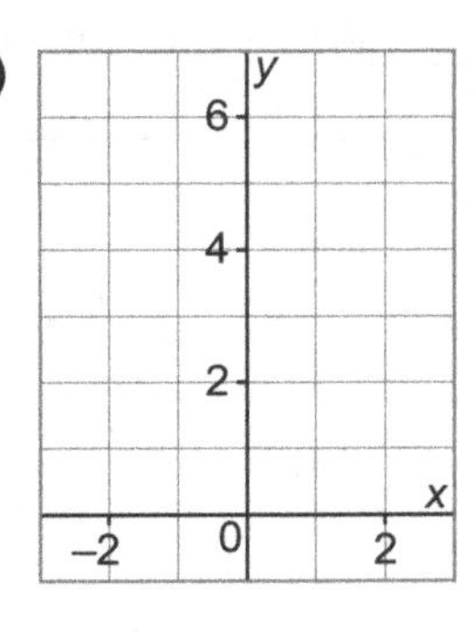

x	y
____	____
____	____
____	____

c)

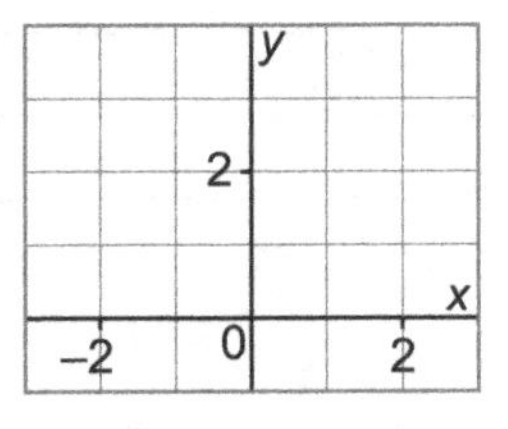

d)

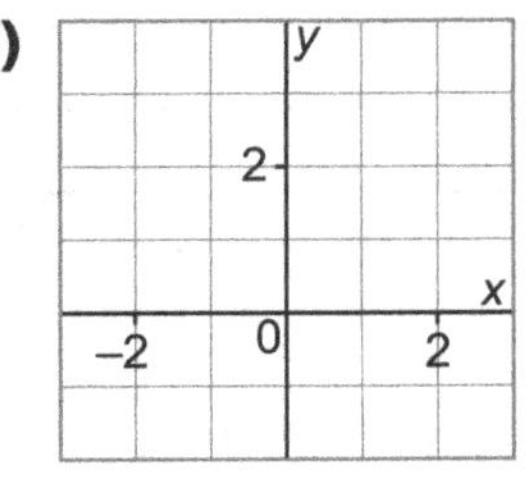

4.4 **7.** Which equation describes the graph?
$y = 2x$ or $y = -2x$
Fill in the tables of values.

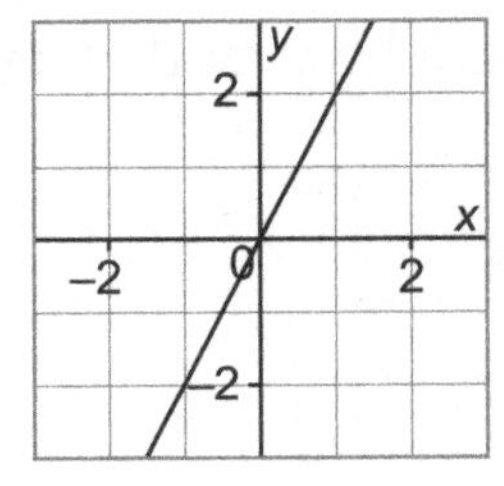

x	$y = 2x$
−1	2(____) = ____
0	2(____) = ____
1	2(____) = ____

x	$y = -2x$
−1	−2(____) = ____
0	________ = ____
1	________ = ____

From the tables:
$y = 2x$ has points (________), (________), and (________).
$y = -2x$ has points (________), (________), and (________).
The graph passes through the points (________), (0, 0), and (________).
So, $y =$ ______ describes the graph.

8. Which graph represents the equation $x - y = 2$?

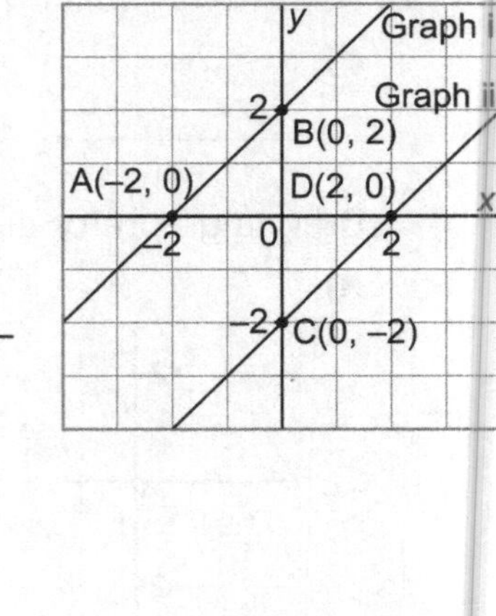

For A(−2, 0):

Left side: $x - y =$ ________ Right side: ________

$=$ ________

The left side ________ equal the right side.

____

For C(0, −2):

Left side: $x - y =$ ________ Right side: ________

$=$ ________

The left side ________ the right side.

For D(2, 0):

____

____

____

So, Graph ______ has equation $x - y = 2$.

4.5 **9.** This graph shows Emma's and Julianna's journey from Saskatoon to Prince Albert.

When Emma and Julianna have travelled 100 km, about how far do they still have to go?

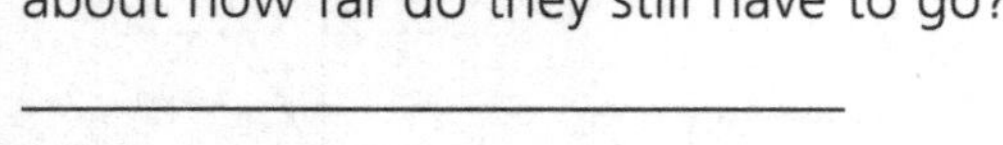

Journey from Saskatoon to Prince Albert

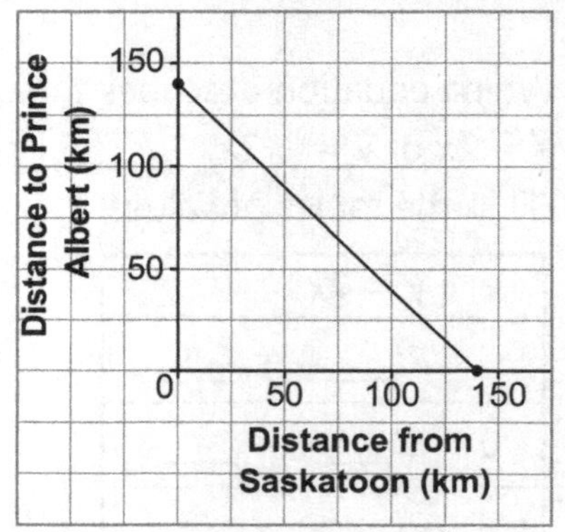

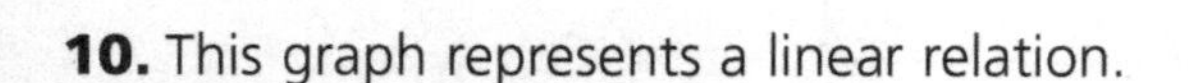

10. This graph represents a linear relation.

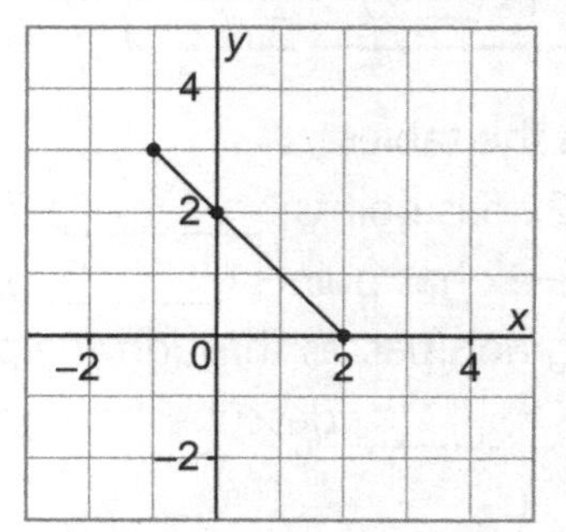

a) Estimate the value of y when:

i) $x = 0$ ________ **ii)** $x = 1$ ________

b) Estimate the value of x when:

i) $y = 4$ ________ **ii)** $y = -2$ ________

UNIT 5

Polynomials

What You'll Learn

- Recognize, write, describe, and classify polynomials.
- Represent polynomials using tiles, pictures, and algebraic expressions.
- Add and subtract polynomials.
- Multiply and divide a polynomial by a monomial.

Why It's Important

Polynomials are used by

- homeowners to calculate mortgage and car payments
- computer technicians to encode information, such as PIN numbers for ATM machines and debit cards

Key Words

term
variable term
constant term
variable
coefficient of the variable
polynomial
degree of a polynomial
monomial
binomial
trinomial
simplify a polynomial
like terms
unlike terms
distributive property

5.1 Skill Builder

Modelling Expressions

We can use algebra tiles to model an expression.
One represents +1. One represents −1.

One represents any variable, such as x or n.

One represents $-x$ or $-n$.

There are 2 . They represent $2x$.

There is 1 . It represents −1.

So, the tiles represent the expression $2x - 1$.

We can use any letter as the variable.

There are 3 . They represent $-3a$.

There are 2 . They represent +2.

So, the tiles represent the expression $-3a + 2$.

Check

1. Which expression does each set of tiles represent?

a) ____________

b) 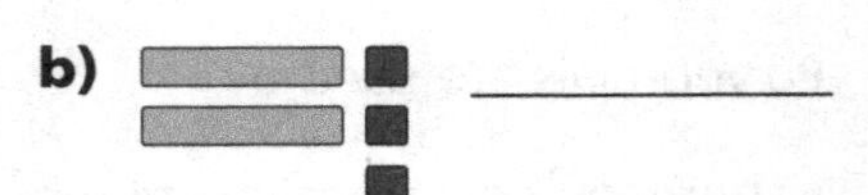 ____________

c) 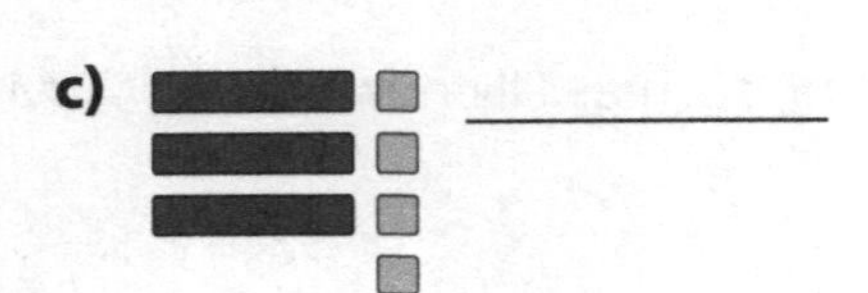 ____________

d) 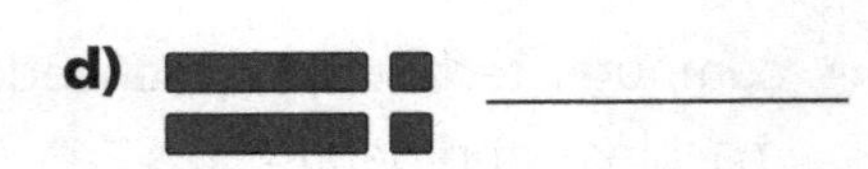 ____________

2. Sketch algebra tiles to model each expression.

a) $s + 4$

b) $5b - 3$

c) $-4n + 5$

d) $-6w - 1$

5.1 Modelling Polynomials

FOCUS **Model, write, and classify polynomials.**

Some expressions contain x^2 terms.

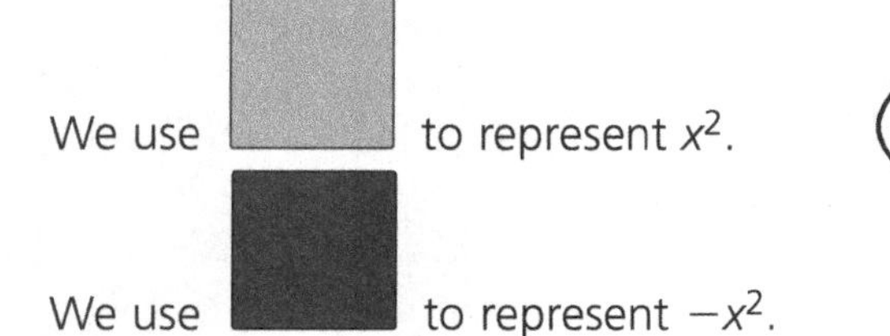

We use [light tile] to represent x^2.

We use [dark tile] to represent $-x^2$.

When the variable is n, the tile is called the n^2-tile.

For the expression $4x^2 + 3x - 1$:

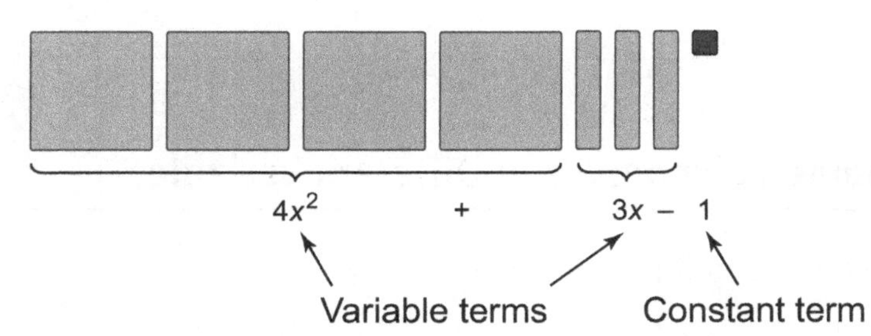

In the term $3x$, the **variable** is x and the **coefficient of the variable** is 3.

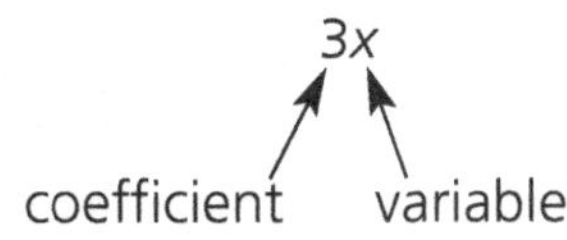

An algebraic expression like this one is also called a **polynomial**.

Example 1 Modelling Polynomials with Algebra Tiles

Use algebra tiles to model each polynomial.

a) $-4t^2$

b) $2n - 5$

Solution

a) To represent $-4t^2$, use 4 [dark tile].

b) To represent $2n$, use 2 [light rectangle].
To represent -5, use 5 [small dark tile].

Check

1. Sketch algebra tiles to model each polynomial.

a) -3

b) $2x + 3$

c) $2e^2 - e + 2$

d) $-3d^2 + 2d - 5$

Example 2 Recognizing the Same Polynomials in Different Variables

Which of these polynomials can be represented by the same algebra tiles?

a) $2x^2 + 7x - 4$

b) $-4 + 2b^2 - 7b$

c) $7s - 4 + 2s^2$

Solution

Select the tiles that match each term.

a) $2x^2 + 7x - 4$

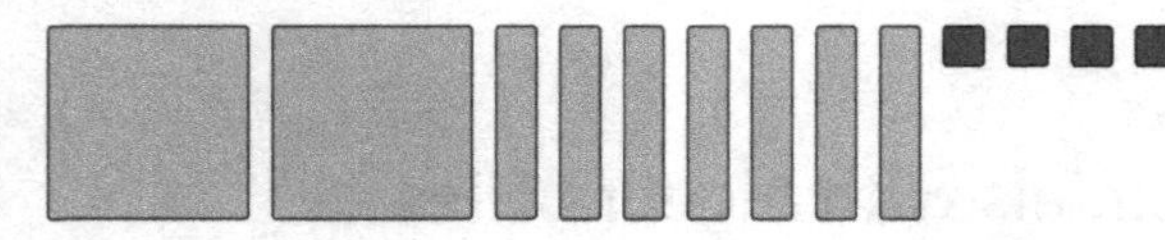

b) $-4 + 2b^2 - 7b$

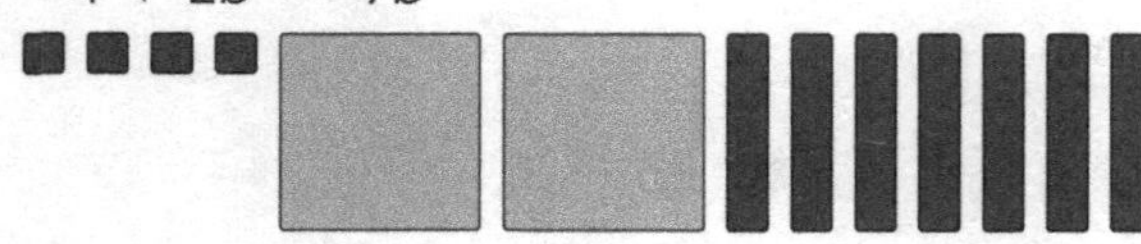

c) $7s - 4 + 2s^2$

The variable used to name a tile does not matter.
In parts a and c, the same algebra tiles are used.
Since $2x^2 + 7x - 4$ and $7s - 4 + 2s^2$ can be represented by the same tiles, the expressions represent the same polynomial.

The order in which the terms are written does not matter.

Check

1. Which of these polynomials can be represented by the same algebra tiles?

a) $3s^2 - 2s + 5$

b) $5 - 3a^2 - 2a$

c) $-2c + 5 - 3c^2$

The same tiles are used in parts _____ and _____.
So, ________________ and ________________ represent the same polynomial.

There are different **types** of polynomials, depending on the number of terms.
The **degree of a polynomial** tells you the greatest exponent of any term.

Type	Number of Terms	Example	Model	Degree
Monomial	1	$2s^2$		2
		$-2n$		1
		4		2
Binomial	2	$x^2 + 3$		2
		$2a - 1$		1
		$-2b^2 + 3b$		2
Trinomial	3	$-c^2 + 4c - 2$		2

A monomial has 1 type of tile.

A constant term has degree 0.

A binomial has 2 different types of tiles.

A trinomial has 3 different types of tiles.

An algebraic expression that contains a term with a variable in the denominator, such as $\frac{5}{n}$, or the square root of a variable, such as $\sqrt{n}$, is not a polynomial.

Practice

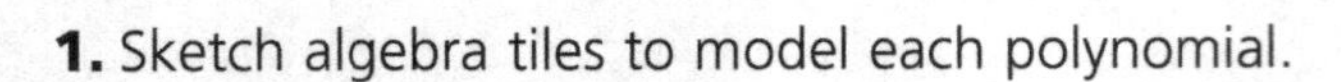

1. Sketch algebra tiles to model each polynomial.

a) $a^2 + 6$

b) $y^2 - y + 3$

______________________ ______________________

c) $-2m^2 + 3m - 4$

d) $2x^2 + 5x + 4$

______________________ ______________________

2. Is the polynomial a monomial, binomial, or trinomial?

a) $-7t$ The polynomial has ___ term, so it is a ______________.

b) $8d^2 + 7$ The polynomial has ___ terms, so it is a ______________.

c) $s^2 + 5s - 6$ The polynomial has ___ terms, so it is a ______________.

d) $4t - 12$ The polynomial has ___ terms, so it is a ______________.

e) -15 The polynomial has ___ term, so it is a ______________.

3. Name the degree of each polynomial.

a) $5a^2 - 3a + 6$ The term with the greatest exponent is $5a^2$.
It has exponent _____.
So, the polynomial has degree _____.

b) $4b - 6$ The term with the greatest exponent is _____.
It has exponent _____.
So, the polynomial has degree _____.

c) $4d^2 - 3d$ The term with the greatest exponent is _____.
It has exponent _____.
So, the polynomial has degree _____.

d) -4 -4 can be written as $-4x$——.
So, the polynomial has degree _____.

4. Write the polynomial represented by each set of tiles.

a) Use the variable f.

____

b) Use the variable n.

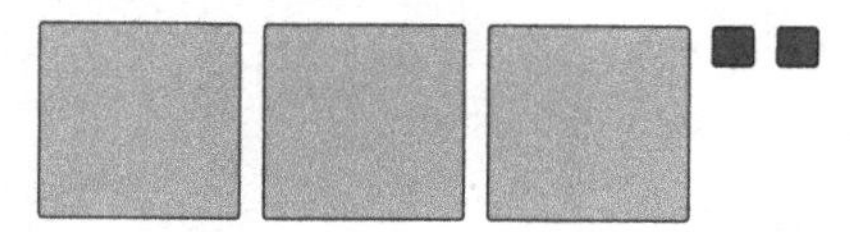

____

c) Use the variable p.

____

5. Choose a set of tiles from question 4.
Write another polynomial that can be represented by the same set of tiles.

____

____

6. Identify the polynomials that can be represented by the same set of algebra tiles.

a) $x^2 + 3x - 1$ 1 , ____ [rectangle tile], and ____ [small square tile]

b) $4r^2 - 5r + 9$ ________________________

c) $9 + 4z^2 - 5z$ ________________________

d) $3s + 1 + s^2$ ________________________

Parts ____ and ____ use the same algebra tiles.
So, ________________ and ________________ both represent the same polynomial.

5.2 Skill Builder

Modelling Integers

One ■ represents +1.
One ■ represents –1.
One ■ and one ■ combine to model 0.

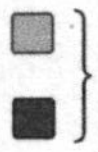

■ } +1
■ } –1
We call this a **zero pair.**

We can model any integer in many ways.

Each set of tiles below models +3.

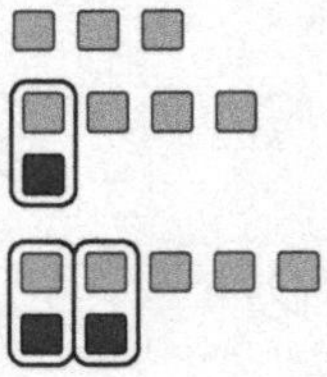

Each pair of 1 ■ and 1 ■ makes a zero pair.

Check

1. Write the integer modelled by each set of tiles.

a)  ■ ■ ■ ■ ■
■ ■ ■

____

b)  ■ ■
■ ■ ■

____

c) ■
■ ■ ■ ■

____

d)  ■ ■ ■
■ ■ ■

____

5.2 Like Terms and Unlike Terms

FOCUS **Simplify polynomials by combining like terms.**

These are all zero pairs:

 and  and 

We can use zero pairs to simplify algebraic expressions.

Example 1 Combining Like Tiles and Removing Zero Pairs

Simplify this tile model.
Write the polynomial that the remaining tiles represent.

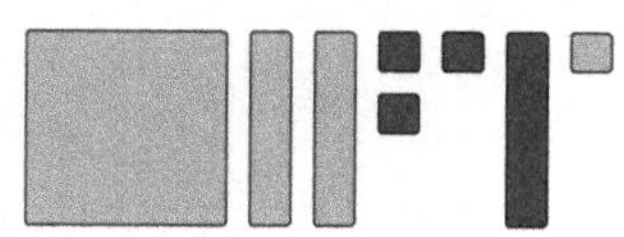

Solution

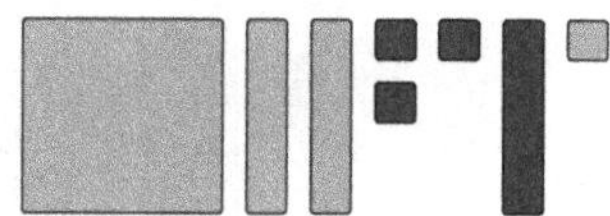

Group like tiles.

Matching tiles have the same size and shape.

Remove zero pairs.

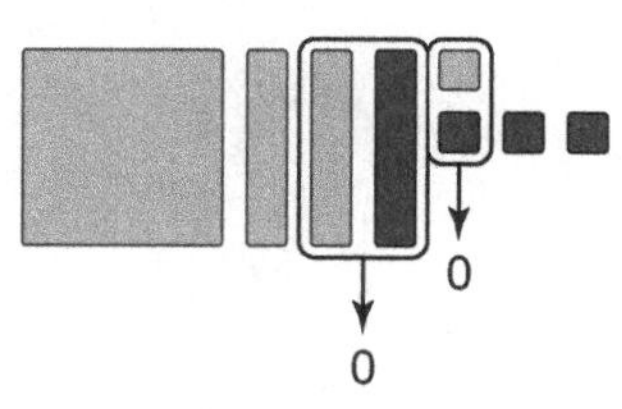

The tiles that remain are:

They represent: $x^2 + x - 2$

When there is only 1 of a type of tile, we omit the coefficient 1.

Check

1. Simplify each tile model.
Write the polynomial that the remaining tiles represent.

a) 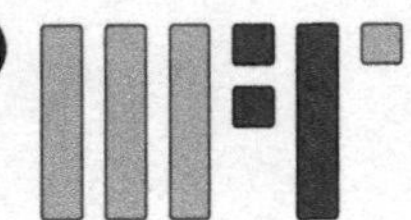

Remaining tiles: ____________________ Polynomial: ______________

b) 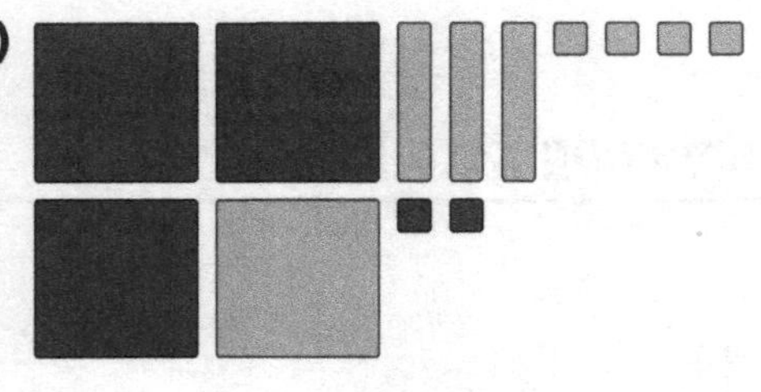

Remaining tiles: ____________________ Polynomial: ______________

c) 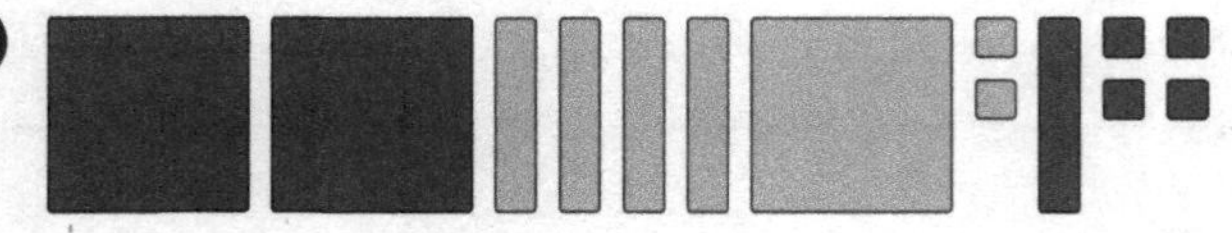

Remaining tiles: ____________________ Polynomial: ______________

Terms that can be represented by matching tiles are called **like terms**.

Like terms:	x^2 and $-2x^2$	$4s$ and $-s$	6 and -2	$5w$ and w
Unlike terms:	$3s$ and s^2	$2x$ and -5	$3d^2$ and 7	

We can **simplify a polynomial** by adding the coefficients of like terms.
To simplify $-5x + 2x$, add the integers: $-5 + 2 = -3$
So, $-5x + 2x = -3x$

Example 2 Simplifying a Polynomial Symbolically

Simplify:

a) $3a + 6 + a - 4$ **b)** $-x^2 + 4x - 5 + 3x^2 - 4x + 1$

Solution

a) $3a + 6 + a - 4$

$= 3a + 1a + 6 - 4$

$= 4a + 2$

Group like terms.
Add the coefficients of like terms.

b) $-x^2 + 4x - 5 + 3x^2 - 4x + 1$

$= -x^2 + 3x^2 + 4x - 4x - 5 + 1$

$= 2x^2 + 0x - 4$

$= 2x^2 - 4$

Group like terms.
Add the coefficients of like terms.

We omit a term when its coefficient is 0.

Check

1. Simplify each polynomial.

a) $5d + 2 + 3d - 1$

$= 5d + 3d + 2 - 1$

$=$ _____d + _____

Group like terms.
Add the coefficients of like terms:
$5 + 3 =$ ____ and $2 + (-1) =$ ____

b) $2a^2 - 3a + 5a^2 + 7a$

= ____________________

= ____________________

Group like terms.
Add the coefficients of like terms:
____ + ____ = ____ and ____ + ____ = ____

c) $-x^2 + 4x - 5 + 2x^2 + x + 3$

= ____________________

____

____

We omit the coefficient when it is 1.

d) $2x^2 + 6x + 7 - 2x^2 + 7x - 11$

= ____________________

____

____

Practice

1. What is the coefficient of each term?

a) $2x^2$ _____ **b)** $6w$ _____ **c)** $-3x$ _____

d) $7t$ _____ **e)** b _____ **f)** $-s$ _____

2. a) Which of these terms are like $3z^2$?

$5z$ $-z^2$ -9 $-6z$ $2z^2$ -11 $-4z^2$

$3z^2$ has variable _____ and exponent _____.

Find all terms with the same variable and exponent: _______________

b) Which of these terms are like $-5x$?

$-4x$ $-3x^2$ -2 $7x$ $5x^2$ 8 $-x$ $-5t$

$-5x$ has variable _____ and exponent _____.

Find all terms with the same variable and exponent: _______________

3. Simplify each tile model.

Write the polynomial that the remaining tiles represent.

a)

Remaining tiles: ____________________ Polynomial: ____________________

b)

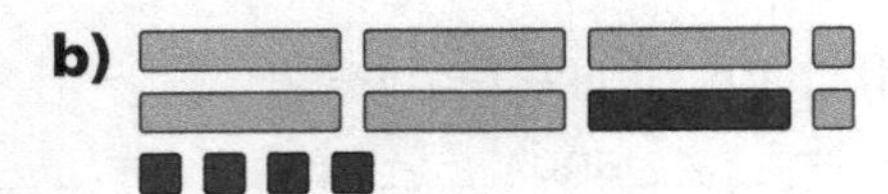

Remaining tiles: ____________________ Polynomial: ____________________

c)

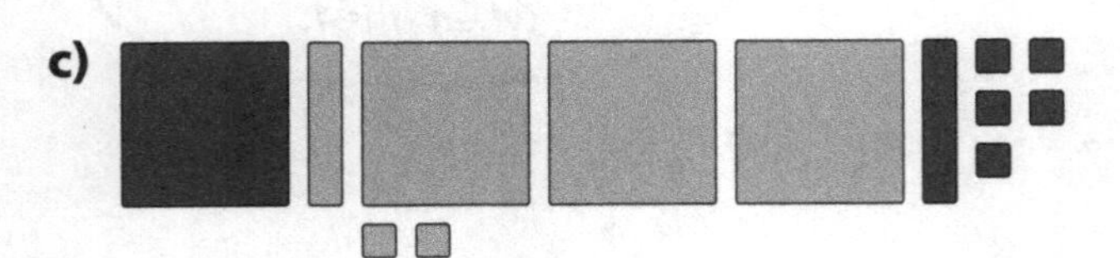

Remaining tiles: ____________________ Polynomial: ____________________

4. Add integers to combine like terms.

a) $-3c + 5c$ $\quad -3 + 5 =$ ____

$-3c + 5c =$ ____

b) $4s - s$ $\quad 4 + (-1) =$ ____

$4s - s =$ ____

c) $-2x^2 + 7x^2$ $\quad$ ____ + ____ = ____

____

d) $8e^2 - 8e^2$ $\quad$ ____

____

5. Simplify each polynomial.

a) $5m + 7 - 2m + 1$

= ____ Group like terms.

= ____ Add the coefficients of like terms.

b) $7c^2 - 6c - 4c^2 + c$

= ____ Group like terms.

= ____ Add the coefficients of like terms.

c) $11 - 9v + v^2 + 2 - v$

= ____

= ____

____

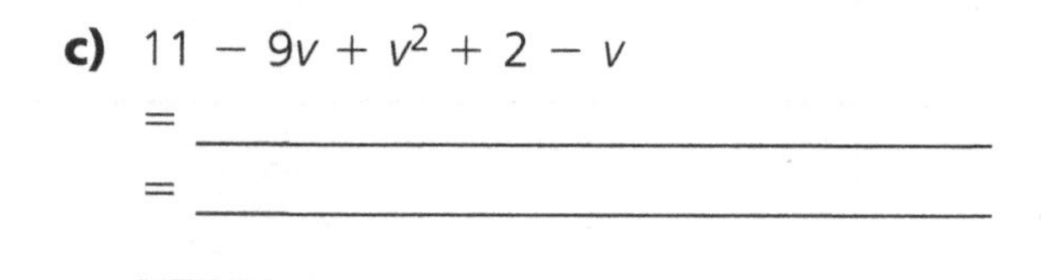

We usually write a polynomial so the exponents of the variable decrease from left to right.

d) $-7f^2 + 12f - 2 -3f^2 - 3f + 5$

= ____

= ____

A polynomial in simplified form is equal to the original polynomial.

6. Identify and explain any errors you find.

a) $3x + 2 = 5x$

____

____

b) $5s + 3s = 8s^2$

____

____

____

____

c) $x^2 - x^2 = 0$

____

____

5.3 Skill Builder

Adding Integers

To add two integers: $3 + (-4)$
We can model each integer with tiles.
Circle zero pairs.

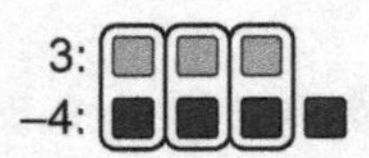

There are 3 zero pairs.
There is 1 tile left.
It models -1.
So, $3 + (-4) = -1$ ← *This is an addition sentence.*

Check

1. Sketch tiles to show the sum of each pair of integers.
Write an addition sentence each time.

a) 4:
5: ____________________

b) 6:
-2: ____________________

c) -3:
-5: ____________________

d) 3:
-3: ____________________

e) 5:
-8: ____________________

5.3 Adding Polynomials

FOCUS **Use different strategies to add polynomials.**

Example 1 Adding Polynomials with Algebra Tiles

Use algebra tiles to model $(3s^2 + 2s - 6) + (-s^2 - 2s + 1)$.
Write an addition sentence.

Solution

Model each polynomial.

$3s^2 + 2s - 6$

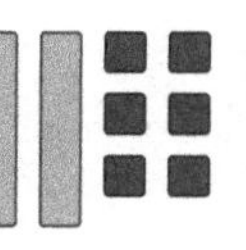

$-s^2 - 2s + 1$

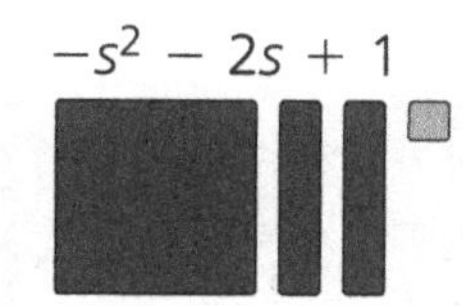

Combine the tiles.

Group matching tiles.

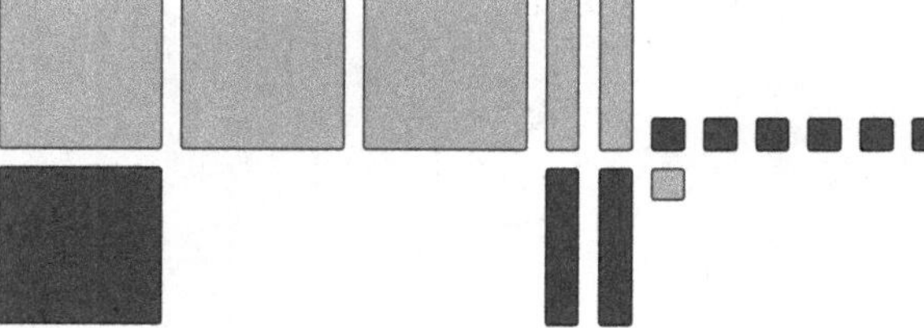

Remove zero pairs.

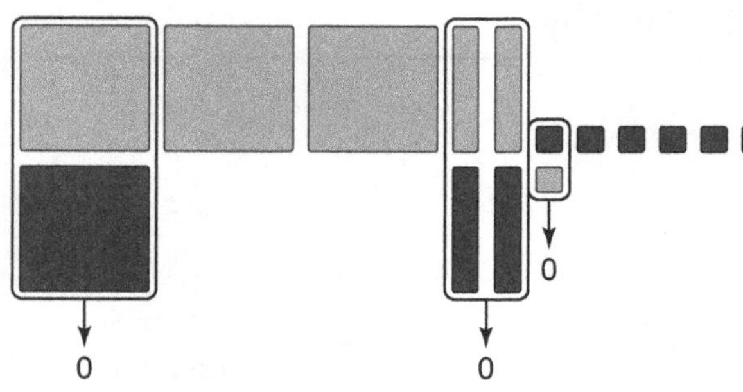

The remaining tiles are:

They represent: $2s^2 - 5$

The addition sentence is: $(3s^2 + 2s - 6) + (-s^2 - 2s + 1) = 2s^2 - 5$

Check

1. Sketch algebra tiles to model each sum.
Then write the sum.

a) $(6p + 4) + (-2p + 1)$

Remaining tiles: ______________________
So, $(6p + 4) + (-2p + 1) =$ ____________

b) $(2x^2 - x + 1) + (x^2 - 3)$

Remaining tiles: ______________________
So, $(2x^2 - x + 1) + (x^2 - 3) =$ ____________

c) $(3e^2 + 6e - 5) + (-4e^2 - 3e + 8)$

Remaining tiles: ______________________
So, $(3e^2 + 6e - 5) + (-4e^2 - 3e + 8) =$ ____________

Algebra tiles are not always available.
To add polynomials without tiles:
- remove the brackets
- add the coefficients of like terms

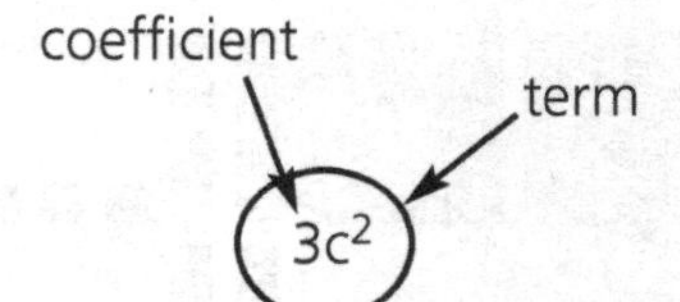

Example 2 Adding Polynomials Symbolically

Add: $(3c^2 + 5c - 6) + (2c^2 - 3c + 4)$

Solution

$(3c^2 + 5c - 6) + (2c^2 - 3c + 4)$	Remove the brackets.
$= 3c^2 + 5c - 6 + 2c^2 - 3c + 4$	Group like terms.
$= \underbrace{3c^2 + 2c^2} + \underbrace{5c - 3c} - \underbrace{6 + 4}$	Add the coefficients of like terms.
$= \quad 5c^2 \quad + \quad 2c \quad -2$	

$3c^2$ and $2c^2$ are like terms.

Check

1. Add.

a) $(7g - 8) + (3g + 1)$ Remove the brackets.
$= 7g - 8 + 3g + 1$ Group like terms.
$= \underbrace{7g + 3g} - \underbrace{8 + 1}$ Add the coefficients of like terms.
$=$ ______________ $7 + 3 =$ ____ and $-8 + 1 =$ ____

b) $(2a^2 - 9a) + (-5a^2 + 12a)$ Remove the brackets.
$=$ ______________ Group like terms.
$=$ ______________ Add the coefficients of like terms.
$=$ ______________ ____ + ____ = ____ and ____ + ____ = ____

c) $(-c^2 + 11c - 3) + (4c^2 + 5)$
$=$ ______________
$=$ ______________
$=$ ______________

Recall: $-c^2$ has coefficient -1.

We can also add 2 polynomials by aligning like terms vertically.

Example 3 Adding Polynomials Vertically

Add: $(2m + 9) + (3m^2 + m - 14)$

Solution

To add the polynomials, remove the brackets and align like terms vertically.

In $3m^2 + m - 14$, the term m has coefficient 1, so write m as $1m$.

$$\begin{array}{r} 2m + 9 \\ + 3m^2 + 1m - 14 \\ \hline 3m^2 + 3m - 5 \end{array}$$

Add the coefficients of like terms.

$$\begin{array}{r} 0 \\ +3 \\ \hline 3 \end{array} \quad \begin{array}{r} 2 \\ +1 \\ \hline 3 \end{array} \quad \begin{array}{r} 9 \\ +(-14) \\ \hline -5 \end{array}$$

So, $(2m + 9) + (3m^2 + m - 14) = 3m^2 + 3m - 5$

Check

1. Add vertically.

a) $(2x + 3) + (4x + 8)$

$$\begin{array}{r} 2x + 3 \\ + \; 4x + 8 \\ \hline \_\_\_x + \_\_\_ \end{array}$$

b) $(5p^2 + 12) + (-2p^2 + 3p - 7)$

$$\begin{array}{rrr} 5p^2 & & + 12 \\ + \; -2p^2 & + \; 3p & - \; 7 \\ \hline & & \end{array}$$

c) $(-6b^2 - 2b + 8) + (9b - b^2 - 19)$

____

____

____

Practice

1. Write the addition sentence modelled by each set of tiles.
Use the variable x.

a) 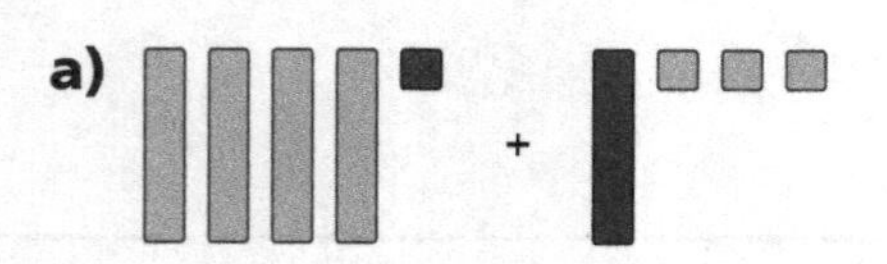

____

b) 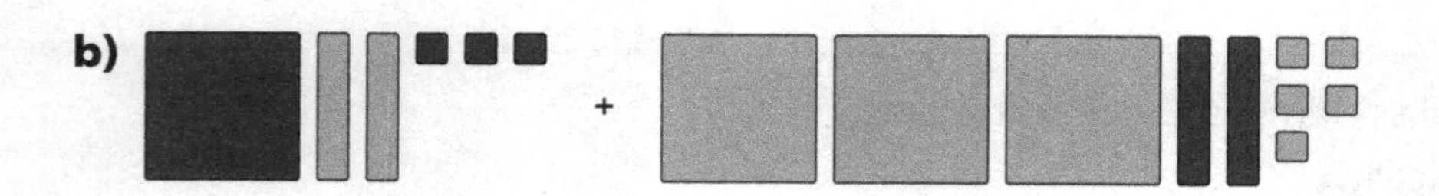

____

2. Sketch algebra tiles to model each sum.
Then write the sum.

a) $(-5w + 8) + (7w - 3) =$ __________

Remaining tiles: ______________________

b) $(-6t^2 - 3t + 2) + (4t^2 - t + 1) =$ ______________

Remaining tiles: ______________________________

3. Add horizontally.

a) $(2r - 3) + (3r - 1)$ — Remove the brackets.

$= 2r - 3 + 3r - 1$ — Group like terms.

$= 2r + 3r - 3 - 1$ — Add the coefficients of like terms.

$= \_\_\_\_r - \_\_\_\_$ — $2 + 3 = \_\_\_\_$ and $-3 + (-1) = \_\_\_\_$

b) $(7h^2 - 2h) + (-4h^2 + 9h - 4)$

= ____________________

= ____________________

= ____________________

c) $(-2y^2 + 6y - 1) + (2y^2 - 6y + 5)$

= ____________________

= ____________________

= ____________________

4. Add vertically.

a) $(9r + 7) + (2r - 3)$

$$\begin{array}{r} 9r + 7 \\ + \; 2r - 3 \\ \hline \_\_\_\_r + \_\_\_\_ \end{array}$$

b) $(-a^2 + 4a) + (-3a^2 + 2a - 5)$

$$\begin{array}{r} -1a^2 + 4a \\ + \; -3a^2 + 2a - 5 \\ \hline \_\_\_\_\_\_\_\_\_\_ \end{array}$$

c) $(8v - 2v^2 - 3) + (9 + 6v^2 - 10v)$

____

____

____

5. Find the perimeter of this triangle.

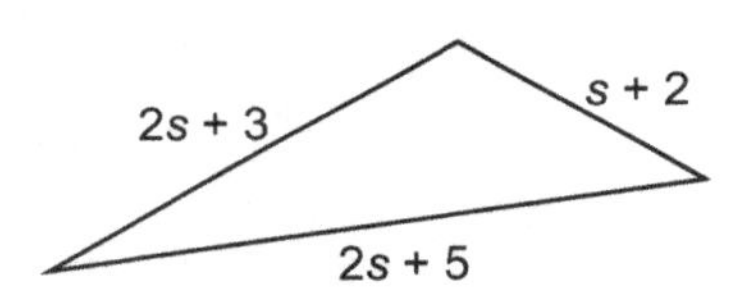

***Perimeter** is the distance around a shape. To find the perimeter, add the side lengths.*

Perimeter = ____________________ — Remove the brackets.

= ____________________ — Group like terms.

= ____________________ — Add coefficients of like terms.

= ____________________

5.4 Skill Builder

Subtracting Integers Symbolically

To subtract an integer without tiles, we add the opposite integer.
-3 and 3, -6 and 6, and -15 and 15 are opposite integers.

To subtract: $(-4) - (-3)$
Add the opposite integer.
The opposite of -3 is 3.
And, $(-4) + 3 = -1$
So, $(-4) - (-3) = -1$

We omit the + sign when the integer is positive.

We can use algebra tiles to check:
Model -4:
■■■■
Take away -3:
■(■■■) → -3
One ■ remains.
So, $(-4) - (-3) = -1$

Check

1. Subtract.

a) $6 - (-2)$:
The opposite of -2 is ____.
Add the opposite: $6 +$ ____ $=$ ____
So, $6 - (-2) =$ ____

b) $3 - (4)$:
The opposite of 4 is ____.
Add the opposite: ________________
So, $3 - (4) =$ ____

c) $(-8) - (-5)$:
The opposite of -5 is ____.
Add the opposite: ________________
So, ________________

d) $(-9) - (4)$:
The opposite of 4 is ____.
Add the opposite: ________________
So, ________________

5.4 Subtracting Polynomials

FOCUS **Use different strategies to subtract polynomials.**

To subtract a polynomial, we subtract each term of the polynomial.

Example 1 Subtracting Polynomials with Algebra Tiles

Use algebra tiles to model $(3b^2 - 2b - 1) - (-2b^2 - b + 2)$.
Write a subtraction sentence.

Solution

Model: $3b^2 - 2b - 1$

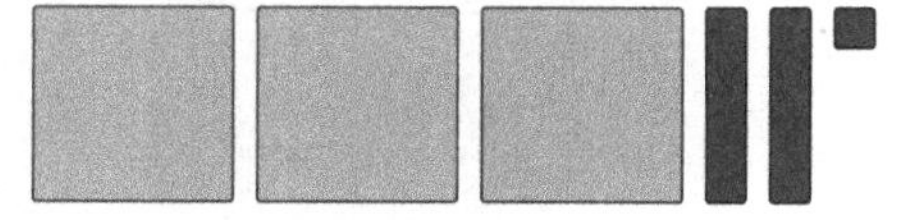

To subtract $-2b^2 - b + 2$, take away 2 , 1 , and 2 .

There are no or to take away.
So, add 2 zero pairs of each tile:

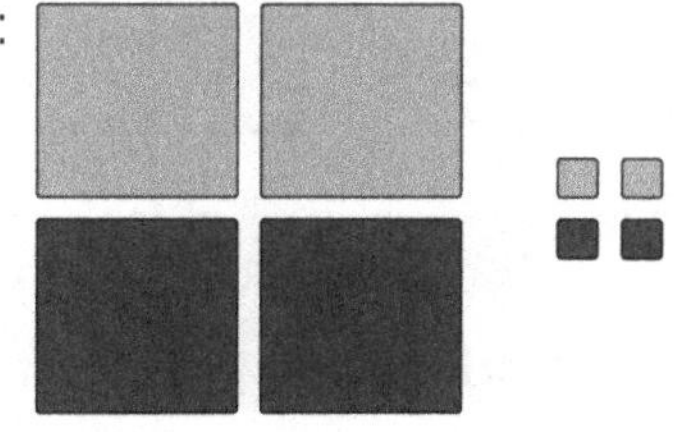

So, these tiles also model $3b^2 - 2b - 1$.

Take away the tiles for $-2b^2 - b + 2$.

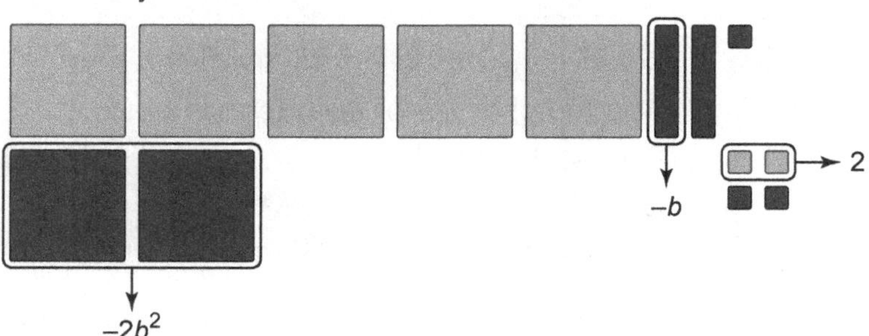

The remaining tiles represent: $5b^2 - b - 3$
The subtraction sentence is: $(3b^2 - 2b - 1) - (-2b^2 - b + 2) = 5b^2 - b - 3$

Check

1. Use algebra tiles to model each difference.
Sketch the tiles that remain, then write the difference.

a) $(4p + 3) - (2p + 1)$

Remaining tiles: ____________________

So, $(4p + 3) - (2p + 1) =$ ____________

b) $(5t + 1) - (-2t + 3)$

Remember to add zero pairs when there are not enough tiles to subtract.

Remaining tiles: ____________________

So, $(5t + 1) - (-2t + 3) =$ ____________

c) $(3e^2 + 2e - 4) - (4e^2 + 3e - 2)$

Remaining tiles: ____________________

So, $(3e^2 + 2e - 4) - (4e^2 + 3e - 2) =$ ____________

To subtract integers without tiles, we can add the opposite integer.
To subtract polynomials without tiles, we can add the opposite terms.

Example 2 Subtracting Polynomials Symbolically

Subtract: $(-5k^2 + 2k - 6) - (3k^2 - 4k + 1)$

Solution

$(-5k^2 + 2k - 6) - (3k^2 - 4k + 1)$	Remove the brackets from the first term.
$= -5k^2 + 2k - 6 - (3k^2 - 4k + 1)$	Add the opposite of each term in brackets.
$= -5k^2 + 2k - 6 + (-3k^2 + 4k - 1)$	Remove the brackets.
$= -5k^2 + 2k - 6 - 3k^2 + 4k - 1$	Group like terms.
$= -5k^2 - 3k^2 + 2k + 4k - 6 - 1$	Add the coefficients of like terms.
$= -8k^2 + 6k - 7$	

Check

1. Subtract.

a) $(8f - 3) - (7f + 5)$ — Remove the brackets from the first term.

$= \_\_\_\_\_\_\_\_\_\_ - (7f + 5)$

The opposite of $7f$ is: ______

The opposite of 5 is: ______

Add the opposites.

$= 8f - 3 + \_\_\_\_\_\_\_\_\_\_\_\_$ — Remove the brackets.

$= \_\_\_\_\_\_\_\_\_\_\_\_$ — Group like terms.

$= \_\_\_\_\_\_\_\_\_\_\_\_$ — Add the coefficients of like terms.

$= \_\_\_\_\_\_\_\_\_\_\_\_$

b) $(2 + 5g - 7g^2) - (9g - 4g^2 + 2)$

$= \_\_\_\_\_\_\_\_\_\_\_\_$

$= \_\_\_\_\_\_\_\_\_\_\_\_$

$= \_\_\_\_\_\_\_\_\_\_\_\_$

$= \_\_\_\_\_\_\_\_\_\_\_\_$

$= \_\_\_\_\_\_\_\_\_\_\_\_$

$= \_\_\_\_\_\_\_\_\_\_\_\_$

Remember to write the polynomial in descending order.

Practice

1. Write the subtraction sentence modelled by each set of tiles.

a)

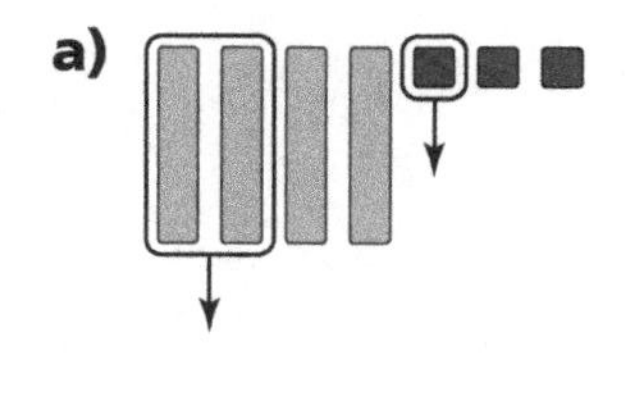

____

____

b)

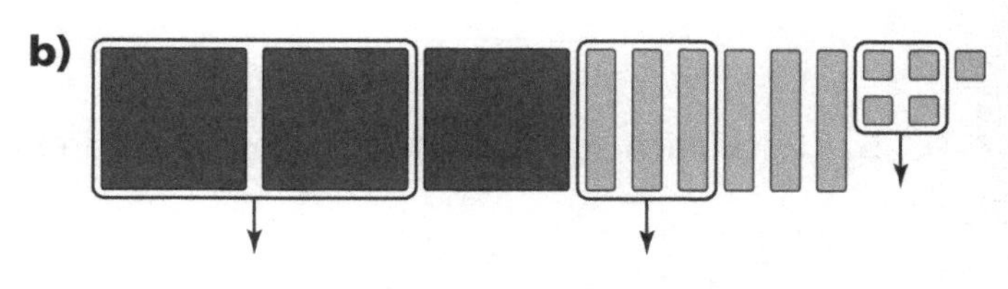

____

____

2. Use algebra tiles to model each difference.
Sketch the tiles that remain, then write the difference.

a) $(3r + 2) - (-2r + 3)$

Remaining tiles: ____________

So, $(3r + 2) - (-2r + 3) =$ __________

b) $(-4v^2 + 5v - 1) - (-3v^2 + 4v - 2)$

Remaining tiles: ____________

So, $(-4v^2 + 5v - 1) - (-3v^2 + 4v - 2)$

$=$ __________

3. Write the opposite of each term.

a) -9: ____ **b)** $3r$: ____ **c)** $-2s^2$: ____ **d)** t: ____

4. Subtract.

a) $(4p + 1) - (p + 10)$ — Remove the brackets from the first term.

$= \_\_\_\_\_\_\_\_ - (p + 10)$ — The opposite of p is: ____
The opposite of 10 is: ____
Add the opposites.

$= 4p + 1 + \_\_\_\_\_\_\_\_$ — Remove the brackets.

$= \_\_\_\_\_\_\_\_$ — Group like terms.

$= \_\_\_\_\_\_\_\_$ — Add the coefficients of like terms.

$= \_\_\_\_\_\_\_\_$

b) $(3h^2 + 5h - 4) - (h^2 - 4h + 6)$ — Remove the brackets from the first term.

$= \_\_\_\_\_\_\_\_$ — Add the opposites.

$= \_\_\_\_\_\_\_\_$ — Remove the brackets.

$= \_\_\_\_\_\_\_\_$ — Group like terms.

$= \_\_\_\_\_\_\_\_$ — Add the coefficients of like terms.

$= \_\_\_\_\_\_\_\_$

c) $(4q^2 + 3) - (3q - q^2 + 3)$

$= \_\_\_\_\_\_\_\_$

$= \_\_\_\_\_\_\_\_$

$= \_\_\_\_\_\_\_\_$

$= \_\_\_\_\_\_\_\_$

$= \_\_\_\_\_\_\_\_$

5. Check each solution. Identify any errors and correct them.

a) $(7x^2 + 3x + 7) - (3x^2 - 4)$
$= 7x^2 + 3x + 7 - 3x^2 - 4$
$= 7x^2 - 3x^2 + 3x + 7 - 4$
$= 4x^2 + 3x + 3$

$(7x^2 + 3x + 7) - (3x^2 - 4)$
$= \_\_\_\_\_\_\_\_$
$= \_\_\_\_\_\_\_\_$
$= \_\_\_\_\_\_\_\_$

b) $(3a^2 - 2a + 4) - (2a^2 + 3)$
$= 3a^2 - 2a + 4 - 2a^2 - 3$
$= 3a^2 - 2a^2 - 2a + 4 - 3$
$= a^2 + 2a - 3$

$(3a^2 - 2a + 4) - (2a^2 + 3)$
$= \_\_\_\_\_\_\_\_$
$= \_\_\_\_\_\_\_\_$
$= \_\_\_\_\_\_\_\_$

Can you ...

- Recognize, write, describe, and classify polynomials?
- Represent polynomials using tiles, pictures, and algebraic expressions?
- Simplify polynomials by combining like terms?
- Add and subtract polynomials?

5.1 **1.** Is the polynomial a monomial, binomial, or trinomial?

a) -9 The polynomial has ____ term, so it is a ____________.

b) $3f - 5$ The polynomial has ____ terms, so it is a ____________.

c) $2s^2 - s + 1$ The polynomial has ____ terms, so it is a ____________.

d) $-a^2 + 2a$ The polynomial has ____ terms, so it is a ____________.

2. Write the polynomial represented by each set of tiles.

a) Use the variable g.

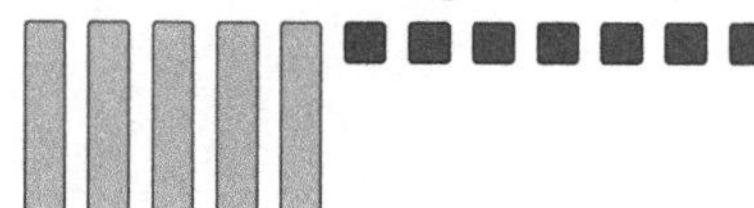

____

b) Use the variable r.

____

c) Use the variable w.

____

5.2 **3.** Simplify each tile model.
Write the polynomial that the remaining tiles represent.

a)

Remaining tiles: ________________ Polynomial: ________________

b) 

Remaining tiles: ________________ Polynomial: ________________

4. Simplify each polynomial.

a) $8e - 9 - 5e + 4$ Group like terms.
$=$ ____________ Add the coefficients of like terms.
$=$ ____________

b) $4d^2 - 3d + 11 - d^2 + 5d - 13$
$=$ ________________________
$=$ ________________________

5.3 **5.** Sketch tiles to model each sum.
Then write the sum.

a) $(4v - 4) + (-2v + 7)$

Remaining tiles: ________________
So, $(4v - 4) + (-2v + 7) =$ ________

b) $(6u^2 - 5u - 7) + (-3u^2 + 3u + 7)$

Remaining tiles: ________________________
So, $(6u^2 - 5u - 7) + (-3u^2 + 3u + 7) =$ ________

6. Add.

a) $(3t + 11) + (-7t - 4)$

$$\begin{array}{r} 3t + 11 \\ + -7t - 4 \\ \hline __t + __ \end{array}$$

b) $(10y^2 - 9) + (-3y^2 + 4y - 2)$

$$\begin{array}{r} 10y^2 \qquad - 9 \\ + -3y^2 + 4y - 2 \\ \hline ________ \end{array}$$

7. Find the perimeter of this rectangle.

Perimeter = ________________________
= ________________________
= ________________________
= ________________________

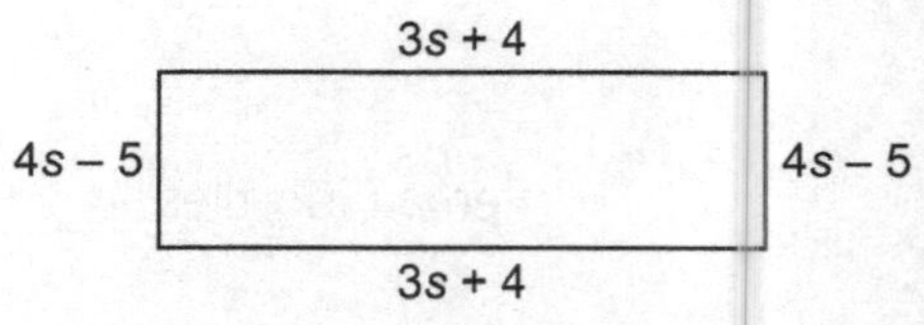

5.4 **8.** Use algebra tiles to model each difference.
Sketch the tiles that remain, then write the difference.

a) $(5n - 6) - (-n - 3)$

Remaining tiles: ____________________
So, $(5n - 6) - (-n - 3) =$ ________

b) $(-v^2 + 3v - 5) - (-v^2 + 4v + 2)$

Remaining tiles: ____________________
So, $(-v^2 + 3v - 5) - (-v^2 + 4v + 2) =$ ________

9. Subtract.

a) $(11h + 3) - (9h - 2)$ — Remove the brackets from the first term.
= ________ $- (9h - 2)$ — Add the opposites.
= ________ + (________) — Remove the brackets.
= ____________________ — Group like terms.
= ____________________ — Add the coefficients of like terms.
= ____________________

b) $(7j^2 - 11j - 7) - (12j^2 - 8j - 3)$
= ____________________
= ____________________
= ____________________
= ____________________
= ____________________

5.5 Skill Builder

The Distributive Property

We can use this diagram to model 5×24.

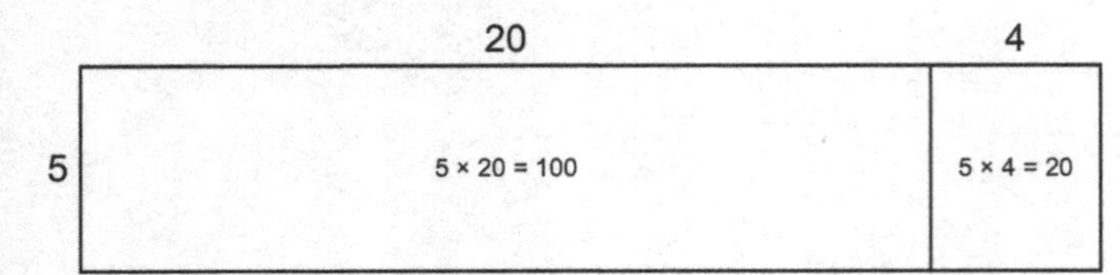

This diagram shows:

$$5 \times 24 = 5 \times (20 + 4)$$
$$= (5 \times 20) + (5 \times 4)$$
$$= 100 + 20$$
$$= 120$$

We multiply the term outside the brackets by each term inside the brackets, then find the sum.

This shows the **distributive property** of multiplication.

Check

1. How does each diagram show the distributive property?

a) 10, 5; 6

____

____

b) 30, 5; 8

____

2. Use the distributive property to multiply.

a) $7 \times 21 = 7 \times (20 + 1)$
$= (7 \times 20) + (7 \times 1)$
= ______________
= ______________

b) $8 \times 43 = 8 \times (40 + 3)$
= ______________
= ______________
= ______________

Multiplying and Dividing Integers

When multiplying or dividing 2 integers, look at the sign of each integer:

- When the integers have the same sign, their product or quotient is positive.
- When the integers have different signs, their product or quotient is negative.

×/÷	(−)	(+)
(−)	(+)	(−)
(+)	(−)	(+)

$7 \times (-4)$
$7 \times (-4) = -28$

These 2 integers have different signs, so their product is negative.

$(-12) \div (-3)$
$(-12) \div (-3) = 4$

These 2 integers have the same sign, so their quotient is positive.

When one number is divided by another number, the result is called the quotient.

Check

1. Will the product be positive or negative?

a) 9×5 __________ **b)** $8 \times (-3)$ __________

c) $(-12) \times 5$ __________ **d)** $(-7) \times (-6)$ __________

2. Multiply.

a) $6 \times 5 =$ __________ **b)** $4 \times (-10) =$ __________

c) $(-7) \times 3 =$ __________ **d)** $(-8) \times (-6) =$ __________

e) $12 \times (-5) =$ __________ **f)** $(-4) \times (-8) =$ __________

3. Will the quotient be positive or negative?

a) $18 \div 3$ __________ **b)** $(-36) \div 6$ __________

c) $72 \div (-9)$ __________ **d)** $(-48) \div (-8)$ __________

4. Divide.

a) $(-49) \div 7 =$ __________ **b)** $(-56) \div (-8) =$ __________

c) $48 \div 6 =$ __________ **d)** $81 \div (-9) =$ __________

e) $(-27) \div (-3) =$ __________ **f)** $(-42) \div 7 =$ __________

5.5 Multiplying and Dividing a Polynomial by a Constant

FOCUS **Use different strategies to multiply and divide a polynomial by a constant.**

To multiply $2(4x)$ with algebra tiles:

Model 2 rows of 4 .

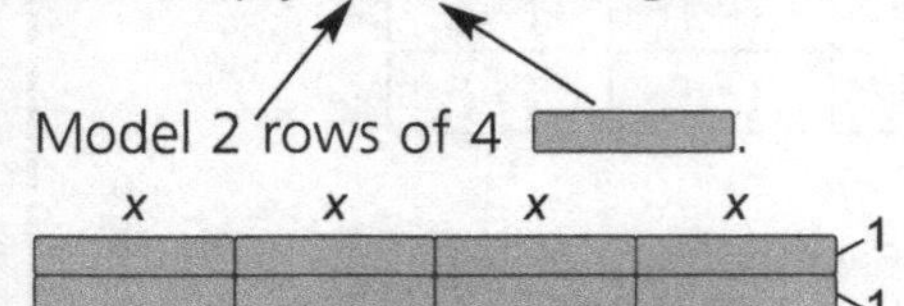

There are 8 x-tiles. So, $2(4x) = 8x$

Recall: $2(4x) = 2 \times 4x$

Example 1 Using Algebra Tiles to Multiply a Polynomial by a Constant

Find the product: $3(2b^2 - 2b + 1)$

Solution

$3(2b^2 - 2b + 1)$

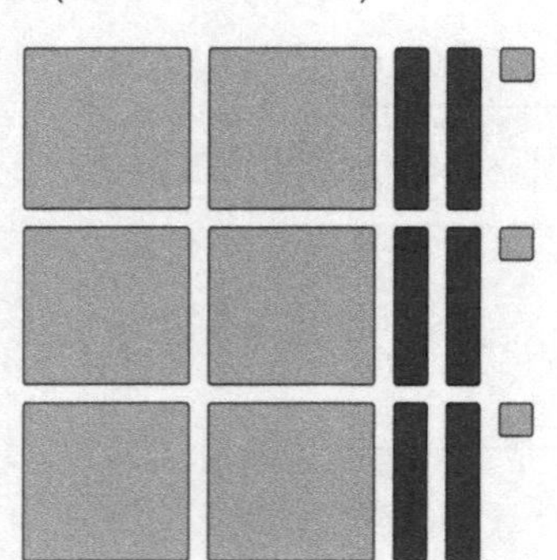

Model 3 rows of 2 , 2 , and 1 .
These tiles represent: $6b^2 - 6b + 3$.
So, $3(2b^2 - 2b + 1) = 6b^2 - 6b + 3$

Check

1. Sketch algebra tiles to multiply. Write the product each time.

a) $3(4p - 3) =$ ____________ **b)** $2(-s^2 + s + 3) =$ ____________

When working symbolically, remember the rules for integer multiplication and division.

Example 2 Using the Distributive Property to Multiply a Polynomial by a Constant

Find the product: $-5(4e^2 - 5e + 3)$

Solution

$-5(4e^2 - 5e + 3)$ Multiply each term in brackets by -5.

$= (-5)(4e^2) + (-5)(-5e) + (-5)(3)$ Multiply.

$= -20e^2 + 25e + (-15)$

$= -20e^2 + 25e - 15$

Check

1. Multiply.

a) $3(7s^2 + 9)$
$= 3(7s^2) + 3(9)$
$=$ ____$s^2 +$ ____

Multiply each term in brackets by 3.
Multiply: $3 \times 7 =$ ____ and $3 \times 9 =$ ____

b) $-4(5e^2 - 8e)$
$=$ ____________________
$=$ ____________________

Multiply each term in brackets by -4.
Multiply.

c) $-5(-2d^2 - 3d + 6)$
$=$ ____________________
$=$ ____________________
$=$ ____________________

d) $7(6y^2 - 8y + 9)$
$=$ ____________________
$=$ ____________________
$=$ ____________________

We can use algebra tiles to divide a polynomial by a constant.
To divide: $(-8x) \div 2$

Arrange 8 [tile] into 2 equal rows.

In each row there are 4 [tile].
So, $(-8x) \div 2 = -4x$

Example 3 Using Algebra Tiles to Divide a Polynomial by a Constant

Find the quotient: $(6s - 9) \div 3$

Solution

$(6s - 9) \div 3$

Arrange 6 [tile] and 9 [tile] into 3 equal rows.

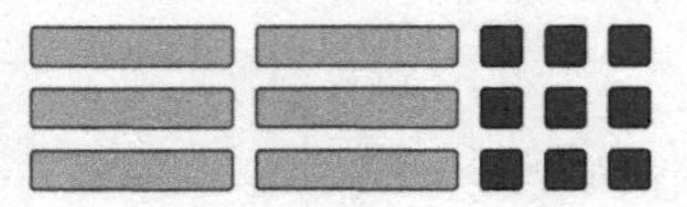

In each row, there are 2 [tile] and 3 [tile].

So, $(6s - 9) \div 3 = 2s - 3$

Check

1. Sketch algebra tiles to divide. Write the quotient each time.

a) $(3g^2 + 12g) \div 3 =$ ____________

b) $(-4b^2 + 6) \div 2 =$ ____________

c) $(4s^2 - 4s + 8) \div 4 =$ ____________

d) $(-6t^2 + 9t - 9) \div 3 =$ ____________

When algebra tiles are not available,
or when the divisor is negative,
we can use what we already know about division.

In the division sentence $6 \div 3 = 2$, the divisor is 3.

We can write $8x \div 4$ as a fraction: $\frac{8x}{4}$

We write the fraction as a product, then simplify each fraction.

$$\frac{8x}{4} = \frac{8}{4} \times x$$
$$= 2 \times x$$
$$= 2x$$

Example 4 Dividing a Polynomial by a Constant Symbolically

Find the quotient: $\frac{-9v^2 + 6}{3}$

Solution

$\frac{-9v^2 + 6}{3}$ Write as the sum of 2 fractions with denominator 3.

$= \frac{-9v^2}{3} + \frac{6}{3}$ Simplify the fractions.

$= \frac{-9}{3} \times v^2 + 2$ When 2 integers have different signs, the quotient is negative.

$= -3 \times v^2 + 2$

$= -3v^2 + 2$

Check

1. Divide.

a) $\frac{12r^2 + 8}{4}$ Write as the sum of 2 fractions with denominator 4.

$= \frac{\_\_\_}{4} + \frac{\_\_\_}{4}$ Simplify the fractions.

$=$ ____________ When 2 integers have the same sign, the quotient is ________.

$=$ ____________

b) $\frac{18v^2 - 6v + 12}{6}$

$= \frac{\_\_\_}{6} + \frac{\_\_\_}{6} + \frac{\_\_\_}{6}$

= ______________________

= ______________________

c) $\frac{-4e^2 - 8e}{2}$

= ______________________

= ______________________

= ______________________

Practice

1. Which multiplication sentence does each rectangle represent?

a)

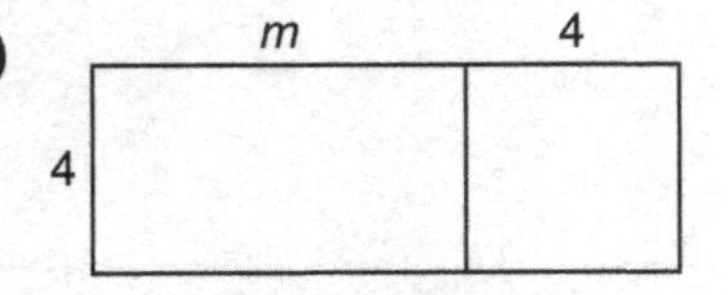

$4(m + 4) = (4 \times \_\_\_) + (4 \times \_\_\_)$

= ____________

b)

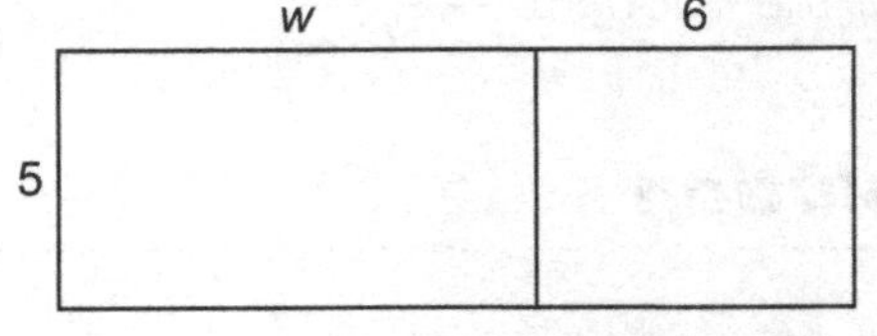

____

= ____________

2. Write the multiplication sentence modelled by each set of tiles.

a)

____

b)

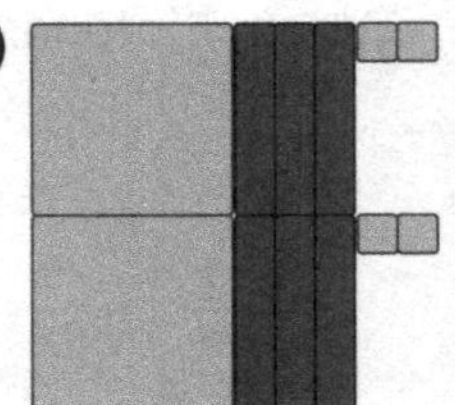

____

3. Sketch algebra tiles to multiply. Write the product each time.

a) $3(6r - 4) =$ ____________

b) $2(-2b^2 - b + 3) =$ ____________

4. Multiply.

a) $6(-4t^2 + 3)$
$= 6(\_\_\_\_) + 6(\_\_\_\_)$
$=$ ____________________

b) $-8(-3k^2 - 2k + 4)$
$=$ ______________________________
____
____

5. Which of these quotients is modelled by the tiles below?

a) $(15x - 9) \div 3$

b) $(-15x - 9) \div 3$

c) $(-15x + 9) \div 3$

6. Sketch algebra tiles to divide. Write the quotient each time.

a) $(3h^2 - 15h) \div 3 =$ ______________

b) $(-2a^2 - 6a + 4) \div 2 =$ ______________

7. Divide.

a) $\frac{-10z^2 + 15}{5}$

$= \frac{\_\_\_\_}{5} + \frac{\_\_\_\_}{5}$

$=$ ________________________

$=$ ________________________

b) $\frac{7x^2 - 7x + 21}{-7}$

$=$ ________________________

$=$ ________________________

$=$ ________________________

5.6 Skill Builder

Multiplying Monomials

The area of this square is:

4 cm

4 cm

$4 \times 4 = 16$

Area $= 16$ cm^2

The area of this rectangle is:

6 cm

4 cm

$4 \times 6 = 24$

Area $= 24$ cm^2

We can use the models above to help us multiply 2 monomials.

The area of this square is:

n

n

$n \times n = n^2$

The area of this rectangle is:

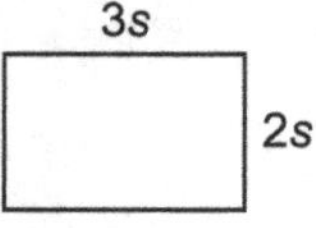

$$\begin{aligned} 3s \times 2s &= 3 \times s \times 2 \times s \\ &= 3 \times 2 \times s \times s \\ &= 6 \times s^2 \\ &= 6s^2 \end{aligned}$$

When one or both of the monomials is negative, we cannot use an area model.
We multiply using the rules for multiplying integers.

$$\begin{aligned} 4v \times (-2v) &= 4 \times v \times (-2) \times v \\ &= 4 \times (-2) \times v \times v \\ &= -8 \times v^2 \\ &= -8v^2 \end{aligned}$$

4 and −2 have different signs, so their product is negative.

Check

1. Multiply.

a) $b \times b =$ _____

b) $c \times (-c) =$ _____

c) $(-f) \times (-f) =$ _____

d) $(-g) \times g =$ _____

2. Multiply.

a) $5r \times 6r = 5 \times r \times 6 \times r$
$= 5 \times 6 \times r \times r$
$=$ ____________
$=$ ____________

b) $(-2d) \times 8d = (-2) \times d \times 8 \times d$
$= (-2) \times 8 \times d \times d$
$=$ ____________
$=$ ____________

c) $4a \times (-7a) =$ ____________
$=$ ____________
$=$ ____________
$=$ ____________

d) $(-5v) \times (-9v) = (-5) \times v \times (-9) \times v$
$=$ ____________
$=$ ____________
$=$ ____________

5.6 Multiplying and Dividing a Polynomial by a Monomial

FOCUS **Use different strategies to multiply and divide a polynomial by a monomial.**

To multiply $2d(3d)$ with algebra tiles:

Draw 2 adjacent sides of a rectangle.
Position [tile] tiles to show side lengths $2d$ and $3d$.

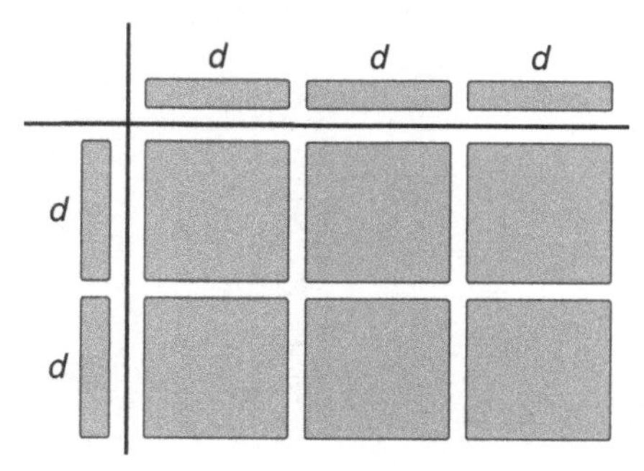

$d \times d = d^2$, so use a d^2-tile.

Then fill the rectangle with tiles.

We used 6 d^2-tiles to fill the rectangle. So, $2d(3d) = 6d^2$

Example 1 Multiplying a Binomial by a Monomial

Find the product: $3e(4e - 2)$

Solution

$3e(4e - 2)$
Draw 2 adjacent sides of a rectangle.
Position tiles to show side lengths $3e$ and $4e - 2$.

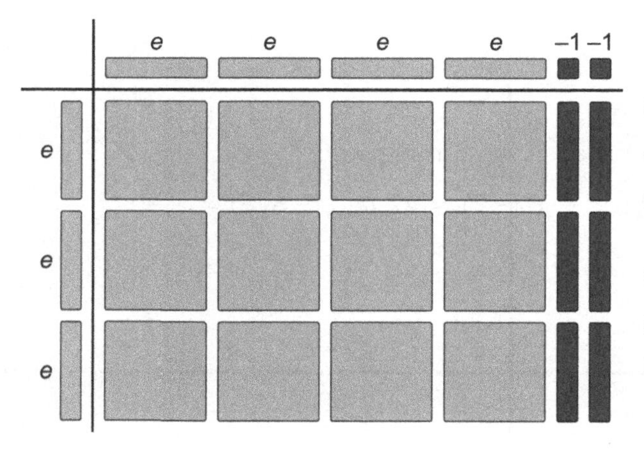

When two tiles have the same colour, use a positive tile in the rectangle.

When two tiles have different colours, use a negative tile in the rectangle.

Fill the rectangle with tiles.
We used 12 e^2-tiles and 6 $-e$-tiles to fill the rectangle.
So, $3e(4e - 2) = 12e^2 - 6e$

Check

1. Sketch algebra tiles to multiply. Write the product each time.

a) $2f(4f)$

Number of f^2-tiles used: ____

So, $2f(4f) =$ ____f^2

b) $3m(-2m + 4)$

$3m(-2m + 4) =$ ________________

Example 2 Multiplying a Binomial by a Monomial Symbolically

Find each product:

a) $3x(9x - 4)$ **b)** $-6x(-7x + 5)$

Solution

a) $3x(9x - 4)$ — Use the distributive property.

Multiply each term in brackets by $3x$.

$= (3x)(9x) + (3x)(-4)$ — Multiply: $3 \times 9 = 27$ and $3(-4) = -12$

$= 27x^2 + (-12x)$

$= 27x^2 - 12x$

b) $-6x(-7x + 5)$ — Multiply each term in brackets by $-6x$.

$= (-6x)(-7x) + (-6x)(5)$ — Multiply: $(-6)(-7) = 42$ and $(-6)(5) = -30$

$= 42x^2 + (-30x)$

$= 42x^2 - 30x$

Check

1. Multiply.

a) $7x(4x + 5)$
$= (7x)(4x) + (7x)(5)$
$=$ ____ $x^2 +$ ____ x

Multiply each term in brackets by $7x$.
Multiply: $7 \times 4 =$ ____ and $7 \times 5 =$ ____

b) $s(-3s + 4)$
$= (s)(-3s) + (s)(4)$
$=$ ________________

Multiply each term in brackets by s.
Multiply: $1 \times (-3) =$ ____ and $1 \times 4 =$ ____

c) $-9r(4r - 5)$
$=$ ________________
$=$ ________________

To divide a polynomial by a monomial, we use what we already know about division.

To divide: $\frac{6a^2}{3a}$

We write the fraction as a product of 2 fractions, then simplify each fraction.

$$\frac{6a^2}{3a} = \frac{6}{3} \times \frac{a^2}{a}$$
$$= 2 \times \frac{a \times \cancel{a}^1}{\cancel{a}_1}$$
$$= 2 \times a$$
$$= 2a$$

a is a common factor of the numerator and the denominator.

$2a$ is the *quotient* of $\frac{6a^2}{3a}$.

Example 3 Dividing a Binomial by a Monomial

Find the quotient: $\frac{-8s^2 + 6s}{-2s}$

Solution

$\frac{-8s^2 + 6s}{-2s}$ — Write as the sum of 2 fractions each with denominator $-2s$.

$= \frac{-8s^2}{-2s} + \frac{6s}{-2s}$ — Simplify the fractions.

$= \frac{-8}{-2} \times \frac{s^2}{s} + \frac{6}{-2} \times \frac{s}{s}$

$= 4 \times s + (-3) \times 1$

$= 4s - 3$

A variable divided by itself is 1.

Check

1. Divide.

a) $\frac{12a^2}{-6a}$ Write as a product of 2 fractions.

$= \frac{\_\_\_}{\_\_\_} \times \frac{a^2}{a}$ Simplify each fraction.

$= \_\_\_ \times \frac{a \times \cancel{a}^{1}}{\cancel{a}_{1}}$

$= \_\_\_\_\_\_$

$= \_\_\_\_\_\_$

b) $\frac{9b^2 + 3b}{3b}$

$= \frac{\_\_\_}{3b} + \frac{\_\_\_}{3b}$

$= \_\_\_\_\_\_$

$= \_\_\_\_\_\_$

$= \_\_\_\_\_\_$

c) $\frac{-14c^2 + 21c}{-7c}$

$= \_\_\_\_\_\_$

$= \_\_\_\_\_\_$

$= \_\_\_\_\_\_$

$= \_\_\_\_\_\_$

Practice

1. Write the multiplication sentence modelled by each set of tiles.

a)

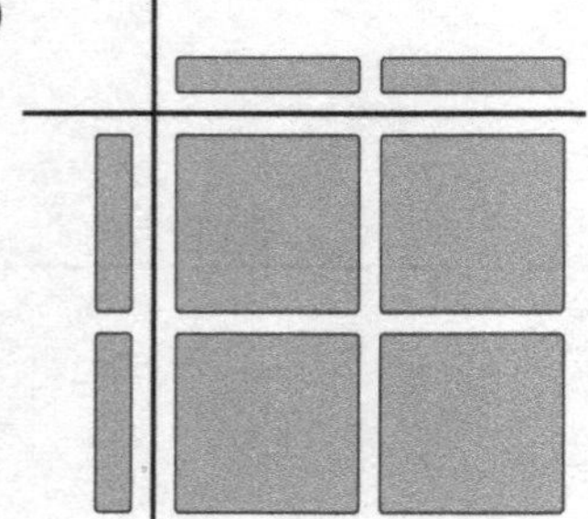

____

b)

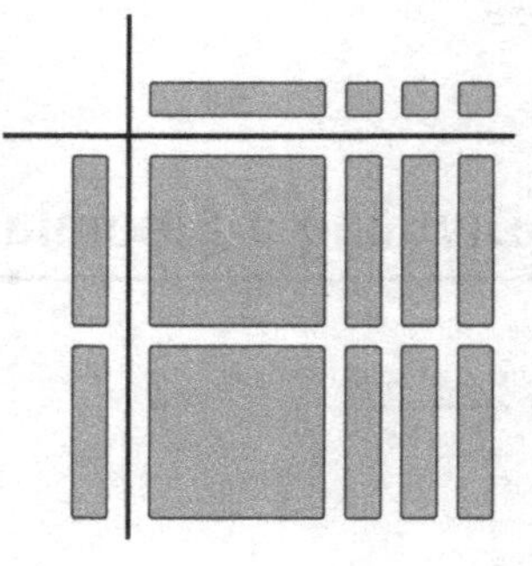

____

2. Sketch algebra tiles to multiply. Write the product each time.

a) $2s(s + 4) =$ ________

b) $t(-2t + 3) =$ ____________

3. Multiply.

a) $4r(5r - 1)$

$= (4r)(\_\_) + (4r)(\_\_)$

$= \_\_r^2 + (\_\_r)$

$=$ ________________

b) $7s(-3s + 6)$

$=$ ________________

$=$ ________________

c) $-6t(t - 3)$

$=$ ________________

$=$ ________________

d) $-8u(-6u + 7)$

$=$ ________________

$=$ ________________

4. Divide.

a) $\frac{12v^2}{4v}$

$= \frac{\_\_}{\_\_} \times \frac{v^2}{v}$

$= \_\_ \times \frac{v \times \cancel{v}^1}{\cancel{v}_1}$

$= \_\_ \times v$

$= \_\_$

b) $\frac{15w^2}{-3w}$

$= \frac{\_\_}{\_\_}$

$=$ ________

$=$ ________

c) $\frac{-28x^2}{-7x}$

$= \frac{\_\_}{\_\_}$

$=$ ________

$=$ ________

5. Divide.

a) $\frac{18y^2 + 12y}{2y}$

$= \frac{\_\_}{2y} + \frac{\_\_}{2y}$

$=$ ________________

$=$ ________________

$=$ ________________

b) $\frac{-32z^2 + 24z}{-8z}$

$=$ ________________

$=$ ________________

$=$ ________________

$=$ ________________

c) $\frac{15n^2 + 21n}{-3n}$

$=$ ________________

$=$ ________________

$=$ ________________

$=$ ________________

Unit 5 Puzzle

Alphabet Soup!

The table below contains 15 polynomial expressions.
Simplify each expression.

1 $2x(x - 3)$ __________	**4** $(3x + 7) - (11 - 4x)$ __________	**3** $\frac{5x^2 + 10x}{5x}$ __________
6 $-4(x^2 + 3x - 1)$ __________	**5** $(5x + 4) + (x^2 - 2x + 1)$ __________	**7** $3(2x - 1)$ __________
9 $3x^2 + 5 - 2x$ $- (5 + 3x^2 - 2x)$ __________	**10** $8 + 7x - 11 - 3x$ __________	**13** $(4x^2 + 9x + 5)$ $- (4x^2 + 8x + 3)$ __________
8 $\frac{36x^2 - 18x}{6x}$ __________	**2** $2x^2 + x + 5 - 7x - 5$ __________	**11** $(16x - 12) \div 4$ __________
0 $(6x + 5) + (-9 + x)$ __________	**12** $2(-2x^2 - 6x + 2)$ __________	**15** $(4x^2 + x + 6)$ $- (3x^2 - 2x + 1)$ __________

Seven pairs of expressions have the same answer. Find the 7 pairs.
For each matching pair, add the numbers in the top left corner of each square.
The sum represents a letter of the alphabet.

1	2	3	4	5	6	7	8	9	10	11	12	13	14	15	16	17	18	19	20	21	22	23	24	25	26
A	B	C	D	E	F	G	H	I	J	K	L	M	N	O	P	Q	R	S	T	U	V	W	X	Y	Z

Write the seven letters below.

___ ___ ___ ___ ___ ___ ___

Unscramble the letters to find a math word used in this unit. __________

Unit 5 Study Guide

Skill	Description	Example
Recognize the different parts of a polynomial.	A polynomial may have variable terms and a constant term. The number in front of a variable is its coefficient.	variable term $3x^2 + 2x + 4$ coefficient constant
Describe and classify polynomials.	A polynomial can be classified by its number of terms and by its term with the greatest degree.	Monomial: $3x$ Binomial: $2x + 5$ Trinomial: $x^2 + 2x - 1$ degree 2
Use algebra tiles to represent a polynomial.	We use these tiles: x^2 $-x^2$ x $-x$ 1 -1 A pair of tiles with the same shape and size, but different colours forms a zero pair. The tiles model 0.	$x^2 + 2x - 1$ 0 0 0
Simplify polynomials by combining like terms.	To simplify a polynomial, add the coefficients of like terms.	Like terms: $4x^2$ and $-2x^2$ Unlike terms: $3x$ and -5 $4x^2 - 2x^2 = 2x^2$
Add polynomials.	To add polynomials, remove the brackets and add the coefficients of like terms.	$(4x^2 + 3x) + (x^2 - 5x)$ $= 4x^2 + 3x + x^2 - 5x$ $= 4x^2 + x^2 + 3x - 5x$ $= 5x^2 - 2x$
Subtract polynomials.	To subtract a polynomial, add the opposite terms.	$(3x^2 + 5x) - (2x^2 - x)$ $= 3x^2 + 5x + (-2x^2 + x)$ $= 3x^2 + 5x - 2x^2 + x$ $= 3x^2 - 2x^2 + 5x + x$ $= x^2 + 6x$
Multiply a polynomial by a monomial.	To multiply a polynomial by a monomial, use the distributive property.	$3x(6x - 5)$ $= 3x(6x) + (3x)(-5)$ $= 18x^2 + (-15x)$ $= 18x^2 - 15x$
Divide a polynomial by a monomial.	To divide a polynomial by a monomial, divide each term of the polynomial by the monomial.	$\frac{24x^2 - 32x}{8x} = \frac{24x^2}{8x} + \frac{-32x}{8x}$ $= 3x - 4$

Unit 5 Review

5.1 **1.** Is the polynomial a monomial, binomial, or trinomial?

a) $-3s^2 + 11$ ____________.

b) $8d$ ____________.

c) $2e^2 - 9e + 7$ ____________.

d) $8h - 1$ ____________.

2. Sketch algebra tiles to model each polynomial.

a) $3k - 4$

b) $2m^2 - m + 3$

c) $-n^2 + 5n - 2$

____________ ____________ ____________

5.2 **3.** Simplify each polynomial.

a) $-7d - 4 + 8d + 2$
= ____________
= ____________

b) $3e^2 - 8e + 2e^2 + 11e$
= ____________
= ____________

c) $13 - 6h + 2h^2 + 7h - 9$
= ____________
= ____________
= ____________

d) $-9k^2 + 15k - 8 - 2k^2 - 4k + 3$
= ____________
= ____________

4. Identify and explain any errors you find.

a) $2x^2 + 5x = 7x^2$ ____________

b) $5s - 7s = -2s$ ____________

5.3 **5.** Sketch algebra tiles to model each sum. Then write the sum.

a) $(-5e + 7) + (4e - 1)$

Remaining tiles: ____________
So, $(-5e + 7) + (4e - 1) =$ ____________

b) $(6f^2 - 2f + 5) + (-4f^2 - f - 3)$

Remaining tiles: ____________
So, $(6f^2 - 2f + 5) + (-4f^2 - f - 3)$
= ____________

6. Add.

a) $(7r + 11) + (-2r + 3)$
= ______
= ______
= ______

b) $(-9s^2 + 5s) + (16s^2 - 9s - 14)$
= ______
= ______
= ______

5.4 **7.** Use algebra tiles to model each difference.
Sketch the tiles that remain, then write the difference.

a) $(-2t + 5) - (-5t + 7)$

Remaining tiles: ______
So, $(-2t + 5) - (-5t + 7) =$ ______

b) $(-7u - 2) - (-u^2 - 3u - 1)$

Remaining tiles: ______
So, $(-7u - 2) - (-u^2 - 3u - 1) =$ ______

8. Subtract.

a) $(6v + 5) - (13v - 3)$
$= 6v + 5 +$ (______)
= ______
= ______
= ______

b) $(10w^2 - 7) - (-2w + 9w^2 + 5)$
= ______
= ______
= ______
= ______

5.5 **9.** Write the multiplication sentence modelled by each set of tiles.

a)

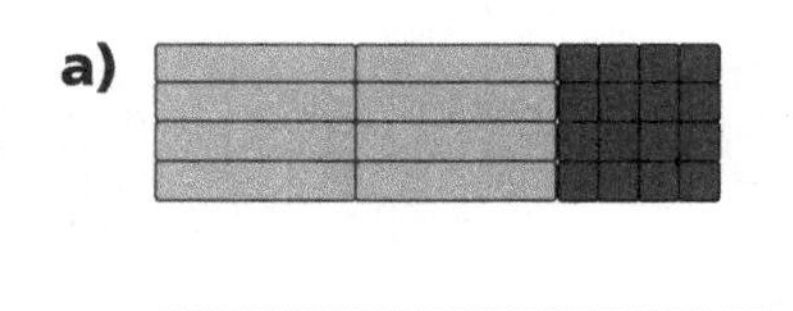

______

b)

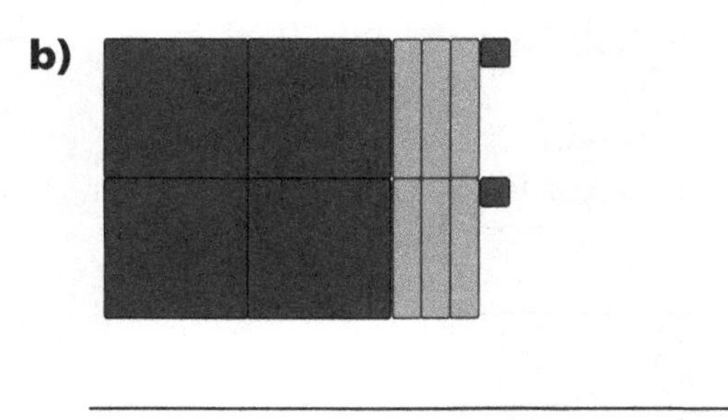

______

10. Multiply.

a) $6(-7y^2 + 1)$
= 6(______) + 6(______)
= ______

b) $-9(-2z^2 - 4z + 5)$

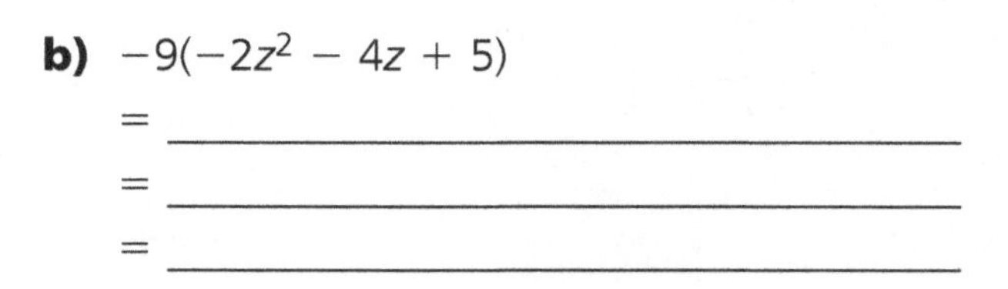

= ______
= ______
= ______

11. Divide.

a) $\frac{16a - 40}{8}$

$= \frac{\_\_}{8} + \frac{\_\_}{8}$

$= \frac{16}{8} \times a + (\_\_)$

= ________

= ________

b) $\frac{27b^2 - 9b + 36}{-9}$

= ________

= ________

= ________

= ________

5.6 **12.** Sketch algebra tiles to multiply. Write the product each time.

a) $2c(c + 5) =$ ________

b) $3d(-d + 4) =$ ________

13. Multiply.

a) $3e(5e - 2)$

$= (3e)(\_\_) + (3e)(\_\_)$

$= \_\_e^2 + (\_\_)e$

= ________

b) $-4f(5f + 2)$

= ________

= ________

= ________

14. Divide.

a) $\frac{-21k^2}{7k}$

$= \frac{-21}{7} \times \frac{k^2}{k}$

$= \_\_ \times \frac{k \times \cancel{k}^1}{\cancel{k}_1}$

$= \_\_ \times k$

= ________

b) $\frac{81m^2 - 45m}{-9m}$

= ________

= ________

= ________

= ________

c) $\frac{-33n^2 + 36n}{-3n}$

= ________

= ________

= ________

= ________

UNIT 6

Linear Equations and Inequalities

What You'll Learn

- Expand your understanding of solving equations.
- Model and solve problems using linear equations.
- Investigate the properties of inequalities.
- Explain and illustrate strategies to solve linear inequalities.

Why It's Important

Linear equations and inequalities are used by

- nurses, home health aides, and medical assistants, to take temperatures and blood pressures, and set up equipment
- purchasing agents and buyers, to find the best merchandise at the lowest price for their employers, and stay aware of changes in the marketplace

Key Words

inverse operations
variable
inequality

6.1 Skill Builder

Order of Operations

We use this order of operations to evaluate expressions with more than one operation.

B Do the operations in **b**rackets first
E Evaluate any **e**xponents
D
M **D**ivide and **m**ultiply in order from left to right
A
S **A**dd and **s**ubtract in order from left to right

$7 - 8 \div 2 + (6 - 1)$	Evaluate brackets first: $(6 - 1)$
$= 7 - 8 \div 2 + 5$	Then divide: $8 \div 2$
$= 7 - 4 + 5$	Then add and subtract from left to right.
$= 3 + 5$	
$= 8$	

Check

1. In each expression, circle what you will do first.

a) $-7 + 2 \times (-3)$	Add	Multiply		
b) $3 \times (-10 \div 2) - (-4)$	Multiply	Divide	Subtract	
c) $19 - 4 \times 3^2 \div 6$	Subtract	Multiply	Power	Divide
d) $-30 \div 5 - 10 \times 2$	Divide	Subtract	Multiply	

2. Evaluate.

a) $-17 + 4 \times 3$
$= -17 +$ ____
$=$ _____

b) $-16 \div 4 + 24 \div (-8)$
$=$ ____ $+ 24 \div (-8)$
$=$ ____ $+$ (____)
$=$ _____

c) $3^2 + 4^2 \div 8 + (-5)$
$=$ ____ $+$ ____ $\div 8 + (-5)$
$= 9 +$ ____ $+ (-5)$
$=$ ____ $+ (-5)$
$=$ ____

The Distributive Property

To multiply $5 \times (3 + 4)$, we can:

- Add $3 + 4$ in the brackets, then multiply the sum by 5:
  $5 \times (3 + 4)$
  $= 5 \times 7$
  $= 35$

OR

- Multiply each number in the brackets by 5, then add:
  $5 \times (3 + 4) = 5 \times 3 + 5 \times 4$
  $= 15 + 20$
  $= 35$

We can use the distributive property to write this expression as a sum of terms:
$7(a + b) = 7a + 7b$

Check

1. Expand.

a) $3(b - 2)$
= ____________
= ____________

b) $6(2 - y)$
= ____________
= ____________

6.1 Solving Equations by Using Inverse Operations

FOCUS **Model a problem with a linear equation, and solve the equation pictorially and symbolically.**

Look at the equation $2x + 5 = 13$.
How was it built?
Start with x. Multiply by 2, then add 5.

Build equation
$\times 2$ $+5$
x → $2x$ → $2x + 5$
4 ← 8 ← 13
$\div 2$ -5
Solve equation

To solve the equation, "undo" the operations, in reverse order.

$2x + 5 = 13$ Subtract 5.
$2x + 5 - 5 = 13 - 5$
$2x = 8$ Divide by 2.
$x = 4$

Do the same operation to both sides of the equation to preserve the equality.

Inverse operations undo each other's results. For example: addition and subtraction are inverse operations.

Example 1 Writing Then Solving One-Step Equations

Write then solve an equation to find each number. Verify the solution.

a) A number plus 5 is 20. **b)** Four times a number is -32.

Solution

a) Let x represent the number. Then, x plus 5 is 20.
The equation is: $x + 5 = 20$
To solve the equation, apply the inverse operations.

$x + 5 = 20$ Undo the addition. Subtract 5 from each side.
$x + 5 - 5 = 20 - 5$
$x = 15$

To verify the solution, substitute $x = 15$ into $x + 5 = 20$.
$15 + 5 = 20$, so the solution is correct.

b) Let n represent the number. Then, 4 times n is -32.
The equation is: $4n = -32$
To solve the equation, apply the inverse operations.

$4n = -32$ Undo the multiplication. Divide each side by 4.
$\frac{4n}{4} = \frac{-32}{4}$
$n = -8$

To verify the solution, substitute $n = -8$ into $4n = -32$.
$4(-8) = -32$, so the solution is correct.

Check

1. Let n represent a number. Two less than a number is 10. What is the number?

$n -$ ________ $=$ ________

$n -$ ________ $=$ ________

$n =$ ________

Check: ____________________

Example 2 Solving a Two-Step Equation

Solve, then verify each equation.

a) $3x + 4 = -5$

b) $2(-2 + w) = 18$

Solution

a) Perform the inverse operations in reverse order.

$3x + 4 = -5$ Subtract 4 from each side.

$3x + 4 - 4 = -5 - 4$

$3x = -9$ Divide each side by 3.

$\frac{3x}{3} = \frac{-9}{3}$

$x = -3$

To verify the solution, substitute $x = -3$ into $3x + 4 = -5$.

Left side $= 3x + 4$ Right side $= -5$

$= 3(-3) + 4$

$= -9 + 4$

$= -5$

Since the left side equals the right side, $x = -3$ is correct.

b) $2(-2 + w) = 18$ Use the distributive property to expand $2(-2 + w)$.

$2(-2) + 2(w) = 18$

$-4 + 2w = 18$ Add 4 to each side.

$-4 + 2w + 4 = 18 + 4$

$2w = 22$ Divide each side by 2.

$\frac{2w}{2} = \frac{22}{2}$

$w = 11$

To verify the solution, substitute $w = 11$ into $2(-2 + w) = 18$.

Left side $= 2(-2 + w)$ Right side $= 18$

$= 2(-2 + 11)$

$= 2(9)$

$= 18$

Since the left side equals the right side, $w = 11$ is correct.

Check

1. What operations would you use to solve each equation?

a) $-5h + 4 = 6$

First ______________, then ______________.

b) $2 + 5p = -3$

First ______________, then ______________.

2. Solve, then verify the equation.

$2(t - 1) = 12$

$2(\_) - 2(\_) = 12$

Use the distributive property to expand 2(t − 1).

____

____

____

____

____

Substitute $t =$ ___ into the equation.

Left side $= 2(t - 1)$ Right side = ___

$= 2(\_\_\_\_)$

= ____

= ____

Since the left side equals the right side, $t =$ ___ is correct.

Practice

1. Solve each equation.

a) $z + 9 = 10$

____

____

b) $s - 4 = -12$

____

____

c) $6 + c = 2$

____

____

d) $5 = v - 2$

____

____

2. For each statement, write then solve an equation to find the number.
Verify the solution.

a) A number divided by 4 is -3.

$\_\_\_ = -3$

____

____

Left side = __________

____

____

Right side = ____

$n =$ ____ is correct.

b) Three times a number is 15.

$\_\_\_ = 15$

____

____

Left side = __________

____

____

Right side = ____

$x =$ ____ is correct.

3. Emma tried to solve the equation $4x = 16$ by subtracting 4 from each side.
Show the correct way to solve the equation.
$4x = 16$

______

______

4. Solve each equation. Verify the solution.

a) $5k - 6 = 24$

____

____

____

____

Left side $= 5k - 6$
$= 5(\_) - 6$
= __________

____

Right side = ____
$k =$ ____ is correct.

b) $3 + 4y = -9$

____

____

____

____

Left side $= 3 + 4y$
= __________
= __________

____

Right side = ______
$y =$ ____ is correct.

5. a) Tuyen tried to solve the equation $3x - 6 = 15$ like this:

$$\frac{3x}{3} - 6 = \frac{15}{3}$$
$$x - 6 = 5$$
$$x - 6 + 6 = 5 + 6$$
$$x = 11$$

Where did she make a mistake?

____

b) Show the correct way to solve $3x - 6 = 15$.
Verify the solution.

Left side = $3x - 6$ Right side = ___
= ____________
= ____________
= ____________

Since the left side equals the right side, x = ___ is correct.

6. A rectangle has length 4 cm and perimeter 12 cm.

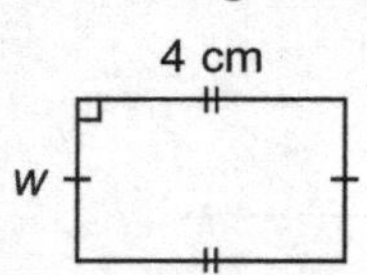

The perimeter is the sum of all the sides.

a) Write an equation that can be used to determine the width of the rectangle.
________________ = 12

b) Solve the equation.
____
____
____
____

The width is ____ cm.

c) Verify the solution.
Left side = ________________   Right side = ___
____
____
____

____
____

6.2 Skill Builder

Solving Equations Using Models

We can use a balance-scales model to solve an equation.
Keep the scales balanced by doing the same operation on both sides.
For example, we can add or remove the same mass:

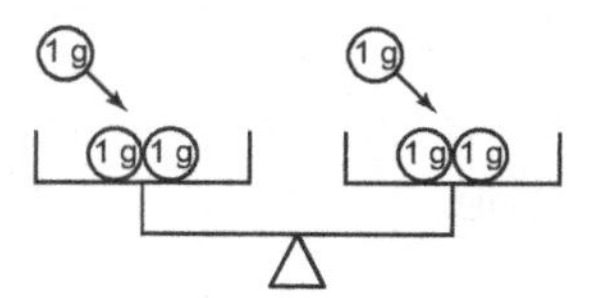

The scales remain balanced.

We can also use algebra tiles to solve an equation.
Rearrange the tiles so the variable tiles are on one side,
and the unit tiles are on the other side.
For example, to solve $3x - 2 = 7$:

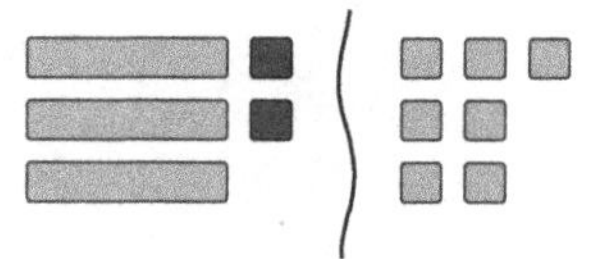

Isolate the x-tiles.
Add 2 positive unit tiles to make zero pairs.

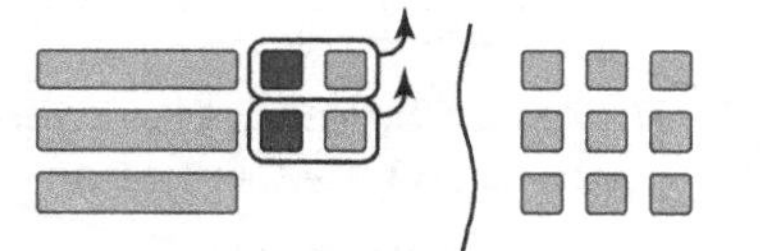

There are 3 x-tiles.
Arrange the unit tiles into 3 equal groups.

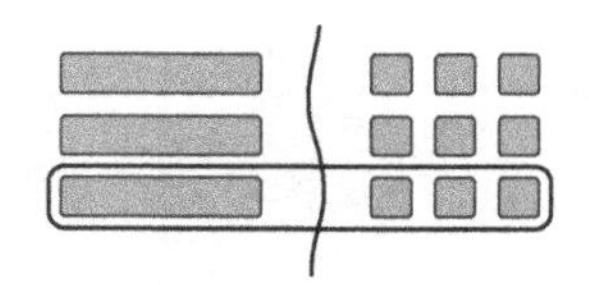

The solution is $x = 3$.

Check

1. Use algebra tiles to solve: $4m + 6 = -2$
Record your steps algebraically.

____

____

____

____

6.2 Solving Equations by Using Balance Strategies

FOCUS **Model a problem with a linear equation, use balance strategies to solve the equation pictorially, and record the process symbolically.**

To solve an equation, isolate the variable on one side of the equation.

We can use balance scales to model this.

Everything we do to one side of the equation must be done to the other side.
This way, the scales remain balanced.

Example 1 Modelling Equations with Variables on Both Sides

a) Solve: $3x + 2 = 6 + x$

b) Verify the solution.

Solution

a) Isolate x to solve the equation.

There can be more than one way to solve an equation.

Pictorial Solution | **Algebraic Solution**

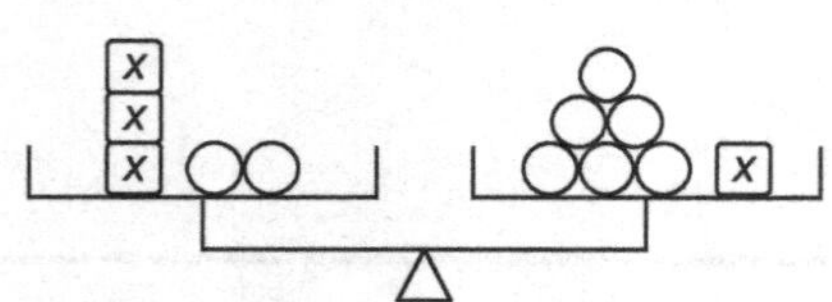

$$3x + 2 = 6 + x$$

Each ○ has a mass of 1 g.

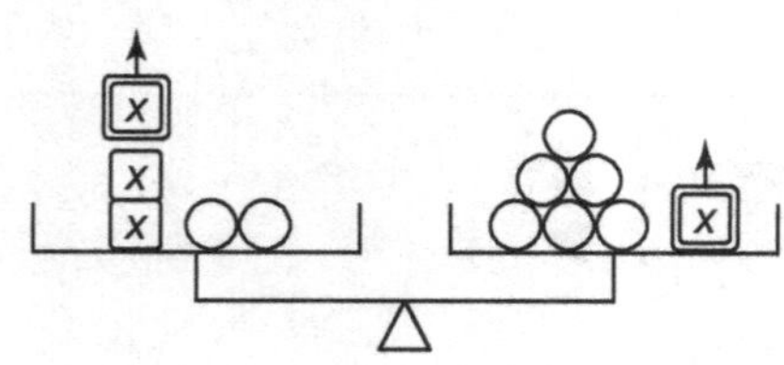

$$3x + 2 - x = 6 + x - x$$
$$2x + 2 = 6$$

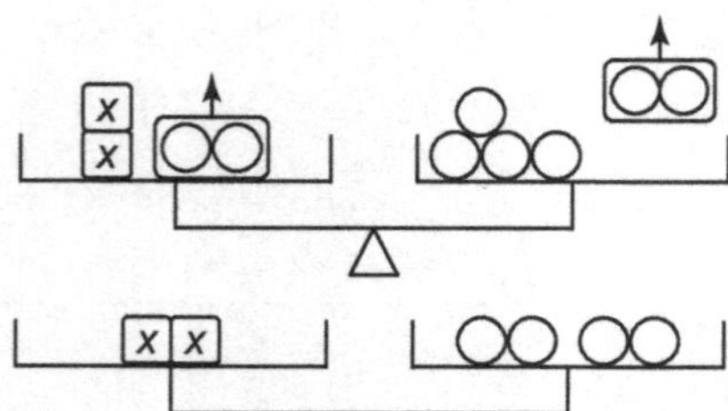

$$2x + 2 - 2 = 6 - 2$$
$$2x = 4$$

$$\frac{2x}{2} = \frac{4}{2}$$
$$x = 2$$

b) Check: Substitute $x = 2$ in each side of the equation.

Left side $= 3x + 2$
$= 3(2) + 2$
$= 6 + 2$
$= 8$

Right side $= 6 + x$
$= 6 + 2$
$= 8$

Since the left side equals the right side, $x = 2$ is correct.

Check

1. Write the equation given by the picture.
Solve, and record your steps algebraically.

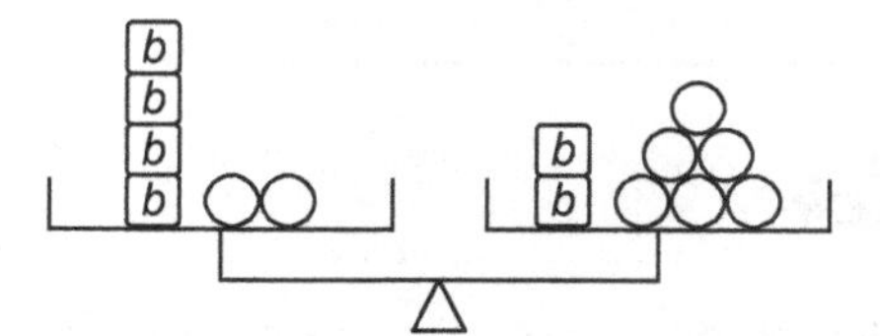

$4b$ + _____ = _____ + _____

$4b$ + _____ − _____ = _____ + _____ − _____

____

____

____

____

____

If we have an equation with negative terms, it is easier to use algebra tiles to model and solve the equation.

We add the same tiles to each side or subtract the same tiles from each side to keep the equation balanced.

Example 2 Using Algebra Tiles to Solve an Equation

Solve: $-2n + 5 = 3n - 5$

Solution

Algebra Tile Model | **Algebraic Solution**

$-2n + 5 = 3n - 5$

Add 2 n-tiles to each side.
Remove zero pairs.

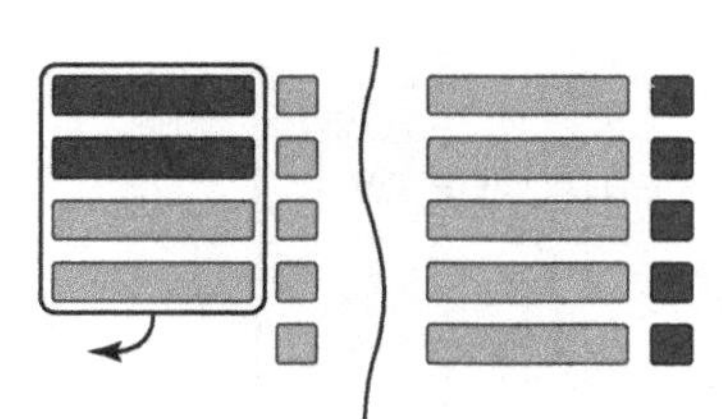

$-2n + 5 + 2n = 3n - 5 + 2n$

$5 = 5n - 5$

Add five 1-tiles to each side.
Remove zero pairs.

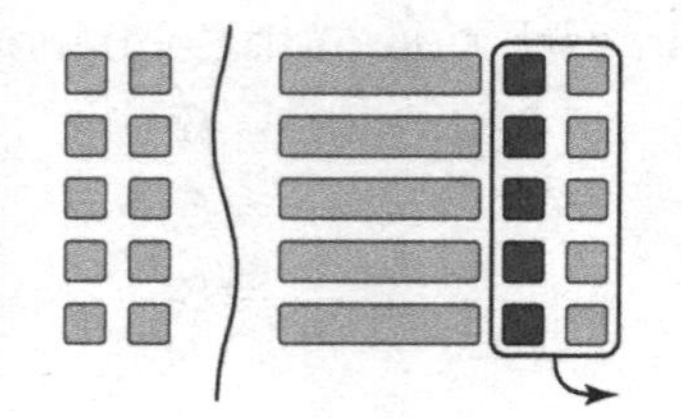

$5 + 5 = 5n - 5 + 5$

$10 = 5n$

You can solve with the variable on either side of the equal sign. The answer is the same.

Arrange the remaining tiles on each side into 5 groups. One n-tile is equal to 2.

$\frac{10}{5} = \frac{5n}{5}$

$2 = n$

Or, $n = 2$

Check

1. Use algebra tiles to model and solve the equation.
Record your work algebraically.

$-c + 5 = 2c - 4$

____

____

____

____

____

____

Example 3 Solving Equations with Rational Coefficients

Solve the equation, then verify the solution.

$\frac{2a}{3} = 6$

Solution

Create an equivalent equation without fractions.
To clear the fraction, multiply each side by the denominator.

$\frac{2a}{3} = 6$	Multiply each side by 3.
$\frac{2a}{3} \times 3 = 6 \times 3$	$\frac{2a}{3} \times \frac{3}{1} = \frac{2a}{1} = 2a$
$2a = 18$	Divide each side by 2.
$\frac{2a}{2} = \frac{18}{2}$	
$a = 9$	

Check: Substitute $a = 9$ in $\frac{2a}{3} = 6$.

Left side $= \frac{2a}{3}$ Right side $= 6$

$= \frac{2(9)}{3}$

$= \frac{18}{3}$

$= 6$

Since the left side equals the right side, $a = 9$ is correct.

Check

1. Solve. Verify the solution.

a) $\frac{x}{4} = 5$

Clear the fraction. Multiply each side by the denominator, 4.

____

____

Check: Substitute $x =$ ____ in $\frac{x}{4} = 5$.

Left side $=$ ____ Right side $=$ ____

$=$ ____

Since ____________________________ $x =$ ____ is correct.

b) $\frac{x}{4} + \frac{7}{4} = \frac{5}{4}$

____

____

____

____

Check: Substitute $x =$ ____ in $\frac{x}{4} + \frac{7}{4} = \frac{5}{4}$.

Left side $=$ ____________ Right side $=$ ____

____

____

Since ____________________________ $x =$ ____ is correct.

Practice

1. Write the equation represented by the picture.
Solve, and record your steps algebraically.

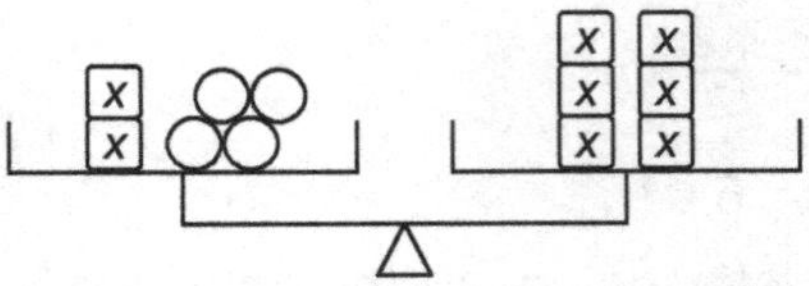

____

____

____

____

____

2. Solve each equation.

a) $3w - 2 = w + 4$

$3w - 2$ _____ $= w + 4$ _____

b) $2 - x = -2 - 3x$

$2 - x$ _____ $= -2 - 3x$ _____

c) $y - 4 = -2 - y$

d) $2 - j = -8 + 4j$

3. Solve each equation. Verify the solution.

a) $\frac{t}{6} + 2 = 4$

$\frac{t}{6} + 2 -$ ____ $= 4 -$ ____

$\frac{t}{6} =$ ___

____

____

Left side $= \frac{t}{6} + 2$

= __________

= __________

= __________

Right side = _____

$t =$ ____ is correct.

b) $5 + \frac{w}{5} = 2$

____

____

____

____

Left side = $5 + \frac{w}{5}$

= __________

= __________

____

Right side = ________

w = ____ is correct.

4. Jake tried to solve $4c - 3 = c + 3$ like this:

$4c - 3 - 3 = c + 3 - 3$

$4c = c + 0$

$4c - c = c - c + 0$

$3c = 0$

$c = 0$

a) Where did he make a mistake?

____

b) Show the correct way to solve $4c - 3 = c + 3$.
Verify the solution.

$4c - 3 = c + 3$

____

____

____

____

____

____

Left side = $4c - 3$

____

____

____

Right side = $c + 3$

____

____

Since the left side equals the right side, c = __ is correct.

Can you ...

- Model a problem with a linear equation, and solve the equation pictorially and symbolically?
- Model a problem with a linear equation, use balance strategies to solve the equation pictorially, and record the process symbolically?

6.1 **1.** For each equation, what is the first operation you would do to isolate the variable?

a) $3k = 9$

b) $m - 2 = 5$

c) $2x - 3 = 4$

d) $5 = 3y - 4$

2. For each statement, write then solve an equation to find the number. Verify the solution.

a) Two times a number is 10.

b) Three less than a number is 15.

3. Solve each equation.

a) $x + 7 = -2$

b) $4c = 20$

c) $4 = y - 2$

d) $\frac{m}{6} = 3$

4. Solve each equation. Verify the solution.

a) $3q - 1 = 17$

b) $2(3 + p) = -4$

6.2 **5.** Write the equation represented by the picture.
Solve, and record your steps algebraically.

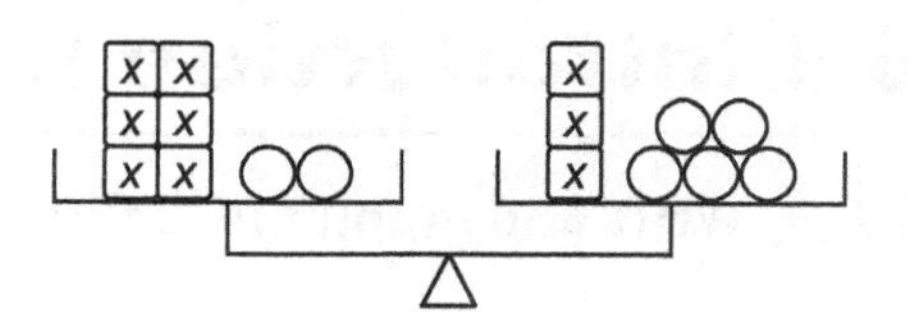

____x + ____ = ____x + ____

____

____

____

____

____

6. Solve each equation. Verify the solution.

a) $3a - 2 = a - 6$

$3a - 2 + 2 = a - 6 + 2$

____

____

____

____

____

Left side $= 3a - 2$

$=$ ______________________

____

Right side $= a - 6$

$=$ ______________________

____

$a =$ ____ is correct.

b) $4 + h = 1 - 2h$

$4 + h - 4 = 1 - 2h - 4$

____

____

____

____

____

Left side $= 4 + h$

$=$ ______________________

____

Right side $= 1 - 2h$

$=$ ______________________

____

$h =$ ____ is correct.

c) $\frac{5a}{6} = 10$

____

____

____

____

Left side $= \frac{5a}{6}$

$=$ ______________________

____

Right side $=$ ______________________

$a =$ ___ is correct.

6.3 Introduction to Linear Inequalities

FOCUS **Write and graph inequalities.**

Less than	$<$	below, under
Less than or equal to	$\leq$	up to, at most, no more than, maximum
Greater than	$>$	over, more than
Greater than or equal to	$\geq$	at least, minimum

Example 1 Writing an Inequality to Describe a Situation

Define a variable and write an inequality to describe the situation.

a) SPEED
LIMIT
60

b) You must be at least 16 years old to get a driver's licence.

Solution

a) Let s represent the speed.
You can go up to 60 km/h, but not faster.
So, s can equal 60 or be any number less than 60.

The inequality is $s \leq 60$.

b) Let a represent the age to get a driver's licence.
"At least 16" means that you must be 16, or older.
You cannot be less than 16.
So, a can equal 16 or be greater than 16.

The inequality is $a \geq 16$.

*$a \geq 16$ is read as **a is greater than or equal to 16.***

Check

1. Let t represent the temperature in degrees Celsius.
Write an inequality to describe each situation:

a) For temperatures less than 0°C, make sure to wear warm clothing. t ____ 0

b) The highest temperature we've had this week was 12°C. t ____ 12

Linear Inequalities
A linear inequality may be true for many values of the variable.

Example 2 Determining Whether a Number Is a Solution of an Inequality

Is each number a solution of the inequality $x \leq 3$? Justify the answers.

a) 5 **b)** 3 **c)** 0 **d)** -2

Solution

Use a number line to show all the numbers.
The solution of $x \leq 3$ is all numbers that are less than or equal to 3.

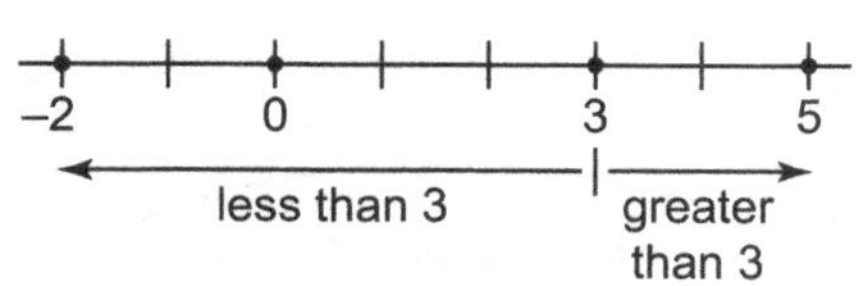

For a number to be less than 3, it must lie to the left of 3.
0 and -2 are to the left of 3, so they are solutions.
3 is equal to itself, so it is a solution.
5 is to the right of 3, so it is not a solution.

Check

1. a) Is 8 a solution of the inequality $x > 0$? Use the number line to help.

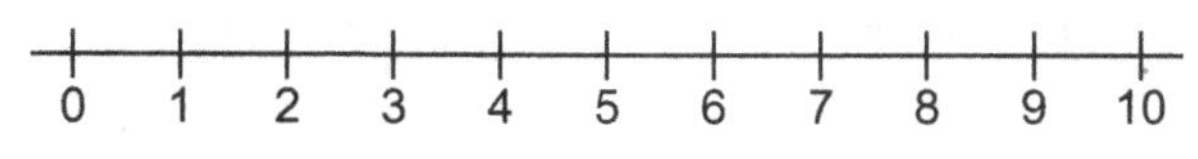

$x > 0$ is read as x is greater than 0.

8 is to the ________ of 0, so 8 _____ a solution.

b) What are 3 other numbers that are solutions of $x > 0$?

____

The solutions of an inequality can be graphed on a number line.
For example:

$a > 0$
a is greater than 0, so 0 is not included in the solution.
This is shown by an open circle at 0.

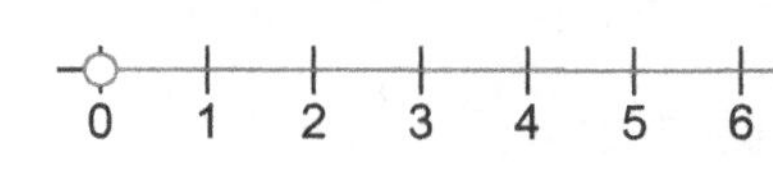

$z \leq 0$
z is less than or equal to 0, so 0 is included in the solution.
This is shown by a shaded circle at 0.

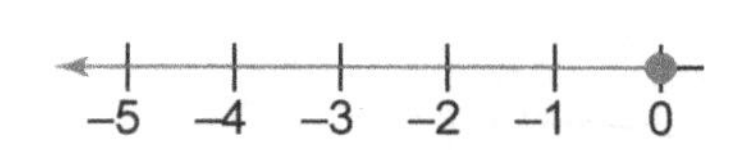

Example 3 Graphing Inequalities on a Number Line

Graph each inequality on a number line.
Write 3 numbers that are possible solutions of the inequality.

a) $b > 5$ **b)** $y \leq -1$ **c)** $-4 \geq n$ **d)** $-1 < r$

Solution

a) $b > 5$

Any number greater than 5 satisfies the inequality.

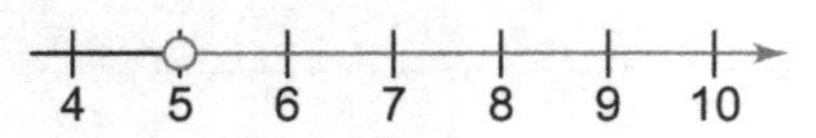

Draw an open circle at 5, because 5 is not part of the solution.

3 possible solutions are: 6, 7, 8

b) $y \leq -1$

Any number less than or equal to -1 satisfies the inequality.

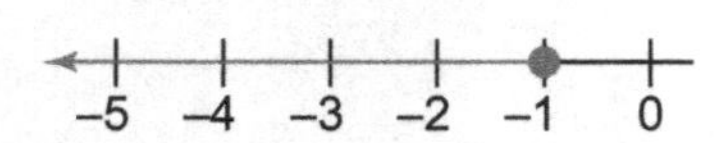

Draw a shaded circle at -1, because -1 is part of the solution.

3 possible solutions are: -1, -2, -5

c) $-4 \geq n$ means -4 is greater than or equal to n, or n is less than or equal to -4.
$-4 \geq n$ is the same as $n \leq -4$.

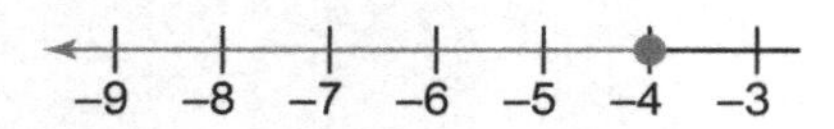

3 possible solutions are: -4, -5, -6

d) $-1 < r$ means -1 is less than r, or r is greater than -1.
$-1 < r$ is the same as $r > -1$.

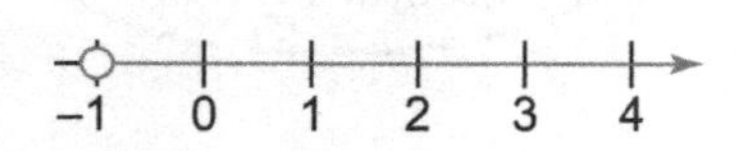

3 possible solutions are: 0, 2, 4

Check

1. Graph each inequality on a number line.
Write 3 numbers that are possible solutions for each inequality.

a) $h \leq 4$

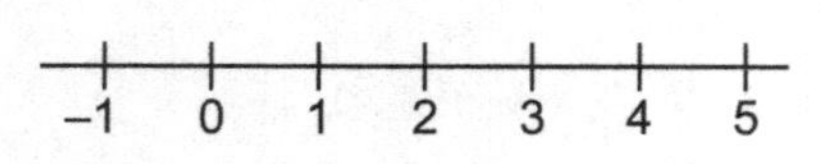

____, ____, ____

b) $-3 < x$

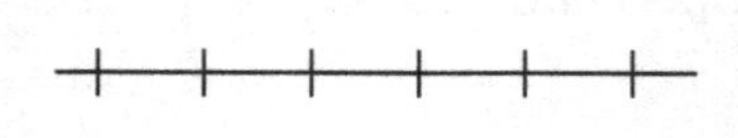

____, ____, ____

Practice

1. Is each inequality true or false?
If it is false, change the sign to write a true inequality.

a) $3 < 10$ ______ **b)** $3 < -10$ ______ **c)** $0 \leq 1$ ______ **d)** $1 \geq 1$ ______

2. Is each number a solution of $x \geq 5$?

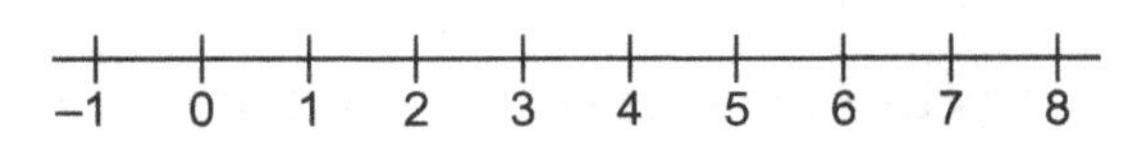

a) 5 ______ **b)** -1 ______ **c)** 0 ______ **d)** 8 ______ **e)** 6 ______

3. a) Graph each inequality on the number line.

i) $m > 3$

2 3 4 5 6 7 8

ii) $x < 2$

iii) $y \geq -5$

b) Write 3 numbers that are possible solutions of each inequality above.

i) ______ **ii)** ______ **iii)** ______

4. Write an inequality to model each situation.

a) The maximum speed is 100 km/h. Let s represent the speed, in km/h. ______

b) The elevator can hold no more than 12 people. Let n represent the number of people the elevator can hold. ______

c) This year, the price of gas has always been at least 70 cents per litre. Let p represent the price of gas, in cents. ______

d) This pass card is good for up to 10 entries to the amusement park. Let n represent the number of entries. ______

5. Match each inequality with the graph of its solution below.

a) $x > 1$ **b)** $x \leq -2$ **c)** $x < 1$ **d)** $x \geq 0$

i) –3 –2 –1 0 1 2

______

ii) –1 0 1 2 3 4 5

______

iii) 0 1 2 3 4 5 6

______

iv) –6 –5 –4 –3 –2 –1

______

6. Write an inequality whose solution is graphed on the number line.

a) 4 5 6 7 8 9

______

b) –6 –5 –4 –3 –2

______

6.4 Solving Linear Inequalities by Using Addition and Subtraction

FOCUS **Use addition and subtraction to solve inequalities.**

Consider the inequality $-1 < 3$.

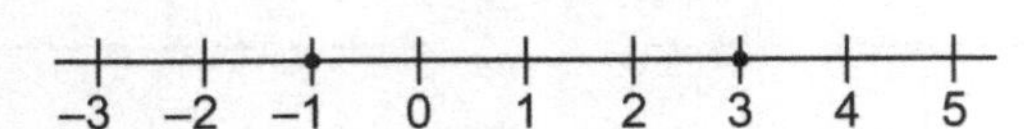

What happens to an inequality if we add the same number to each side?

$-1 < 3$ Add 2 to each side.

Left side: $-1 + 2 = 1$

Right side: $3 + 2 = 5$

The resulting inequality is still true: $1 < 5$

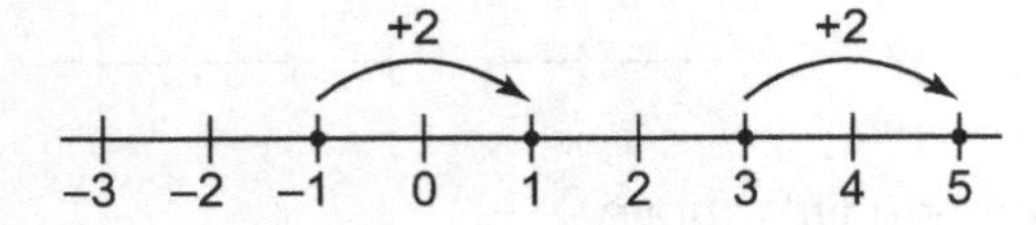

What happens to an inequality if we subtract the same number from each side?

$-1 < 3$ Subtract 2 from each side.

Left side: $-1 - 2 = -3$

Right side: $3 - 2 = 1$

The resulting inequality is still true: $-3 < 1$

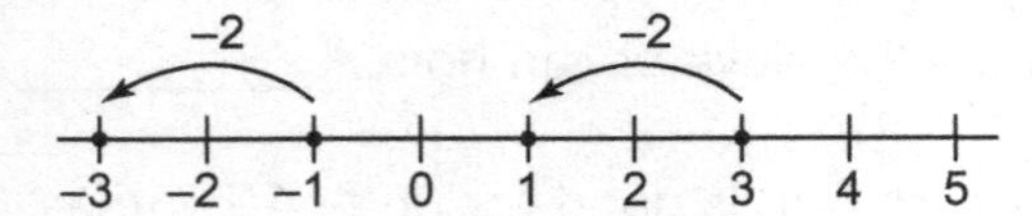

Property of Inequalities

When the same number is added to or subtracted from each side of an inequality, the resulting inequality is still true.

The strategy that we used to solve an equation can be used to solve an inequality. Isolate the variable to solve.

Equation	**Inequality**
$r - 6 = -2$	$r - 6 < -2$
$r - 6 + 6 = -2 + 6$	$r - 6 + 6 < -2 + 6$
$r = 4$	$r < 4$
There is only 1 solution: $r = 4$	Any number less than 4 is part of the solution. The solution includes 3, 2, and 1, for example.

Example 1 Solving an Inequality

a) Solve the inequality $6 \leq x - 4$.

b) Graph the solution on a number line.

Solution

a)
$$6 \leq x - 4$$
$$6 + 4 \leq x - 4 + 4$$
$$10 \leq x$$

Add 4 to each side to isolate x.

This is the same as $x \geq 10$.

b)

Place a shaded circle on 10 because 10 is part of the solution.

Check

1. Solve each inequality. Graph the solution on the number line.

a) $p + 3 \leq 4$

$p + 3$ _____ ≤ 4 _____

$p \leq$ ____________________

____

–3 –2 –1 0 1 2

b) $-5 > 2 + a$

____

____

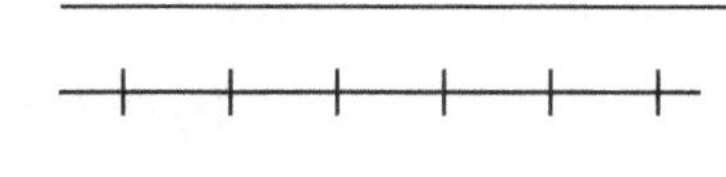

Example 2 Solving an Inequality with Variables on Both Sides

a) Solve the inequality $3d + 2 < 2d - 2$.

b) Graph the solution on a number line.

Solution

a)
$$3d + 2 < 2d - 2$$
Subtract $2d$ from each side.
$$3d + 2 - 2d < 2d - 2 - 2d$$
$$d + 2 < -2$$
Subtract 2 from each side.
$$d + 2 - 2 < -2 - 2$$
$$d < -4$$

b)

–9 –8 –7 –6 –5 –4 –3

Place an open circle on –4 because –4 is not part of the solution.

Check

1. Solve each inequality. Graph the solution on a number line.

a) $3z + 1 \leq 2z - 2$

____

____

____

____

−7 −6 −5 −4 −3 −2

b) $4 - 4x > 6 - 5x$

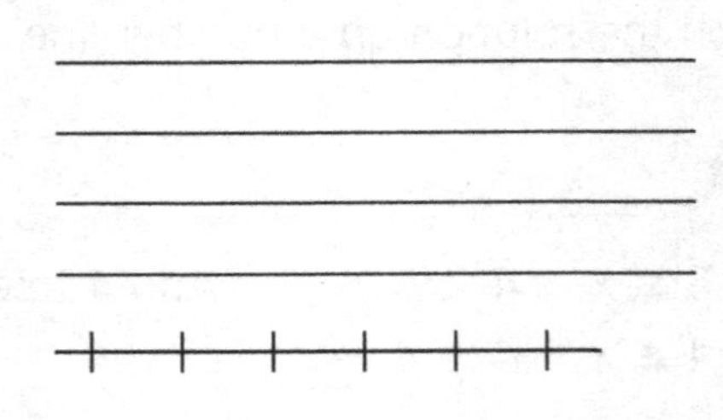

Practice

1. Which operation will you perform to each side of the inequality to isolate the variable?

a) $a + 1 > 3$

____

b) $2 < m - 3$

____

c) $x - 4 \geq 5$

____

d) $6 > 1 - z$

____

2. Fill in the missing steps to get to the solution.

a) $x + 5 > 10$

$x + 5$ _____ > 10 _____

$x >$ __

b) $12 \leq x - 4$

12 _____ $\leq x - 4$ _____

__ $\leq x$

3. Solve each inequality.
Match each inequality with the graph of its solution, below.

a) $n - 4 > -2$

____

____

b) $p + 6 < -2$

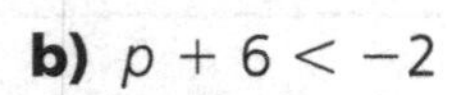

____

____

c) $u - 3 \geq -4$

____

____

d) $2 + y > -2$

____

____

i) −2 −1 0 1 2 3

____

ii)

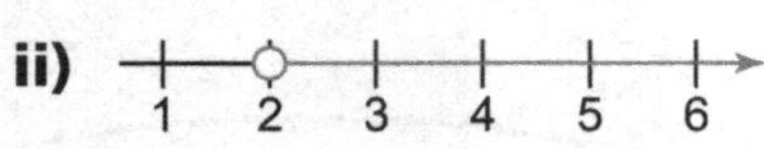

____

iii)

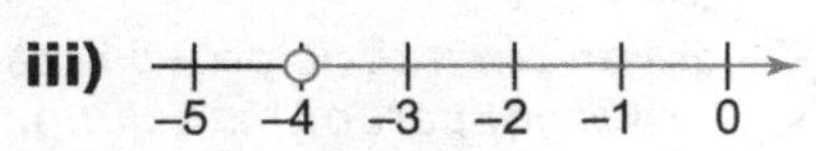

____

iv)

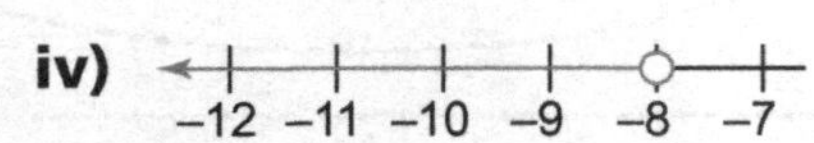

____

4. a) Solve each inequality. Graph the solution on a number line.

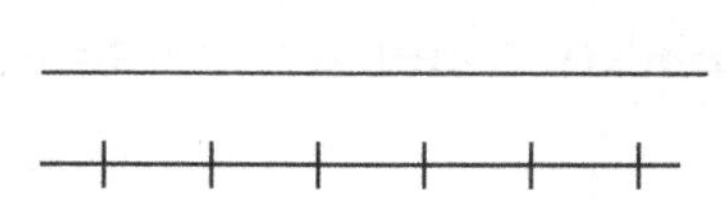

i) $y + 3 \leq -2$

ii) $2 + b < 5$

$-9 \quad -8 \quad -7 \quad -6 \quad -5 \quad -4$

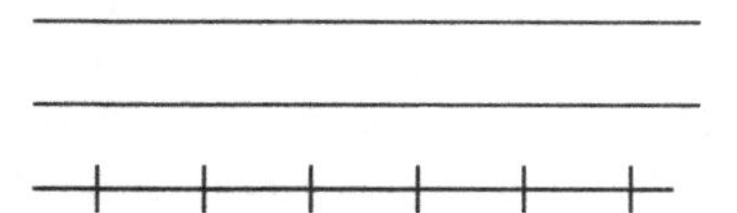

iii) $4 \geq n - 2$

iv) $3 < t - 3$

b) Write 3 numbers that are possible solutions for each inequality.

i) ____________

ii) ____________

iii) ____________

iv) ____________

c) Write a number that is NOT a solution of each inequality.

i) ________

ii) ________

iii) ________

iv) ________

5. Solve, then graph each inequality.

a) $6a + 2 \geq 5a + 1$

b) $3 + 2s > s - 3$

$-2 \quad -1 \quad 0 \quad 1 \quad 2 \quad 3$

6. a) Solve the equation: $4v - 6 = 3v - 3$

b) Solve the inequality: $4v - 6 > 3v - 3$

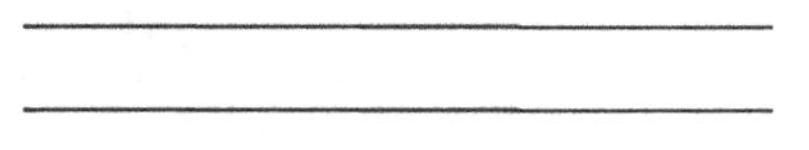

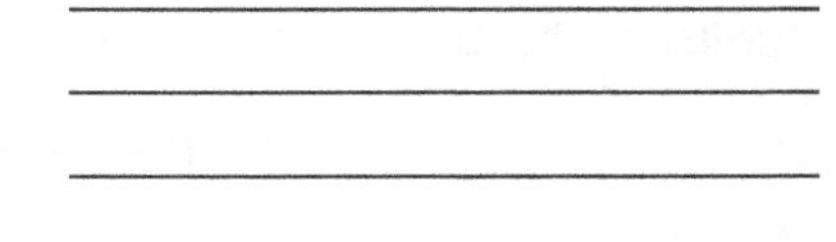

Graph the solution.

$1 \quad 2 \quad 3 \quad 4 \quad 5 \quad 6$

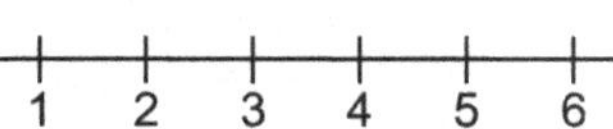

Graph the solution.

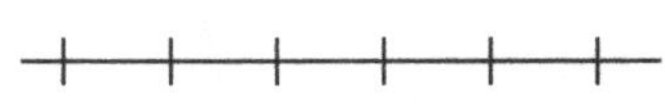

6.5 Solving Linear Inequalities by Using Multiplication and Division

FOCUS **Use multiplication and division to solve inequalities.**

Consider the inequality $-2 < 2$.

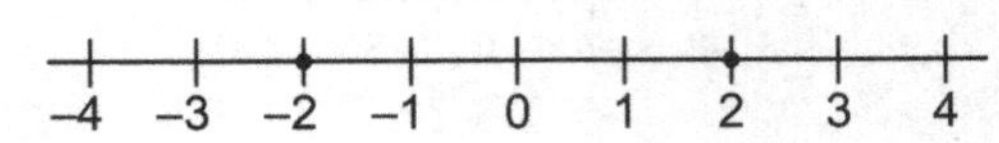

What happens to an inequality when we multiply or divide each side by the same positive number?

$-2 < 2$ Multiply each side by 2.

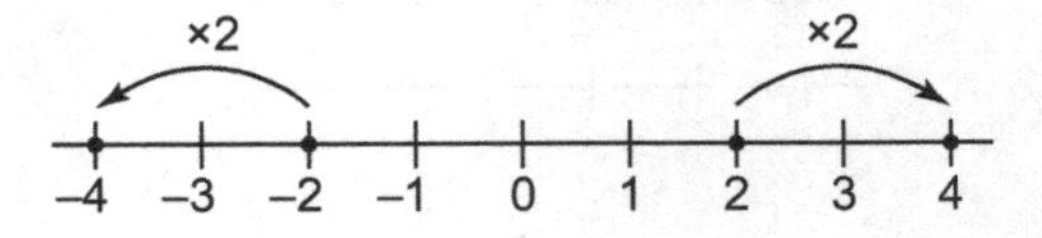

Left side: $(-2)(2) = -4$

Right side: $2(2) = 4$

$-4 < 4$

The resulting inequality is still true.

$-2 < 2$ Divide each side by 2.

Left side: $\frac{-2}{2} = -1$

Right side: $\frac{2}{2} = 1$

$-1 < 1$

The resulting inequality is still true.

Property of Inequalities

When each side of an inequality is multiplied or divided by the same positive number, the resulting inequality is still true.

What happens to an inequality when we multiply or divide each side by the same negative number?

$-2 < 2$ Multiply each side by -2.

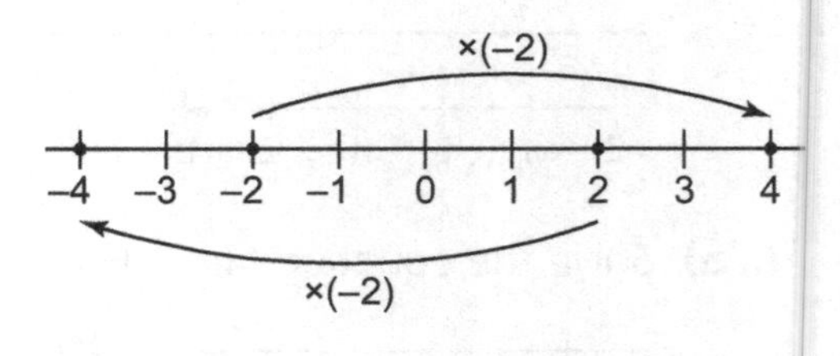

Left side: $(-2)(-2) = 4$

Right side: $2(-2) = -4$

$4 > -4$

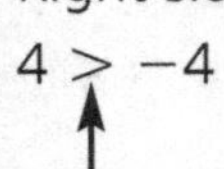

For the inequality to be true, the sign has to be reversed.

$-2 < 2$ Divide each side by -2.

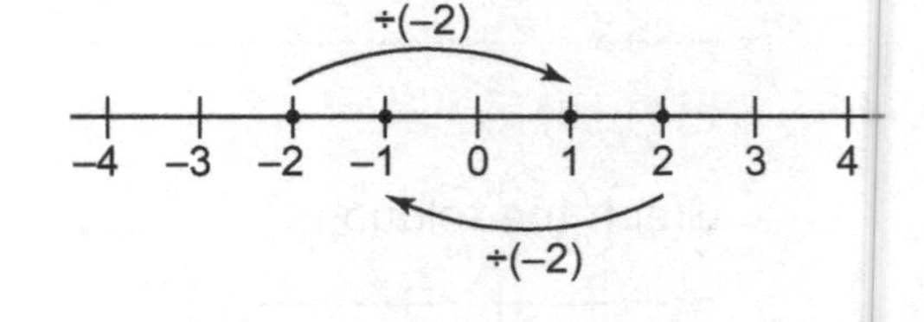

Left side: $\frac{-2}{-2} = 1$

Right side: $\frac{2}{-2} = -1$

$1 > -1$

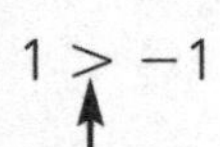

For the inequality to be true, the sign has to be reversed.

Property of Inequalities

When each side of an inequality is multiplied or divided by the same negative number, the inequality sign must be reversed for the inequality to remain true.

Example 1 Solving One-Step Inequalities

Solve each inequality and graph the solution.

a) $4x < -12$ **b)** $-2c \geq 8$ **c)** $\frac{b}{2} \leq 3$ **d)** $\frac{v}{-3} > 4$

Solution

a) $4x < -12$ Divide each side by 4.

$\frac{4x}{4} < \frac{-12}{4}$

When you divide each side by the same positive number, do not reverse the inequality sign.

$x < -3$

The solution of $x < -3$ is all numbers less than -3.

–8 –7 –6 –5 –4 –3 –2

b) $-2c \geq 8$ Divide each side by -2.

$\frac{-2c}{-2} \leq \frac{8}{-2}$

When you divide each side by the same negative number, reverse the inequality sign.

$c \leq -4$

The solution of $c \leq -4$ is all numbers less than or equal to -4.

–9 –8 –7 –6 –5 –4 –3

c) $\frac{b}{2} \geq 3$ Multiply each side by 2.

$2 \times \frac{b}{2} \geq 2(3)$

When you multiply each side by the same positive number, do not reverse the inequality sign.

$b \geq 6$

The solution of $b \geq 6$ is all numbers greater than or equal to 6.

5 6 7 8 9 10 11

d) $\frac{v}{-3} > 4$ Multiply each side by -3.

$(-3)\left(\frac{v}{-3}\right) < (-3)(4)$

When you multiply each side by the same negative number, reverse the inequality sign.

$v < -12$

The solution of $v < -12$ is all numbers less than -12.

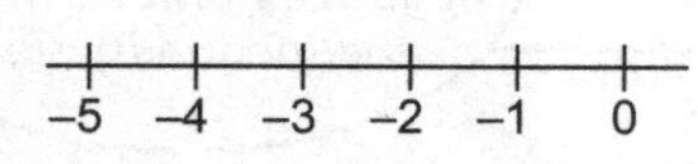

Check

1. State whether you would reverse the inequality sign to solve each inequality.

a) $-2m < 8$ __________

b) $2m \leq 8$ __________

c) $\frac{y}{-2} > 3$ __________

2. Solve the inequalities in question 1. Graph each solution.

a) $-2m < 8$

____

____

–5 –4 –3 –2 –1 0

b) $2m \leq 8$

____

____

c) $\frac{y}{-2} > 3$

____

____

Example 2 Solving a Multi-Step Inequality

a) Solve the inequality: $1 - \frac{2}{3}x > 3$

b) Graph the solution.

Solution

a) $1 - \frac{2}{3}x > 3$ Subtract 1 from each side to isolate x.

$1 - \frac{2}{3}x - 1 > 3 - 1$

$-\frac{2}{3}x > 2$ Multiply each side by –3 to clear the fraction. Reverse the inequality sign.

$(-3)\left(-\frac{2}{3}x\right) < (-3)(2)$

$2x < -6$ Divide each side by 2.

$\frac{2x}{2} < \frac{-6}{2}$

$x < -3$

b) The solution of $x < -3$ is all numbers less than -3.

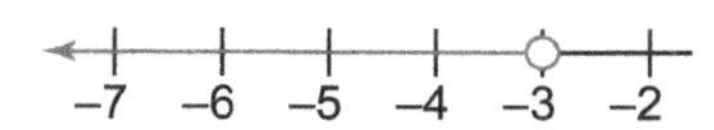

Check

1. Solve the inequality: $-\frac{2f}{5} < 4$

Graph the solution on the number line.

____

____

____

____

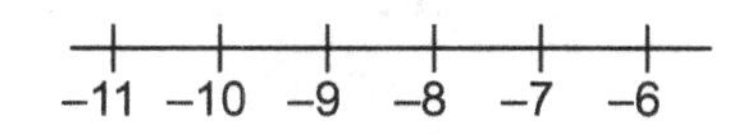

If you multiply or divide by a negative number, remember to reverse the inequality sign.

Practice

1. a) Will the inequality sign change when you perform the indicated operation on each side of the inequality?

i) $3 > -2$; Multiply by 2 __________

ii) $4 \leq 8$; Divide by -4 __________

iii) $-5 < 1$; Multiply by -5 __________

iv) $1 > -4$; Divide by 1 __________

b) Perform each operation above. Write the resulting inequality.

i) __________ **ii)** __________

iii) __________ **iv)** __________

2. a) For the inequality $-2 < 6$, identify which of the following operations will reverse the inequality sign.

i) Multiply both sides by -4 **ii)** Divide both sides by 2

__________ __________

b) Perform each operation above. Write the resulting inequality.

i) __________ **ii)** __________

3. a) What operation do you have to do to solve each inequality?

i) $3x > 9$ **ii)** $-4p < -8$

__________ __________

iii) $-3y \leq 15$ **iv)** $\frac{q}{-2} \leq 5$

__________ __________

b) State whether you would reverse the inequality sign to solve each inequality in part a.

i) ______________ **ii)** ______________

iii) ______________ **iv)** ______________

c) Solve and graph each inequality.

i) $3x > 9$

____

____

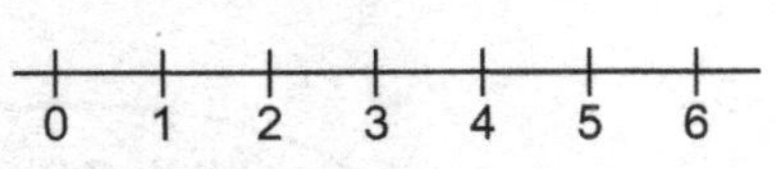

ii) $-4p < -8$

____

____

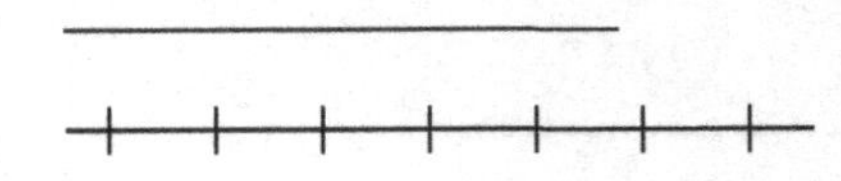

iii) $-3y \leq 15$

____

____

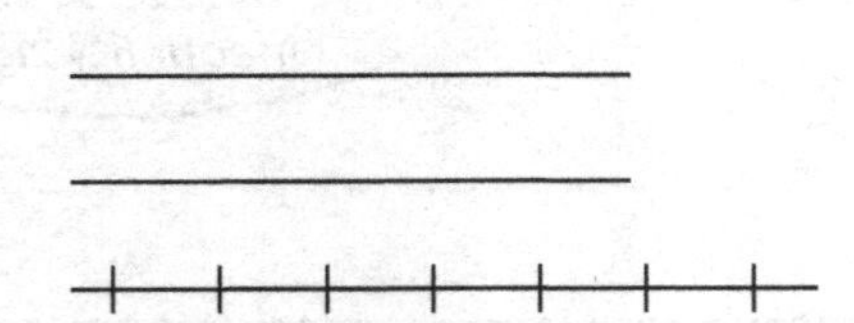

iv) $\frac{q}{-2} \leq 5$

____

____

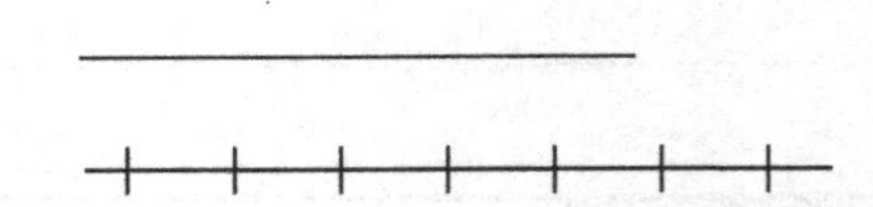

4. Solve each inequality. Graph the solution.

a) $3 - 2r \leq 9$

____

____

____

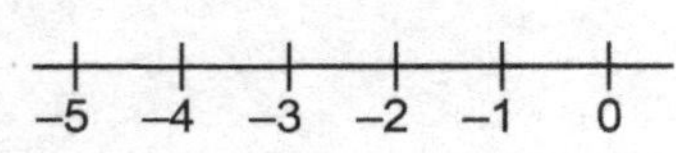

b) $\frac{p}{5} + 2 > -3$

____

____

____

c) $\frac{-s}{6} \geq 3$

____

____

____

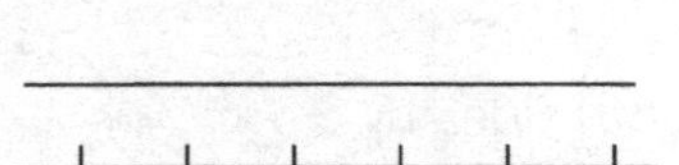

d) $\frac{5w}{8} - 1 < 4$

____

____

____

____

____

____

Unit 6 Puzzle

How Great Is My Number?

You will need

10 red tiles, 10 yellow tiles, 2 die

Label 1 die with the following faces:

$>$	$<$	$=$	$\geq$	$\leq$	$=$

Number of Players

2

Goal of the Game

To get 4 tiles in a row, vertically, horizontally, or diagonally

8	−1	0	2	−2
−4	9	−5	4	1
3	7	6	5	−9
−12	−10	−7	−3	11
−6	−8	10	12	−11

How to Play

1. Roll the number die.
   The player with the greater number goes first.
2. The starting player rolls the die, and covers a number on the board that corresponds to what was rolled. For example, if a player rolls [$\geq$] [•], the player can cover any number that is greater than or equal to 1.
3. Only 1 number can be covered on a turn.
4. Players alternate turns.
5. The first player to get 4 tiles in a row wins.

Unit 6 Study Guide

Skill	Description	Example
Solving Equations	To solve an equation, find the value of the variable that makes the left side of the equation equal to the right side. To solve an equation, isolate the variable on one side of the equation. Use inverse operations or a balance strategy to perform the same operation on both sides of the equation: • Add the same quantity to each side • Subtract the same quantity from each side • Multiply or divide each side by the same non-zero quantity Algebra tiles and balance scales can help model the steps in the solution.	Solve the equation: $3y - 2 = y + 4$ **Solution** $3y - 2 = y + 4$ $3y - 2 + 2 = y + 4 + 2$ $3y = y + 6$ $3y - y = y - y + 6$ $2y = 6$ $\frac{2y}{2} = \frac{6}{2}$ $y = 3$
Solving Inequalities	An inequality is a statement that one quantity is less than (<) another, greater than (>) another, less than or equal to (≤) another, or greater than or equal to (≥) another. The inequality sign reverses when you multiply or divide each side of the inequality by the same negative number. A linear inequality may be true for many values of the variable. We can graph the solutions on a number line.	Solve the inequality and graph the solution: $-2s - 2 \leq s - 5$ **Solution** $-2s - 2 + 2 \leq s - 5 + 2$ $-2s \leq s - 3$ $-2s - s \leq s - 3 - s$ $-3s \leq -3$ $\frac{-3s}{-3} \geq \frac{-3}{-3}$ $s \geq 1$ Since we divide each side by the same negative number, the inequality sign is reversed. 0 1 2 3 4 5

Unit 6 Review

6.1 **1.** Solve each equation. Verify the results.

a) $f + 6 = 3$

$f =$ ____ is correct.

b) $g - 5 = -2$

$g =$ ____ is correct.

c) $5h = 25$

$h =$ ____ is correct.

d) $-2k = 6$

$k =$ ____ is correct.

2. Solve each equation. Verify the solution.

a) $4x - 2 = 6$

$x =$ ____ is correct.

b) $2 - 3c = -7$

$c =$ ____ is correct.

c) $2v - 3 = -9$

$v =$ ____ is correct.

d) $-2(2 + w) = -20$

$w =$ ____ is correct.

6.2 **3.** Write the equation modelled by each set of algebra tiles.
Solve the equation.

a)

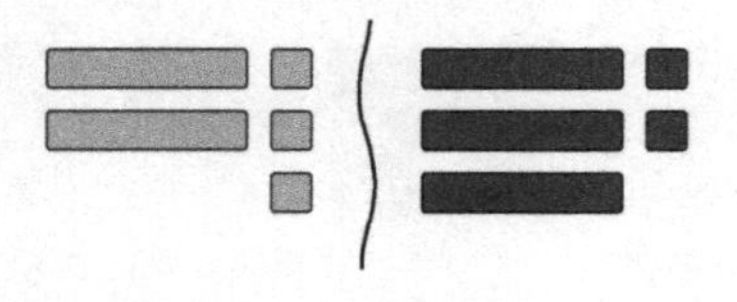

b)

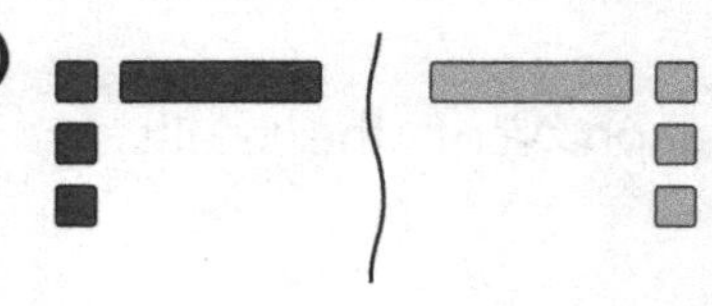

4. Solve each equation.

a) $9 - 2w = w - 6$

b) $e - 6 = 6 - e$

c) $3n + 1 = 3 + n$

d) $m - 2 = 3m + 4$

5. Solve each equation. Verify the solution.

a) $6 + \frac{s}{2} = 7$

Left side $= 6 + \frac{s}{2}$

Right side = ____________

$s =$ ____ is correct.

b) $4 + \frac{2x}{3} = 2$

Left side = $4 + \frac{2x}{3}$

____

____

____

Right side = ______________

x = _____ is correct.

6.3 **6.** Graph each inequality.
Write 3 numbers that are possible solutions for each inequality.

a) $q > -3$ (number line: –4, –3, –2, –1, 0)

____

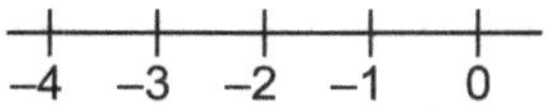

b) $w \leq 0$ (number line)

____

c) $t \geq -1$ (number line)

____

d) $r < 6$ (number line)

____

7. Write an inequality whose solution is graphed on the number line.

a) 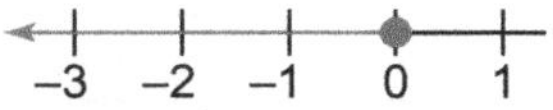 (number line: –3, –2, –1, 0, 1)

____

b) 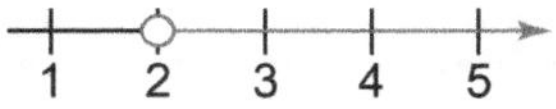 (number line: 1, 2, 3, 4, 5)

____

6.4 **8.** Solve each inequality. Graph the solution.

a) $d - 6 > 4$

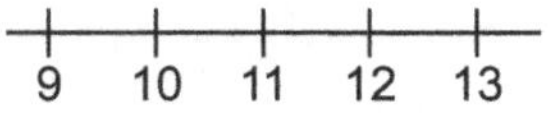

____

____

(number line: 9, 10, 11, 12, 13)

b) $2f + 1 < -3$

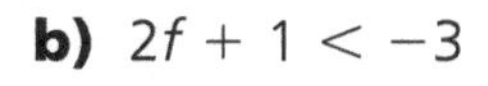

____

____

______________ (number line)

____

9. Solve each inequality. Graph the solution.

a) $4j - 1 \geq 2j + 3$

____

____

____

____

____

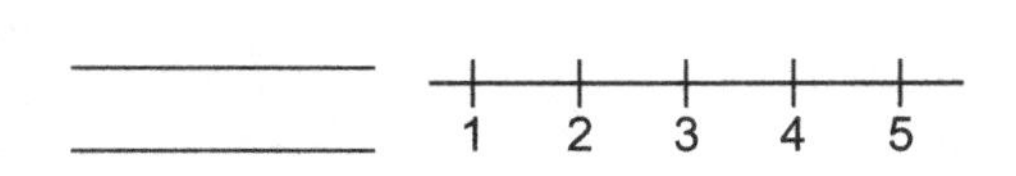

____

b) $k - 2 < 2 - k$

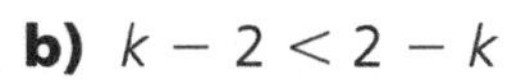

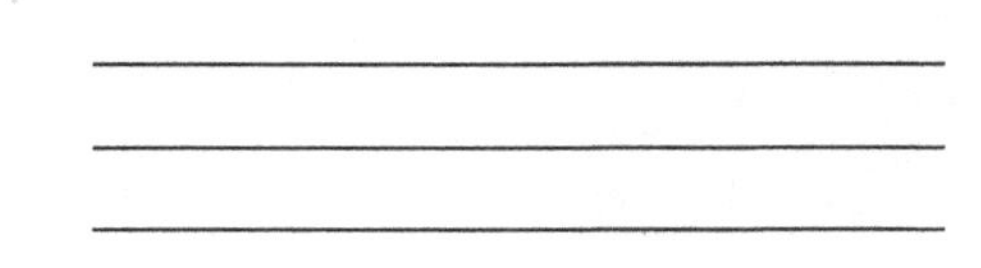

____

____

____

____

____

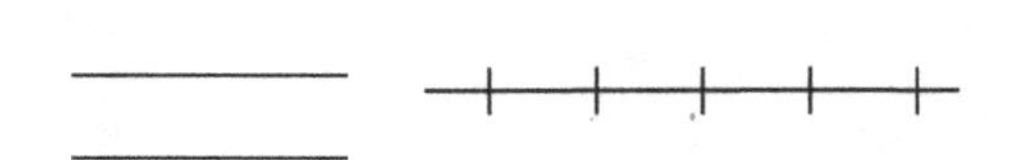

____

6.5 **10.** State whether you would reverse the inequality sign to solve each inequality.

a) $2z < -4$ ____________________ **b)** $-2x \geq 4$ ____________________

c) $\frac{c}{-2} < 4$ ____________________ **d)** $\frac{v}{2} \geq -4$ ____________________

11. Solve each inequality in question 10. Graph the solution.

a) $2z < -4$

−5 −4 −3 −2 −1

b) $-2x \geq 4$

c) $\frac{c}{-2} < 4$

d) $\frac{v}{2} \geq -4$

12. Solve each inequality and graph the solution.

a) $-3b + 4 \geq -5$

0 1 2 3 4

b) $n + 2 < 2n - 2$

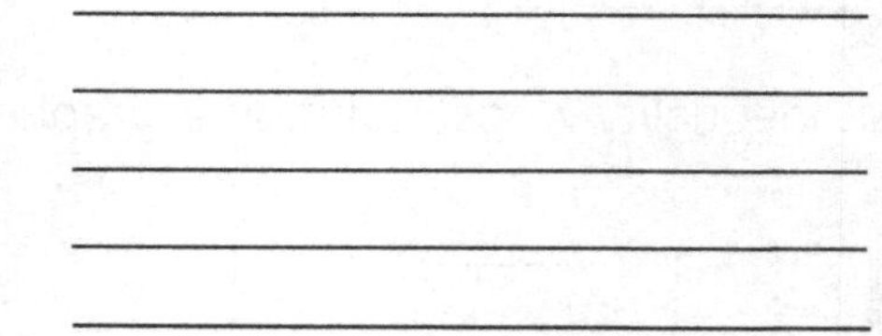

c) $-5 - m < 3 + m$

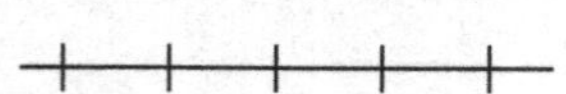

d) $2 - \frac{x}{2} > 1$

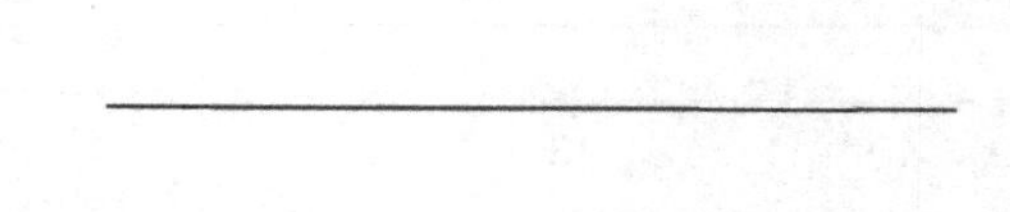

UNIT 7

Similarity and Transformations

What You'll Learn

- Draw and interpret scale diagrams.
- Apply properties of similar polygons.
- Identify and describe line symmetry and rotational symmetry.

Why It's Important

Similarity and scale diagrams are used by

- construction workers when they construct buildings and bridges
- motorists when they use maps to get around a city

Symmetry is used by

- interior designers when they arrange furniture and accessories in a room

Key Words

enlargement
reduction
scale diagram
scale factor
polygon
non-polygon
similar polygons
proportional
line symmetry
congruent
reflection
line of reflection
tessellation
rotation
rotational symmetry
order of rotation
angle of rotation symmetry
translation

7.1 Skill Builder

Converting Between Metric Units of Length

This table shows the relationships among some of the units of length.

1 m = 100 cm 1 m = 1000 mm
1 cm = 0.01 m 1 cm = 10 mm
1 mm = 0.001 m 1 mm = 0.1 cm

To convert 2.3 m to centimetres:
1 m = 100 cm
So, to convert metres to centimetres, multiply by 100.
2.3 m = 2.3(100 cm)
= 230 cm

To convert 255 cm to metres:
1 cm = 0.01 m
So, to convert centimetres to metres, multiply by 0.01.
255 cm = 255(0.01 m)
= 2.55 m

Check

1. Convert each measure to centimetres.

a) 7 m
1 m = ______ cm
So, 7 m = 7(__________)
= __________

b) 21 mm
1 mm = __________
So, 21 mm = __________
= __________

2. Convert each measure to metres.

a) 346 cm
1 cm = ______ m
So, 346 cm = 346(__________)
= __________

b) 1800 mm
1 mm = __________
So, 1800 mm = __________
= __________

3. Convert each measure to millimetres.

a) 6.5 cm
1 cm = _____ mm
So, 6.5 cm = 6.5(__________)
= __________

b) 3.8 m
1 m = __________
So, 3.8 m = __________
= __________

7.1 Scale Diagrams and Enlargements

FOCUS **Draw and interpret scale diagrams that represent enlargements.**

A diagram that is an **enlargement** or a **reduction** of another diagram is called a **scale diagram.** The **scale factor** is the relationship between the matching lengths on the two diagrams.

To find the scale factor of a scale diagram, we divide:

$\frac{\text{length on scale diagram}}{\text{length on original diagram}}$

Example 1 Using Matching Lengths to Determine the Scale Factor

Here is a scale diagram of a pin.
The actual length of the pin is 13 mm.
Find the scale factor of the diagram.

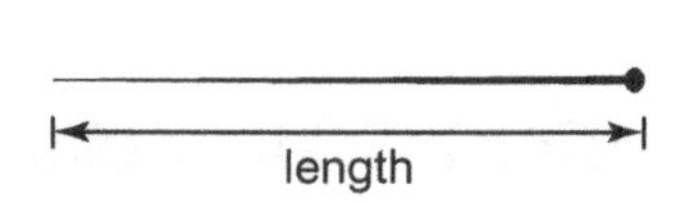

Solution

Measure the length of the pin in the diagram.
The length is 3.9 cm, or 39 mm.

The scale factor is: $\frac{\text{length on scale diagram}}{\text{length of pin}} = \frac{39\text{ mm}}{13\text{ mm}}$

$= 3$

The units of length must be the same.

The scale factor is 3.

When the drawing is an enlargement, the scale factor is greater than 1.

Check

1. Find the scale factor for each scale diagram.

a) The actual length of the ant is 6 mm.

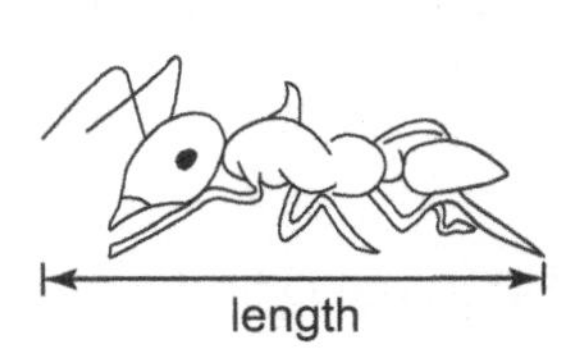

Measure the length of the ant in the diagram.
Length = _______ cm, or _______ mm

Scale factor = $\frac{\text{length on scale diagram}}{\text{length of ant}}$

$= \frac{\_\_\_\_\_\_}{\_\_\_\_\_\_}$

$=$ _______

The scale factor is _______.

b) Length of rectangle in scale diagram: ________
Length of original rectangle: ________

$$\text{Scale factor} = \frac{\text{length on scale diagram}}{\text{length on original diagram}}$$

$= \frac{\_\_\_\_\_\_}{\_\_\_\_\_\_}$

$= \_\_\_\_\_\_$

The scale factor is ______.

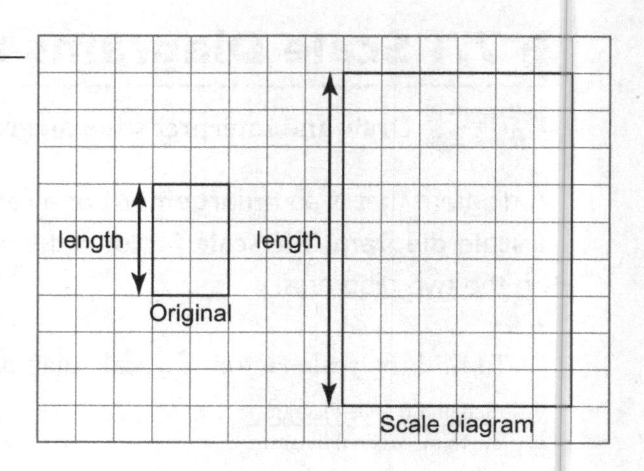

To find the dimensions of a scale diagram, multiply each length on the original diagram by the scale factor.

Example 2 Using a Scale Factor to Determine Dimensions

This cylinder is to be enlarged by a scale factor of $\frac{5}{2}$.
Find the dimensions of the enlargement.

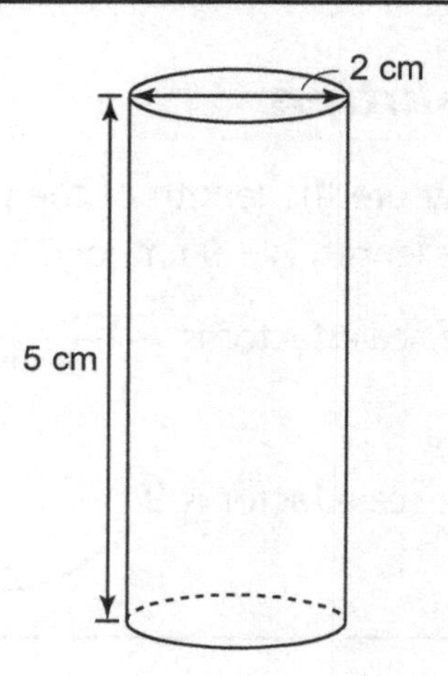

Solution

Write the scale factor as a decimal.

$\frac{5}{2} = 5 \div 2$

$= 2.5$

To write a fraction as a decimal, divide the numerator by the denominator.

Diameter of original cylinder: 2 cm

Diameter of enlargement: 2.5×2 cm = 5 cm

Height of original cylinder: 5 cm

Height of enlargement: 2.5×5 cm = 12.5 cm

The enlargement has diameter 5 cm and height 12.5 cm.

Check

1. A photo has dimensions 10 cm by 15 cm.
Enlargements are to be made with each scale factor below.
Find the dimensions of each enlargement.

a) Scale factor 4

Length of original photo: ________
Length of enlargement: 4 × ________ = ________

The length of a rectangle is always the longer dimension.

Width of original photo: ________
Width of enlargement: 4 × ________ = ________

The enlargement has dimensions ____________________.

b) Scale factor $\frac{13}{4}$

Write the scale factor as a decimal.

____

Length of original photo: ____________________
Length of enlargement: ______________ = ____________

Width of original photo: ____________________
Width of enlargement: ______________ = ____________

The enlargement has dimensions ____________________.

Practice

1. Find the scale factor for each scale diagram.

a) The actual length of the cell phone button is 9 mm.

Measure the length of the button in the diagram.
Length = ________ cm, or ________ mm

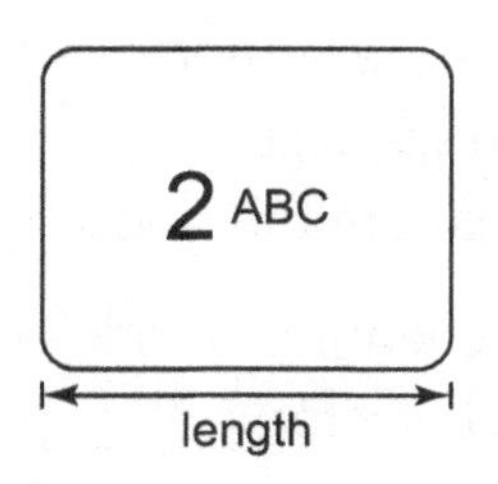

Scale factor $= \frac{\text{length on scale diagram}}{\text{length of button}} = \frac{\_\_\_\_}{\_\_\_\_} =$ ______

The scale factor is ______.

b) The actual width of the paperclip is 6 mm.

The width of the paperclip in the diagram is: Width = ______ cm, or ______ mm

Scale factor $= \frac{\text{width on scale diagram}}{\text{width of paperclip}}$

$= \frac{\_\_\_\_}{\_\_\_\_} =$ ______

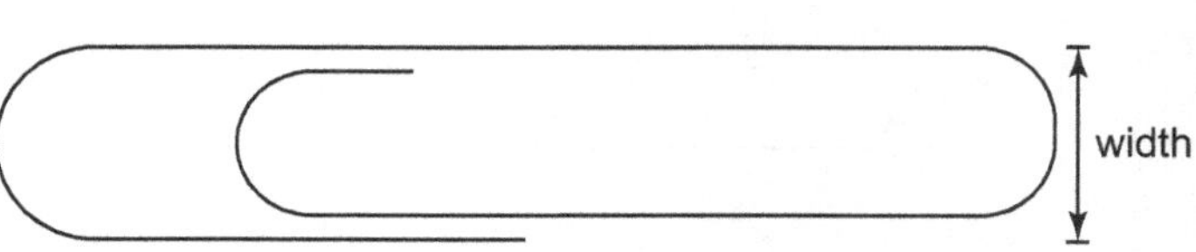

The scale factor is ______.

2. Find the scale factor for this scale diagram.

Original length: ________
Length on scale diagram: ________

Scale factor $= \dfrac{\text{length on scale diagram}}{\text{length on original diagram}}$

$= \dfrac{\_\_\_\_}{\_\_\_\_}$

$=$ ______

The scale factor is ______.

Original diagram

Scale diagram

3. Enlargements of a photo are to be placed in different catalogues.
The original photo has side length 4 cm.
Find the side length for each enlargement of this photo.

a) Enlargement with scale factor 2.5

Side length of original photo: ________
Side length of enlargement: $2.5 \times$ ________ $=$ ________

The enlargement has side length ________.

b) Enlargement with scale factor $\frac{7}{4}$

Write the scale factor as a decimal:

____

____

Side length of original photo: ________
Side length of enlargement: ______________ $=$ ________

The enlargement has side length ________.

4. Suppose you draw a scale diagram of this triangle.
You use a scale factor of 2.75.
What are the side lengths of the enlargement?

Side lengths of original triangle: ____________________
Scale factor: ________

Side lengths of enlargement:

____

____

____

____

7.2 Scale Diagrams and Reductions

FOCUS **Draw and interpret scale diagrams that represent reductions.**

A scale diagram can be smaller than the original diagram.
This type of scale diagram is called a **reduction.**
A reduction has a scale factor that is less than 1.

Example 1 Using Matching Lengths to Determine the Scale Factor

Find the scale factor for this reduction.

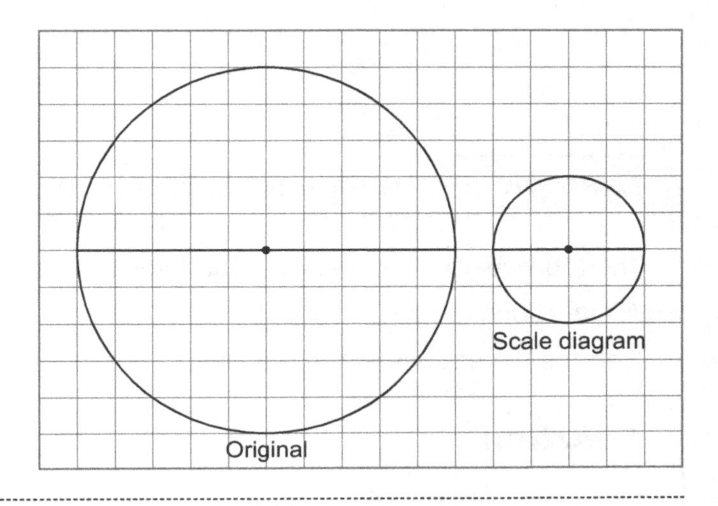

Solution

Measure the diameter of the original circle. The diameter is 5 cm.
Measure the diameter of the scale diagram. The diameter is 2 cm.

The scale factor is: $\frac{\text{diameter on scale diagram}}{\text{diameter on original diagram}} = \frac{2\text{ cm}}{5\text{ cm}} = \frac{2}{5}$

The scale factor is $\frac{2}{5}$. The scale factor is less than 1.

Check

1. Find the scale factor for each reduction.

a) Measure the length of the original line segment.
Length = ______ cm

Original   Scale diagram

Measure the length of the line segment in the scale diagram.
Length = ______ cm

Scale factor = $\frac{\text{length on scale diagram}}{\text{length on original diagram}}$

= $\frac{\_\_\_\_\_\_}{\_\_\_\_\_\_}$ = ______

The scale factor is ______.

b) Length of original rectangle: ________
Length of rectangle in scale diagram: ________

$$\text{Scale factor} = \frac{\text{length on scale diagram}}{\text{length on original diagram}}$$

= ―――

= ________

The scale factor is ______.

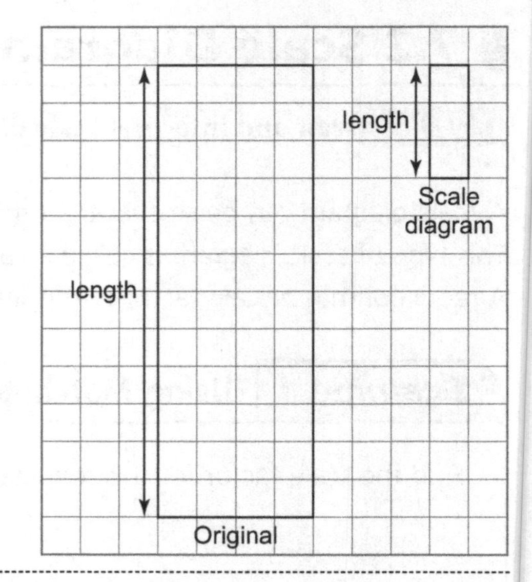

Example 2 Using a Scale Factor to Determine Dimensions

The top view of a rectangular patio table has length 165 cm and width 105 cm.

A reduction is to be drawn with scale factor $\frac{1}{5}$.

Find the dimensions of the reduction.

Solution

Write the scale factor as a decimal.

$\frac{1}{5} = 1 \div 5 = 0.2$

Length of original table: 165 cm

Length of reduction: 0.2×165 cm = 33 cm

Width of original table: 105 cm

Width of reduction: 0.2×105 cm = 21 cm

The reduction has dimensions 33 cm by 21 cm.

Check

1. A window has dimensions 104 cm by 89 cm.

A reduction is to be drawn with scale factor $\frac{1}{20}$.

Find the dimensions of the reduction.

Write the scale factor as a decimal. $\frac{1}{20}$ = ________

Length of original window: ________

Length of reduction: ________________ = ________

Width of original window: ________

Width of reduction: ________________ = ________

The reduction has dimensions ____________________

2. The top view of a rectangular swimming pool has dimensions 10 m by 5 m.
A reduction is to be drawn with scale factor $\frac{1}{50}$.
Find the dimensions of the reduction.

Write the scale factor as a decimal.

____

Length of pool: _______
Length of reduction: ________________
Convert this length to centimetres:
1 m = 100 cm
So, ____________________

____

Width of pool: _______
Width of reduction: ________________
Convert this width to centimetres:

____

____

The reduction has dimensions ______________.

Practice

1. Find the scale factor for each reduction.

a) Diameter of original circle: _____ cm
Diameter of reduction: _____ cm

Scale factor $= \frac{\text{diameter on scale diagram}}{\text{diameter on original diagram}}$

$= \frac{\_\_\_\_}{\_\_\_\_}$

$=$ _____

The scale factor is _____.

b) Length of original line segment: _____
Length of reduction: _____

Scale factor $= \frac{\text{length on scale diagram}}{\text{length on original diagram}}$

$= \frac{\_\_\_\_}{\_\_\_\_}$

$=$ _____

The scale factor is _____.

2. A line segment has length 36 cm.
A reduction is to be drawn with scale factor $\frac{3}{20}$.
Draw a line segment with the new length.

Write the scale factor as a decimal.

____

Original length: __________

Length of reduction: ________________ = __________

Draw the line segment:

3. A reduction of each object is to be drawn with the given scale factor.
Find the matching length in centimetres on the reduction.

a) A water ski has length 170 cm.
The scale factor is 0.04.
Length of water ski: __________
Length of reduction: ________________ = __________

b) A canoe has length 4 m.
The scale factor is $\frac{3}{50}$.
Write the scale factor as a decimal.

____

Length of canoe: __________

Length of reduction: ________________ = __________

Convert this length to centimetres: ________________________

____

4. Suppose you draw a scale diagram of this triangle.
You use a scale factor of $\frac{1}{4}$.
What are the side lengths of the reduction?

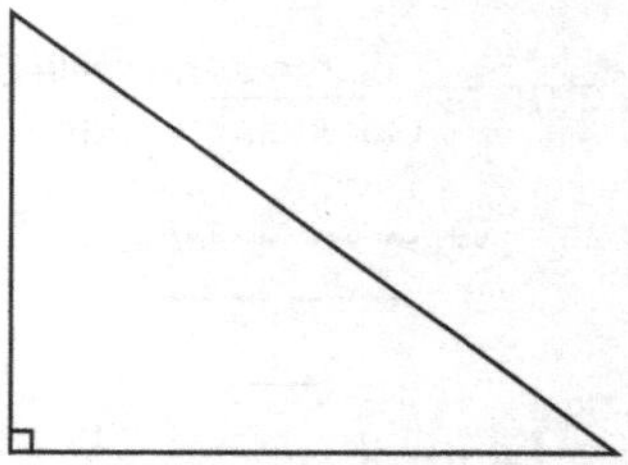

Side lengths of original triangle: ________________
Write the scale factor as a decimal.

____

Side lengths of reduction:

____

____

____

____

____

7.3 Skill Builder

Polygons

A **polygon** is a closed shape with straight sides.
Exactly 2 sides meet at a vertex.

This shape is a polygon.

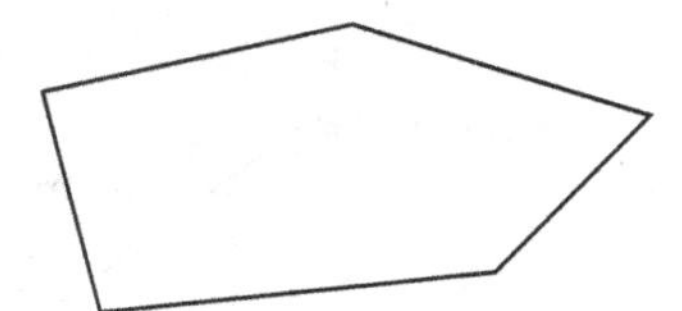

These shapes are **non-polygons.**

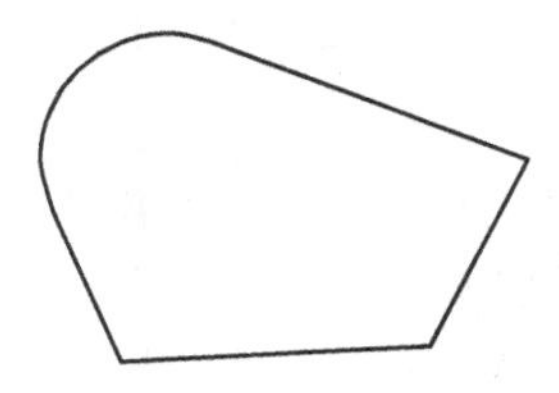

This shape has a curved side.

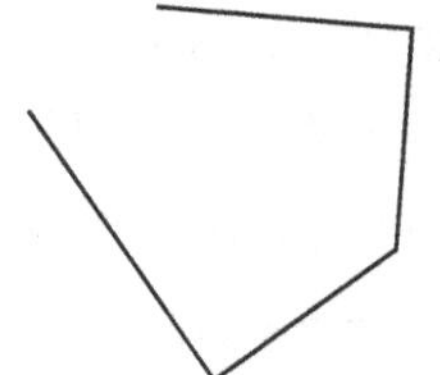

This shape is not closed.

Check

1. Is each shape a polygon or a non-polygon?

a)

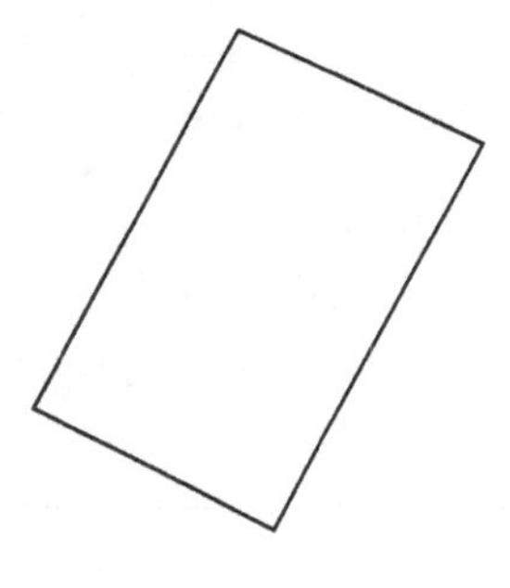

____

b)

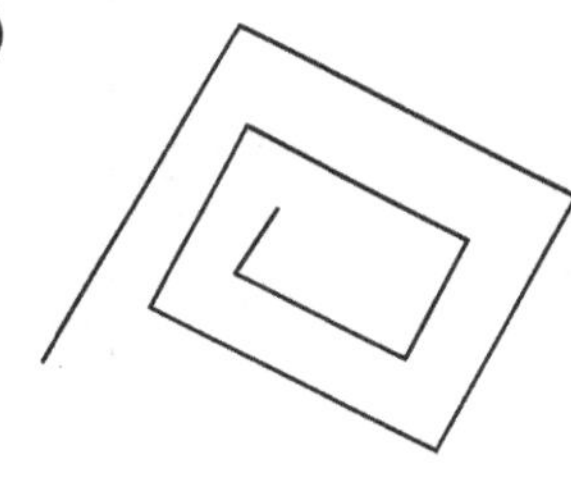

____

c)

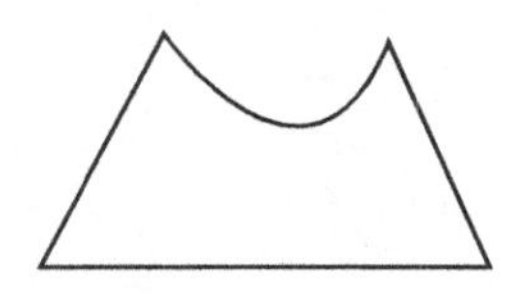

____

d)

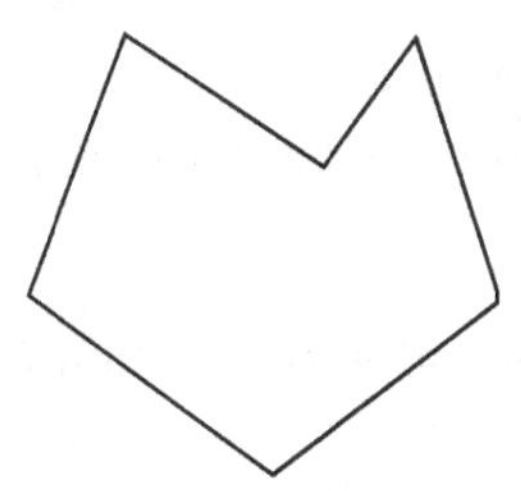

____

e)

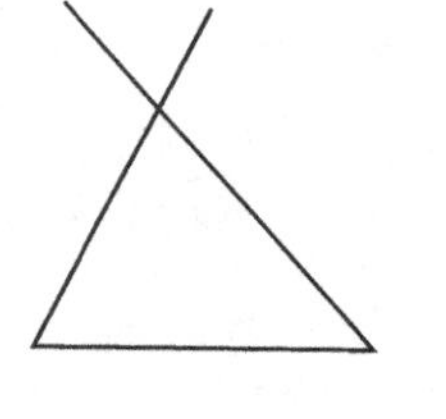

____

f)

____

7.3 Similar Polygons

FOCUS **Recognize similar polygons, then use their properties to solve problems.**

When one polygon is an enlargement or reduction of another polygon, we say the polygons are **similar.**
Similar polygons have the same shape, but not necessarily the same size.

When two polygons are similar:
- matching angles are equal **AND**
- matching sides are proportional

*When all pairs of matching sides have the same scale factor, we say matching sides are **proportional.***

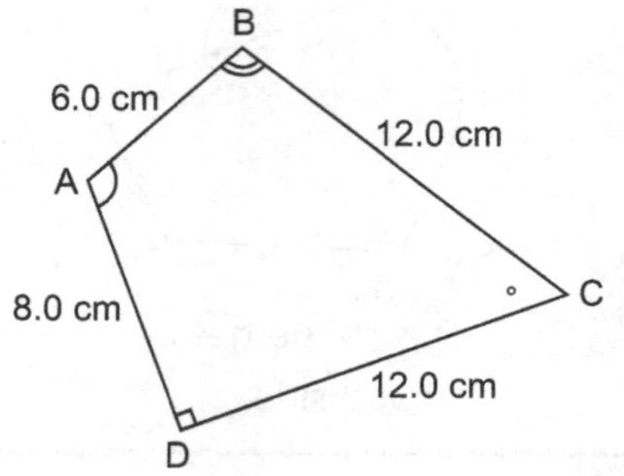

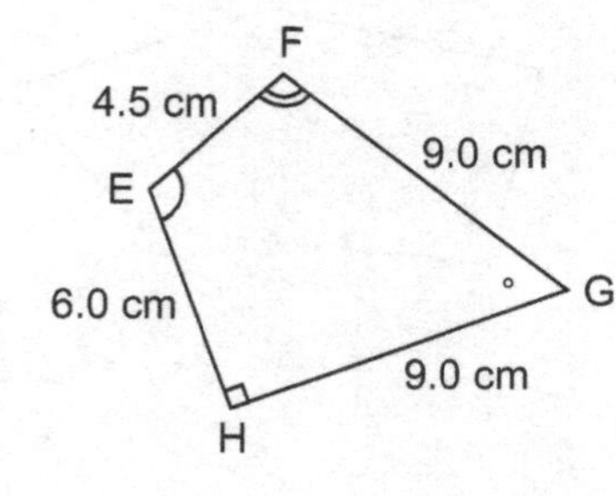

Example 1 Identifying Similar Polygons

Are these quadrilaterals similar?
Explain.

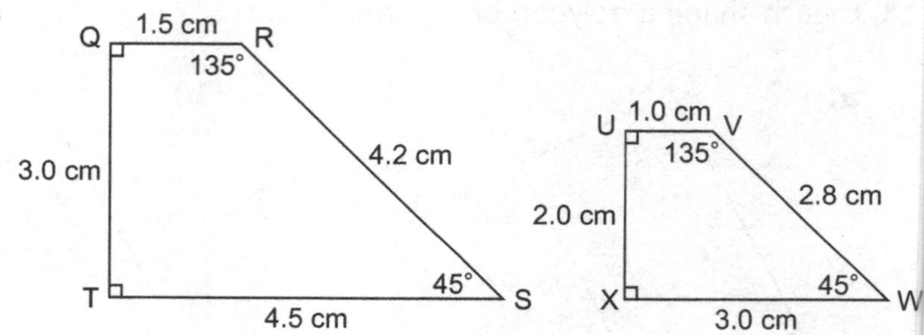

Solution

Check matching angles: $\angle Q = \angle U = 90°$ $\quad \angle R = \angle V = 135°$

$\angle S = \angle W = 45°$ $\quad \angle T = \angle X = 90°$

All matching angles are equal.
So, the first condition for similar polygons is met.

Check matching sides.
The matching sides are: QR and UV, RS and VW, ST and WX, and TQ and XU.
Find the scale factors.

$\frac{\text{length of QR}}{\text{length of UV}} = \frac{1.5\text{ cm}}{1.0\text{ cm}}$
$= 1.5$

$\frac{\text{length of RS}}{\text{length of VW}} = \frac{4.2\text{ cm}}{2.8\text{ cm}}$
$= 1.5$

$\frac{\text{length of ST}}{\text{length of WX}} = \frac{4.5\text{ cm}}{3.0\text{ cm}}$
$= 1.5$

$\frac{\text{length of TQ}}{\text{length of XU}} = \frac{3.0\text{ cm}}{2.0\text{ cm}}$
$= 1.5$

All scale factors are equal, so matching sides are proportional.
Since matching angles are equal and matching sides are proportional, the quadrilaterals are similar.

Check

1. Are these rectangles similar?

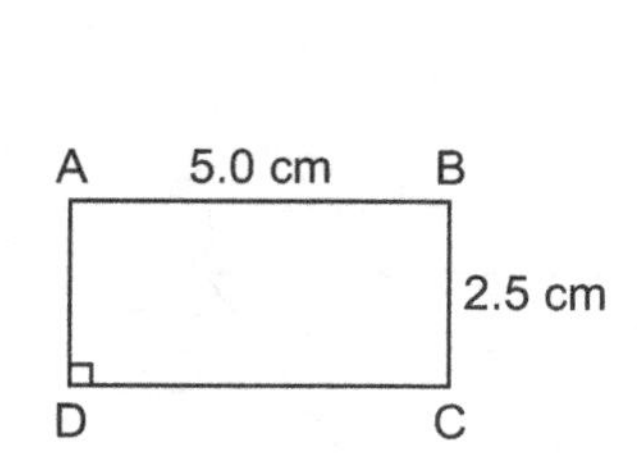

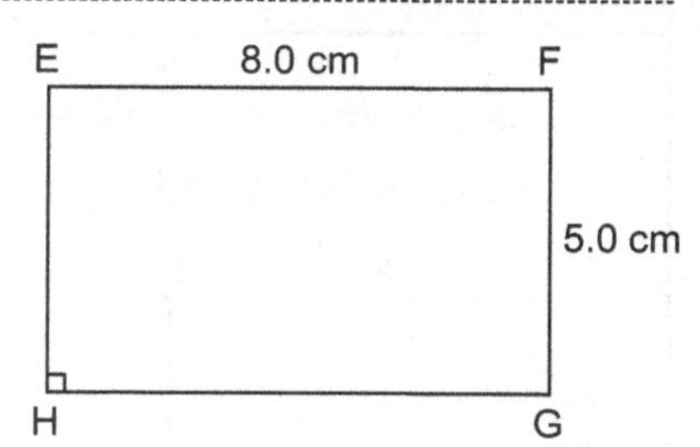

Check matching angles.
The measure of each angle in a rectangle is ______.
So, matching angles are ______.

Check matching sides.
The matching sides are: ____ and ____, and ____ and ____.
Find the scale factors.

Since opposite sides of a rectangle are equal, check only one pair of matching lengths and one pair of matching widths.

$\frac{\text{length of } \_\_\_\_}{\text{length of } \_\_\_\_} = \frac{\_\_\_\_\_\_}{\_\_\_\_\_\_}$ $= \_\_\_\_\_\_$ $\frac{\text{length of } \_\_\_\_}{\text{length of } \_\_\_\_} = \frac{\_\_\_\_\_\_}{\_\_\_\_\_\_}$ $= \_\_\_\_\_\_$

The scale factors ______ equal.
So, the sides ______ proportional.
The rectangles ______ similar.

2. Are these parallelograms similar?

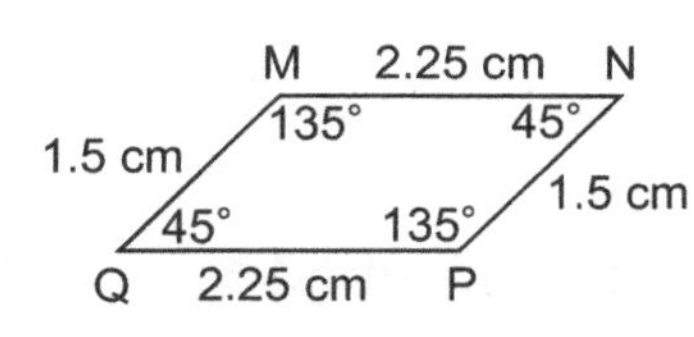

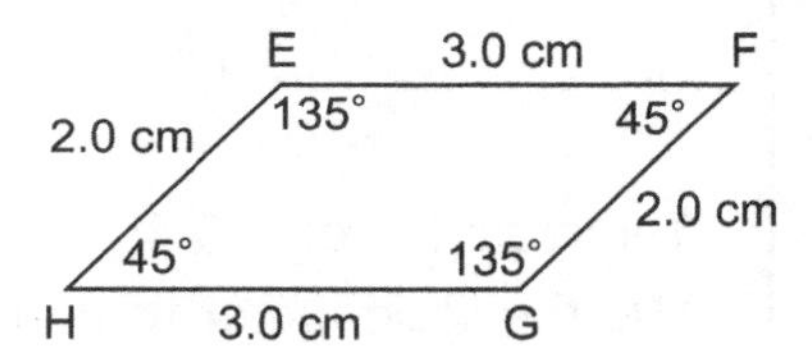

Check matching angles. $\angle M$ = _____ = _____
$\angle N$ = _____ = _____
____
____

All matching angles ______ equal.

Check matching sides.
The matching sides are: _____ and _____, and _____ and _____.
Find the scale factors.

Since opposite sides of a parallelogram are equal, check only two pairs of matching sides.

$\frac{\text{length of } \_\_\_\_}{\text{length of } \_\_\_\_} = \frac{\_\_\_\_\_\_}{\_\_\_\_\_\_}$ $= \_\_\_\_\_\_$ $\frac{\text{length of } \_\_\_\_}{\text{length of } \_\_\_\_} = \frac{\_\_\_\_\_\_}{\_\_\_\_\_\_}$ $= \_\_\_\_\_\_$

The scale factors ______ equal.
So, the sides ______ proportional.
The parallelograms ______ similar.

Example 2 Determining Lengths in Similar Polygons

These two quadrilaterals are similar.
Find the length of JM.

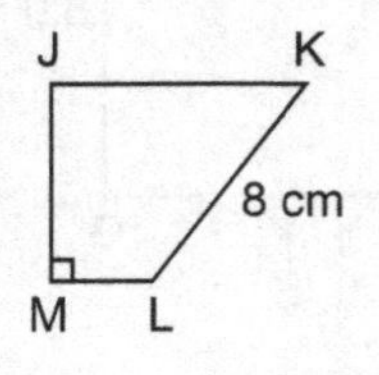

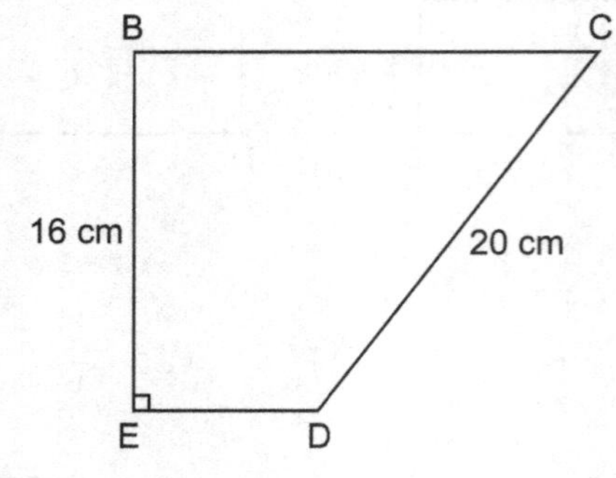

Solution

Consider the polygon with the unknown length as a reduction or enlargement of the other polygon.

Quadrilateral JKLM is a reduction of quadrilateral BCDE.
To find the scale factor of the reduction,
choose a pair of matching sides whose lengths are both known:
CD = 20 cm and KL = 8 cm

$$\text{Scale factor} = \frac{\text{length on reduction}}{\text{length on original}}$$
$$= \frac{8 \text{ cm}}{20 \text{ cm}}$$
$$= 0.4$$

The scale factor is 0.4.
Use the scale factor to find the length of JM.
JM and BE are matching sides.
Length of BE: 16 cm
Scale factor: 0.4
Length of JM: $0.4 \times 16 \text{ cm} = 6.4 \text{ cm}$

So, JM has length 6.4 cm.

Check

1. These two polygons are similar.
   Find the length of JK.

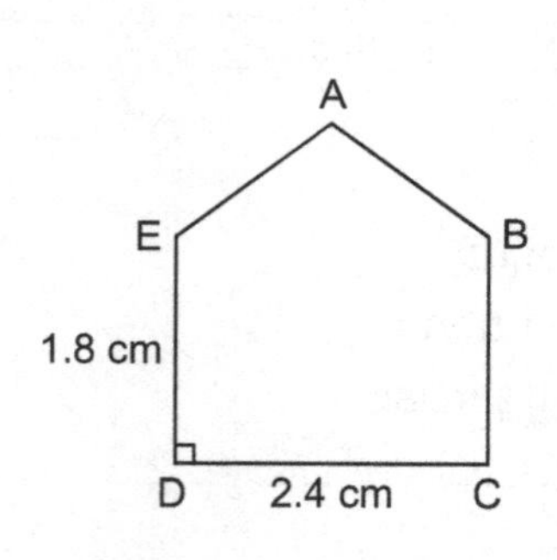

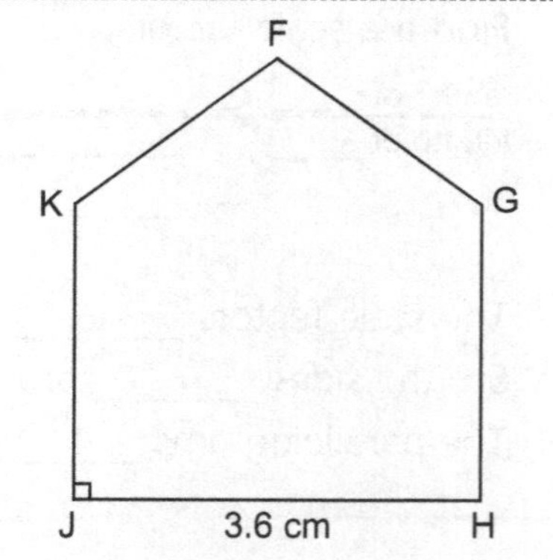

Polygon FGHJK is an enlargement of polygon ABCDE.
To find the scale factor, choose a pair of matching sides whose lengths are both known:

____

Scale factor $= \frac{\text{length on enlargement}}{\text{length on original}}$

$= \frac{\_\_\_\_\_\_}{\_\_\_\_\_\_}$

$= \_\_\_\_$

The scale factor is _____.
Use the scale factor to find the length of JK.
JK and DE are matching sides.
Length of DE: ________
Scale factor: _____
Length of JK: __________________________

So, JK has length ________.

2. These two polygons are similar.
Find the length of YZ.

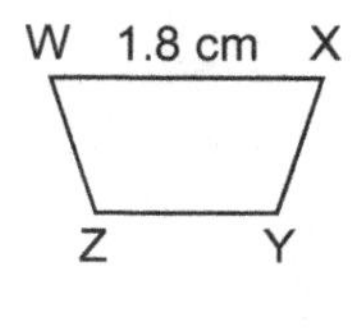

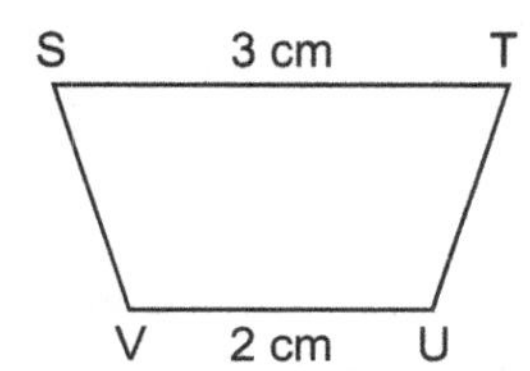

Polygon WXYZ is a ____________ of polygon STUV.
To find the scale factor, choose a pair of matching sides whose lengths are both known:

____

Scale factor $= \frac{\text{length on \_\_\_\_\_\_}}{\text{length on original}}$

$= \frac{\_\_\_\_\_\_}{\_\_\_\_\_\_}$

$= \_\_\_\_$

The scale factor is _____.
Use the scale factor to find the length of YZ.
UV and YZ are matching sides.
Length of UV: ________
Scale factor: _____
Length of UV: __________________________
So, UV has length ________.

Practice

1. Are these quadrilaterals similar?

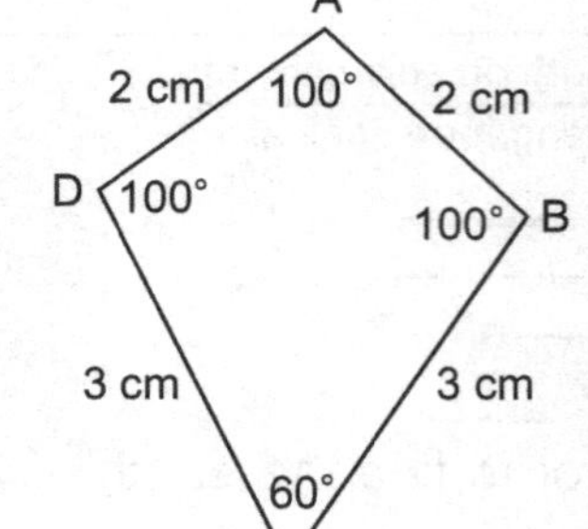

Check matching angles. ∠A = _____ = _____

∠B = _____ = _____

____

____

All matching angles _______ equal.

Check matching sides.

The matching sides are: AB and ____, and BC and ____.

Find the scale factors.

Since adjacent sides of the kites are equal, check only two pairs of matching sides.

$\frac{\text{length of \_\_\_\_}}{\text{length of \_\_\_\_}} = \frac{\_\_\_\_\_\_}{\_\_\_\_\_\_}$ $= \_\_\_\_\_\_$

$\frac{\text{length of \_\_\_\_}}{\text{length of \_\_\_\_}} = \frac{\_\_\_\_\_\_}{\_\_\_\_\_\_}$ $= \_\_\_\_\_\_$

The scale factors _______ equal.

So, the sides _______ proportional.

The quadrilaterals _______ similar.

2. Are any of these rectangles similar?

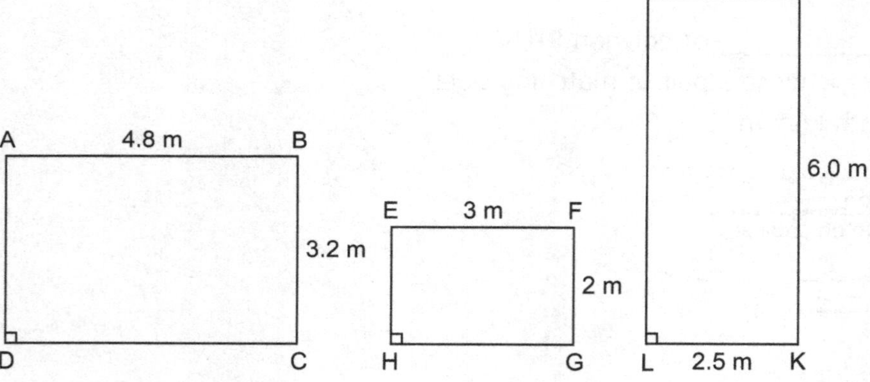

The measure of each angle in a rectangle is ____.

So, for any two rectangles, matching angles are _______.

Check matching lengths and widths in pairs of rectangles.

For rectangles ABCD and EFGH, the scale factors are:

$\frac{\text{length of \_\_\_\_}}{\text{length of \_\_\_\_}} = \frac{\_\_\_\_\_\_}{\_\_\_\_\_\_}$ $= \_\_\_\_\_\_$

$\frac{\text{length of \_\_\_\_}}{\text{length of \_\_\_\_}} = \frac{\_\_\_\_\_\_}{\_\_\_\_\_\_}$ $= \_\_\_\_\_\_$

The scale factors _______ equal.

So, the sides _______ proportional.

The rectangles _______ similar.

For rectangles ABCD and JKLM, the scale factors are:

$\frac{\text{length of ____}}{\text{length of ____}} = \frac{\text{______}}{\text{______}}$ $\frac{\text{length of ____}}{\text{length of ____}} = \frac{\text{______}}{\text{______}}$

= ______   = ______

The scale factors ____________ equal.
So, the sides ____________ proportional.
The rectangles ____________ similar.

Is rectangle EFGH similar to rectangle JKLM?
Use what we know to find out.
We know that rectangle ABCD ________________ to rectangle EFGH.
We know that rectangle ABCD ________________ to rectangle JKLM.
So, we know rectangle EFGH ________________ to rectangle JKLM.

3. These two polygons are similar.
Find the length of UV.

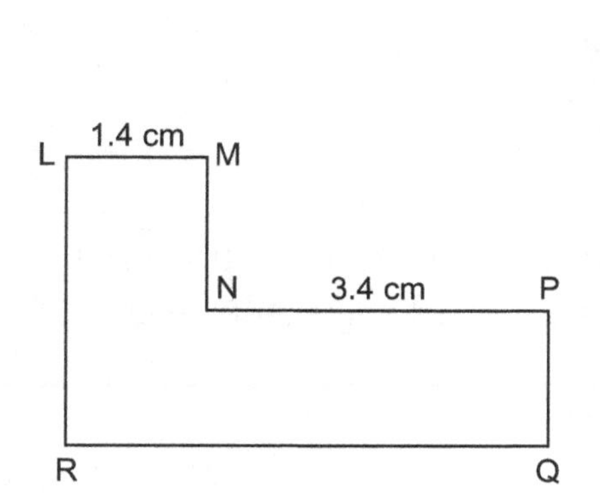

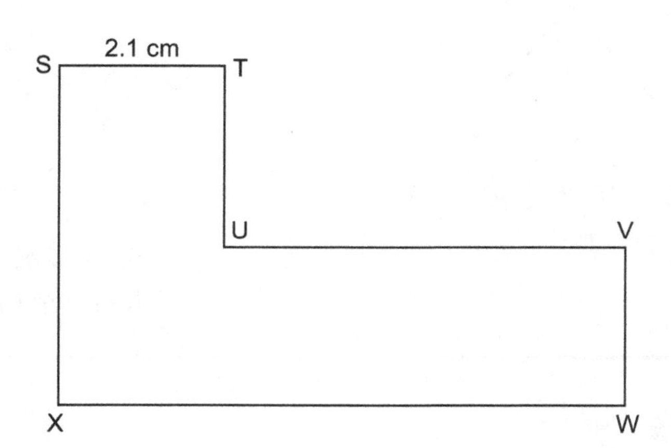

Polygon STUVWX is an enlargement of polygon LMNPQR.
To find the scale factor, choose a pair of matching sides whose lengths are both known:

____

Scale factor $= \frac{\text{length on enlargement}}{\text{length on original}}$

$= \frac{\text{______}}{\text{______}}$

= ______

The scale factor is _________.
Use the scale factor to find the length of UV.
UV and NP are matching sides.
Length of NP: _________
Scale factor: _________
Length of UV: ______________________
So, UV has length _________.

7.4 Skill Builder

Sum of the Angles in a Triangle

In any triangle, the sum of the angle measures is 180°.

So, to find an unknown angle measure:

- start with 180°
- subtract the known measures

An isosceles triangle has 2 equal sides and 2 equal angles.

To find the measure of the third angle, subtract the measure of the equal angles twice.

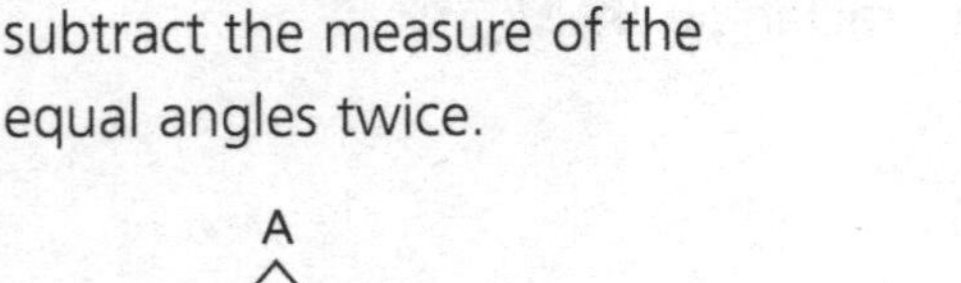

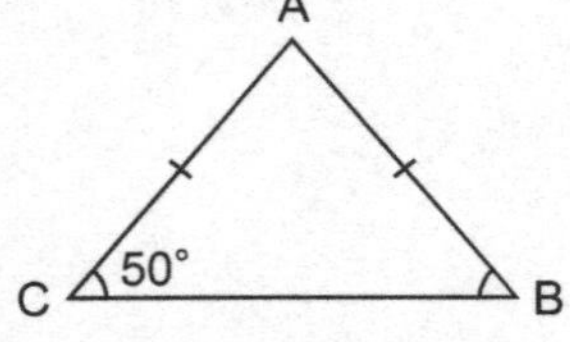

$\angle A = 180° - 50° - 50°$
$= 80°$

To find the measure of each equal angle, subtract the known angle from 180°, then divide by 2.

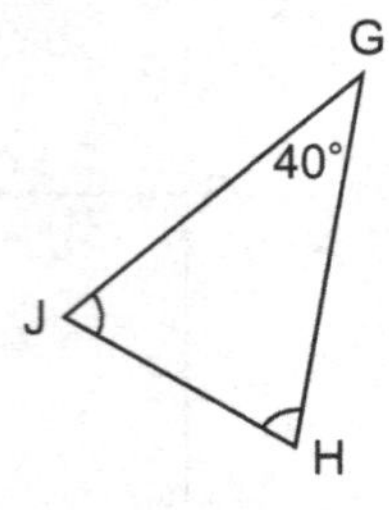

Sum of equal angles is: $180° - 40° = 140°$
Measure of each equal angle: $140° \div 2 = 70°$

Check

1. Find the measure of the third angle.

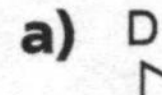

a)

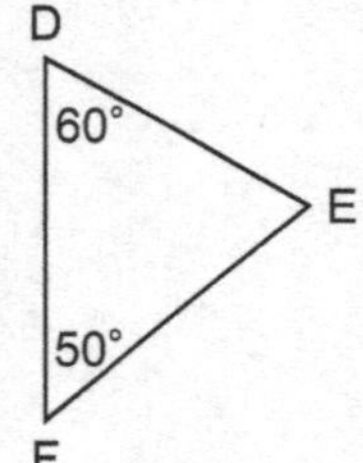

$\angle E = 180° -$ _____ $-$ _____
$=$ _____

b)

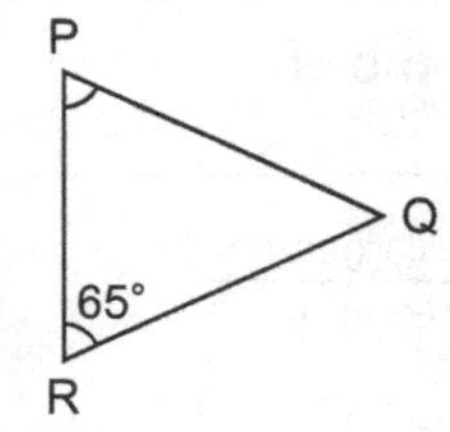

$\angle Q =$ ____________________
$=$ _____

2. Find the measure of each equal angle.

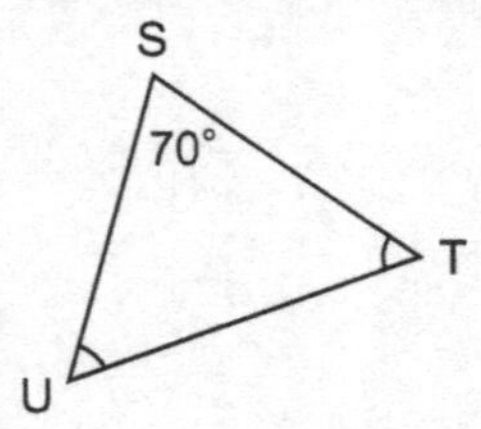

Sum of equal angles is:
$180° -$ _____ $=$ _____
Measure of each equal angle:
_____ $\div 2 =$ _____

7.4 Similar Triangles

FOCUS **Use the properties of similar triangles to solve problems.**

A triangle is a special polygon.
When two triangles are similar:

- matching angles are equal **OR**
- matching sides are proportional

The order in which similar triangles are named gives a lot of information.

The symbol ~ means "is similar to."

Then, ∠A = ∠D, ∠B = ∠E, and ∠C = ∠F
Similarly, AB matches DE, BC matches EF, and AC matches DF.

Example 1 Identifying Similar Triangles

Name the similar triangles.

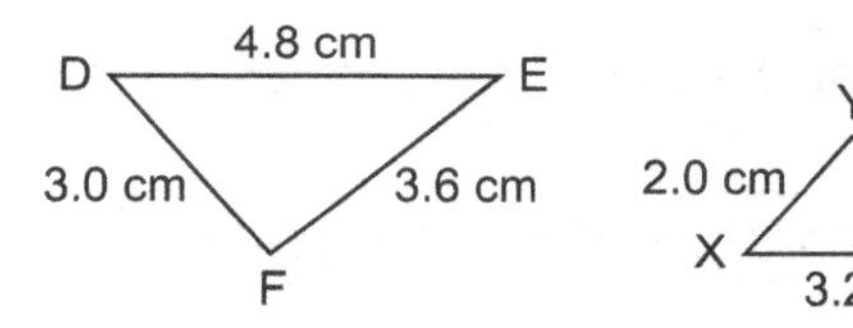

Solution

Angle measures are not given.
So, find out if matching sides are proportional.
In △DEF, order the sides from shortest to longest: FD, EF, DE
In △XYZ, order the sides from shortest to longest: XY, YZ, ZX
Find the scale factors of matching sides.

$\frac{\text{length of FD}}{\text{length of XY}} = \frac{3.0 \text{ cm}}{2.0 \text{ cm}}$
$= 1.5$

$\frac{\text{length of EF}}{\text{length of YZ}} = \frac{3.6 \text{ cm}}{2.4 \text{ cm}}$
$= 1.5$

$\frac{\text{length of DE}}{\text{length of ZX}} = \frac{4.8 \text{ cm}}{3.2 \text{ cm}}$
$= 1.5$

Since all scale factors are the same, the triangles are similar.

The longest and shortest sides meet at vertices: D and X
The two longer sides meet at vertices: E and Z
The two shorter sides meet at vertices: F and Y
So, △DEF ~ △XZY

Read the letters down the columns.

Check

1. In each diagram, name two similar triangles.

a) Two angles in each triangle are given.
The measure of the third angle
in each triangle is:
180° – ____________

List matching angles:
∠A = ____ = ____
∠B = ____ = ____
∠C = ____ = ____
Matching angles ______ equal.
So, the triangles ______ similar.

To name the triangles, order the letters so matching angles correspond.
△ABC ~ △______

b) Find out if matching sides are proportional.
In △DEF, order the sides from shortest to longest:

____

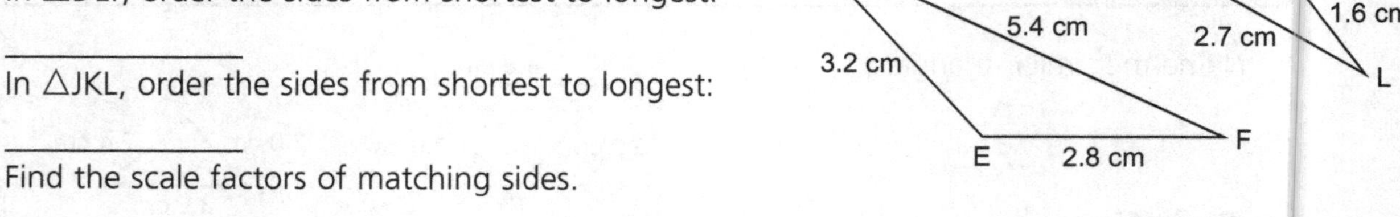

In △JKL, order the sides from shortest to longest:

____

Find the scale factors of matching sides.

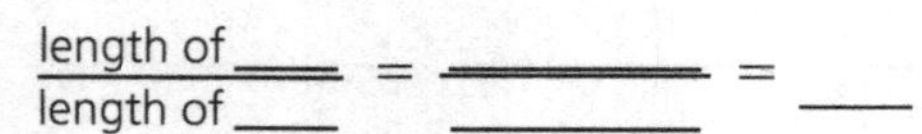

$\frac{\text{length of ____}}{\text{length of ____}}$ = $\frac{\text{______}}{\text{______}}$ = ____

$\frac{\text{length of ____}}{\text{length of ____}}$ = $\frac{\text{______}}{\text{______}}$ = ____

$\frac{\text{length of ____}}{\text{length of ____}}$ = $\frac{\text{______}}{\text{______}}$ = ____

All scale factors are __________. So, the triangles __________.
The two longer sides meet at vertices: ____ and ____
The two shorter sides meet at vertices: ____ and ____
The longest and shortest sides meet at vertices: ____ and ____
So, △DEF ~ △______

Example 2 Using Similar Triangles to Determine a Length

These two triangles are similar.
Find the length of TU.

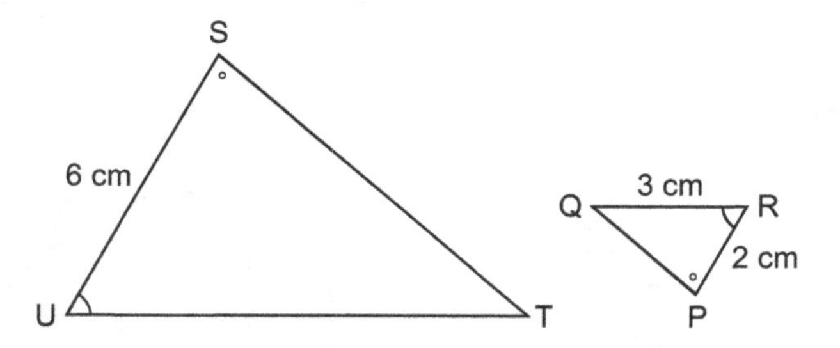

Solution

List matching angles:

$\angle S = \angle P$ $\qquad$ $\angle T = \angle Q$ $\qquad$ $\angle U = \angle R$

So, $\triangle STU \sim \triangle PQR$

$\triangle STU$ is an enlargement of $\triangle PQR$.

Choose a pair of matching sides whose lengths are both known:
SU = 6 cm and PR = 2 cm

Consider the triangle with the unknown length as a reduction or enlargement of the other triangle.

$$\text{Scale factor} = \frac{\text{length on enlargement}}{\text{length on original}}$$
$$= \frac{6 \text{ cm}}{2 \text{ cm}}$$
$$= 3$$

The scale factor is 3.
Use the scale factor to find the length of TU.
TU and QR are matching sides.
Length of QR: 3 cm
Scale factor: 3
Length of TU: 3×3 cm = 9 cm

So, TU has length 9 cm.

Check

1. These two triangles are similar.
Find the length of XV.

F, 20 cm, 10 cm, H, G, V, X, W, 2 cm

List matching angles:

∠F = _____ ∠G = _____ ∠H = _____

So, △FGH ~ △_______

_________ is a reduction of _________.

Choose a pair of matching sides whose lengths are both known:

____

Scale factor = $\frac{\text{length on reduction}}{\text{length on original}}$

= $\frac{\_\_\_\_}{\_\_\_\_}$

= _____

The scale factor is _____.

Use the scale factor to find the length of XV.

XV and FG are matching sides.

Length of FG: ________

Scale factor: ______

Length of XV: __________________________

So, XV has length ______.

Practice

1. In each diagram, name two similar triangles.

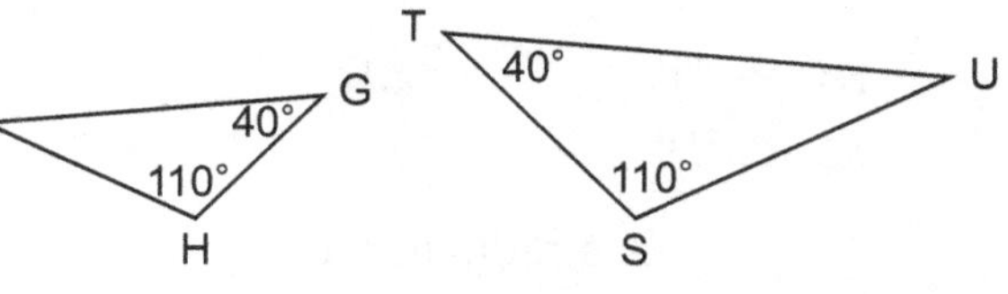

a) Two angles in each triangle are given.
The measure of the third angle
in each triangle is: 180° – ______________

List matching angles:

∠F = _____ = _____ ∠G = _____ = _____ ∠H = _____ = _____

Matching angles _______ equal, so, the triangles _______ similar.

To name the triangles, order the letters so matching angles correspond. △FGH ~ △______

b) Find out if matching sides are proportional.

In △JKL, order the sides from shortest to longest: ____________

In △QRS, order the sides from shortest to longest: ____________

Find the scale factors of matching sides.

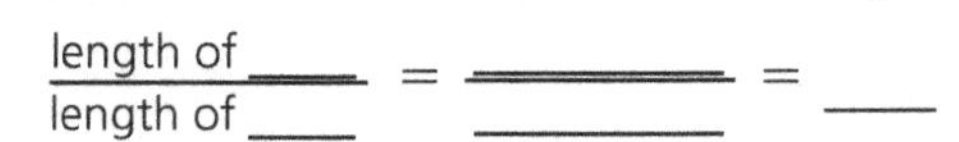

$\frac{\text{length of } \_\_\_}{\text{length of } \_\_\_} = \frac{\_\_\_\_\_}{\_\_\_\_\_} = \_\_\_$

$\frac{\text{length of } \_\_\_}{\text{length of } \_\_\_} = \frac{\_\_\_\_\_}{\_\_\_\_\_} = \_\_\_$

$\frac{\text{length of } \_\_\_}{\text{length of } \_\_\_} = \frac{\_\_\_\_\_}{\_\_\_\_\_} = \_\_\_$

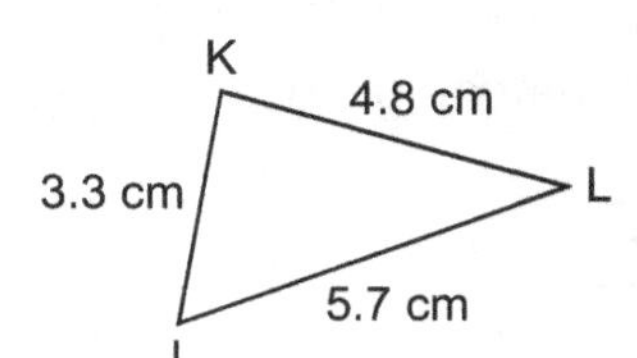

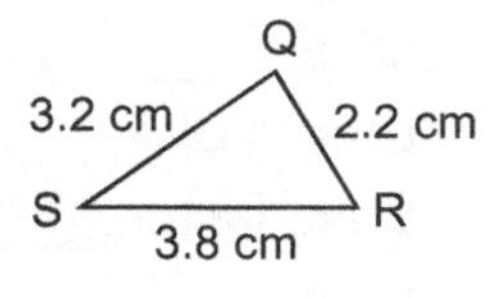

All scale factors are ___________. So, the triangles ___________.

The longest and shortest sides meet at vertices: ____ and ____

The two shorter sides meet at vertices: ____ and ____

The two longer sides meet at vertices: ____ and ____

So, △JKL ~ △_____

2. Are these two triangles similar?

In △PQR, order the sides from shortest to longest:

____

In △BCD, order the sides from shortest to longest:

____

Find the scale factors of matching sides.

$\frac{\text{length of } \_\_\_}{\text{length of } \_\_\_} = \frac{\_\_\_\_\_}{\_\_\_\_\_} = \_\_\_$

$\frac{\text{length of } \_\_\_}{\text{length of } \_\_\_} = \frac{\_\_\_\_\_}{\_\_\_\_\_} = \_\_\_$

$\frac{\text{length of } \_\_\_}{\text{length of } \_\_\_} = \frac{\_\_\_\_\_}{\_\_\_\_\_} = \_\_\_$

All scale factors are ____________. So, the triangles ______________.

3. These two triangles are similar.
Find the length of EC.

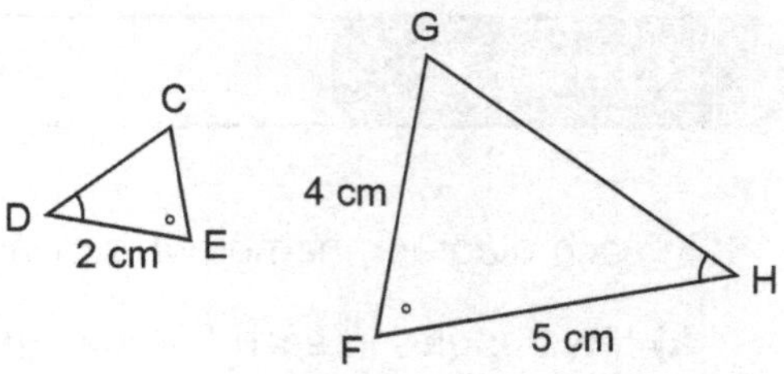

List matching angles:

$\angle C =$ _____ $\angle D =$ _____ $\angle E =$ _____

So, $\triangle CDE \sim \triangle$_____

_____ is a reduction of _____.

Choose a pair of matching sides whose lengths are both known:

____

$$\text{Scale factor} = \frac{\text{length on reduction}}{\text{length on original}}$$

$= \frac{\_\_\_}{\_\_\_}$

$=$ ____

The scale factor is ____.
Use the scale factor to find the length of EC.
EC and ____ are matching sides.
Length of ____: _____
Scale factor: ____
Length of EC: _______________
So, EC has length _____.

4. At a certain time of day, two trees cast shadows.
Find the height of the taller tree.

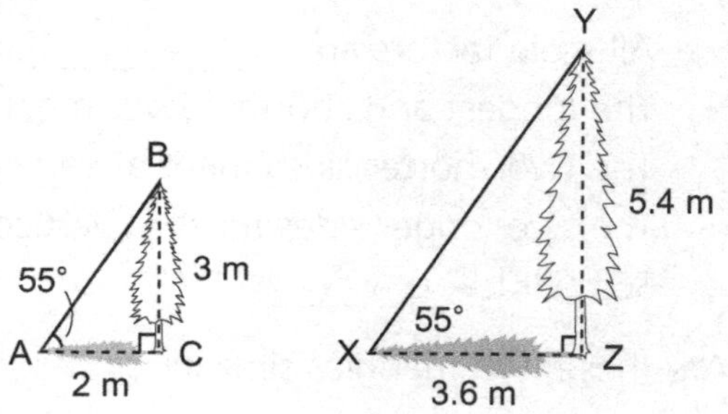

Matching angles are _____.
So, $\triangle ABC \sim \triangle$_____
$\triangle XYZ$ is an __________ of $\triangle ABC$.
Use sides _______________
to find the scale factor.

$$\frac{\text{length on enlargement}}{\text{length on original}} = \frac{\_\_\_}{\_\_\_}$$

$=$ ____

The scale factor is 1.8.
Use the scale factor to find the height of the taller tree, YZ.
BC and YZ are matching sides.
Length of BC: ____ Scale factor: ____
Length of YZ: _______________
So, the height of the taller tree is _________.

5. The two triangles in this diagram are similar.
Find the length of DE.

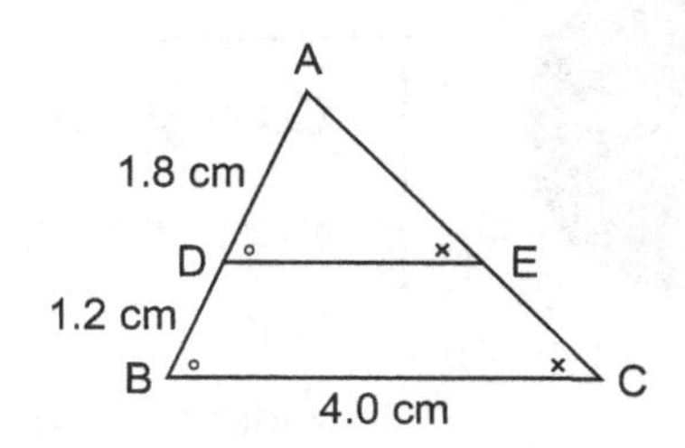

To better see the individual triangles,
we draw the triangles separately.

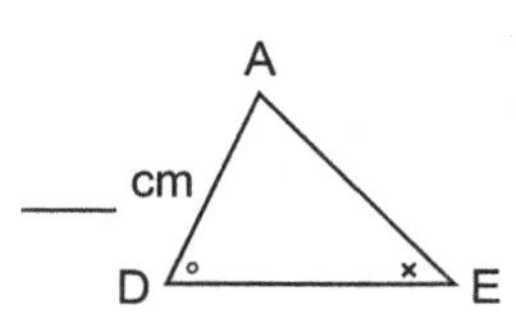

1.8 cm + 1.2 cm = _____ cm

A

B C

___ cm

∠A = _____ ∠B = _____ ∠C = _____

So, △ABC ~ △_____

________ is a reduction of ________.

Choose a pair of matching sides whose lengths are both known:

____

Scale factor = $\frac{\text{length on reduction}}{\text{length on original}}$

= $\frac{\_\_\_\_\_}{\_\_\_\_\_}$

= _____

The scale factor is _____.

Use the scale factor to find the length of DE.

_____ and _____ are matching sides.

Length of _____: __________

Scale factor: _____

Length of DE: ________________________

So, DE has length ________.

Can you ...

- Find the scale factor for a scale diagram?
- Use a scale factor to determine a length?
- Identify similar polygons and triangles?
- Use similar polygons and triangles to determine a length?

7.1 **1.** Find the scale factor for this scale diagram.
The actual diameter of the head of the pushpin is 6 mm.

Measure the diameter of the pushpin in the diagram.
Length = _______ cm, or _______ mm

diameter

Scale factor = $\frac{\text{length on scale diagram}}{\text{length of pushpin}}$

= $\frac{\rule{2cm}{0.4pt}}{\rule{2cm}{0.4pt}}$

= _____

The scale factor is _____.

2. A baby picture is to be enlarged.
The dimensions of the photo are 5 cm by 7 cm.
Find the dimensions of the enlargement with a scale factor of 3.2.
Length of original photo: ______
Length of enlargement: 3.2 × ______ = _________

Width of original photo: ______
Width of enlargement: ______ × ______ = ______

The enlargement has dimensions _______________________.

7.2 **3.** Find the scale factor for this reduction.
Length of original line segment: ______ cm
Length of reduction: ______ cm

Original

Scale diagram

Scale factor = $\frac{\text{length on reduction}}{\text{length on original diagram}}$

= $\frac{\rule{2cm}{0.4pt}}{\rule{2cm}{0.4pt}}$

= _____

The scale factor is _____.

4. A reduction of a lacrosse stick is to be drawn with a scale factor of $\frac{7}{50}$.
The lacrosse stick has length 100 cm.
Find the length of the reduction.

Write the scale factor as a decimal.
$\frac{7}{50}$ = ______
Length of lacrosse stick: ________
Length of reduction: ________ × ________ = ________
The reduction has length ________.

7.3 **5.** These two quadrilaterals are similar.
Find the length of GH.

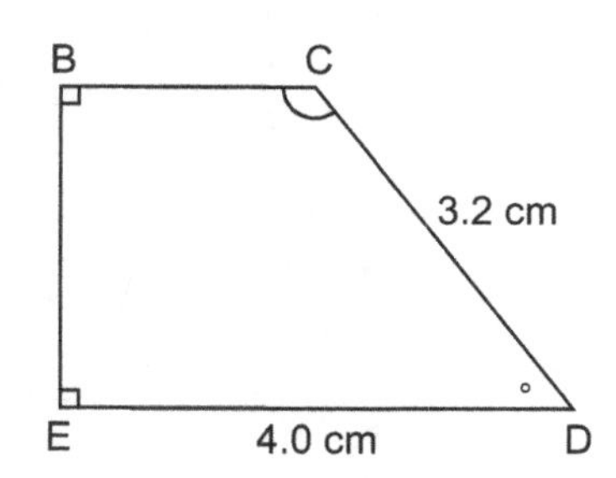

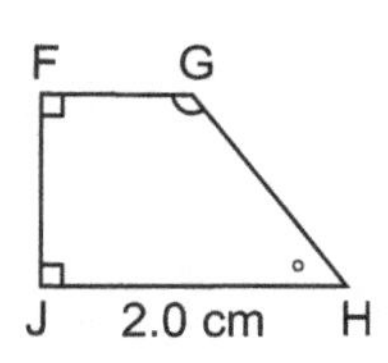

Quadrilateral FGHJ is a _____________ of quadrilateral BCDE.
To find the scale factor, choose a pair of matching sides whose lengths are both known:

____

Scale factor = $\frac{\text{length on \_\_\_\_\_\_\_\_}}{\text{length on original}}$

= $\frac{\_\_\_\_\_\_\_\_}{\_\_\_\_\_\_\_\_}$

= _____

The scale factor is _____.
Use the scale factor to find the length of GH.
GH and _____ are matching sides.
Length of _____: __________
Scale factor: _____
Length of GH: _________________________
So, GH has length _______.

7.4 **6.** Are these 2 triangles similar?

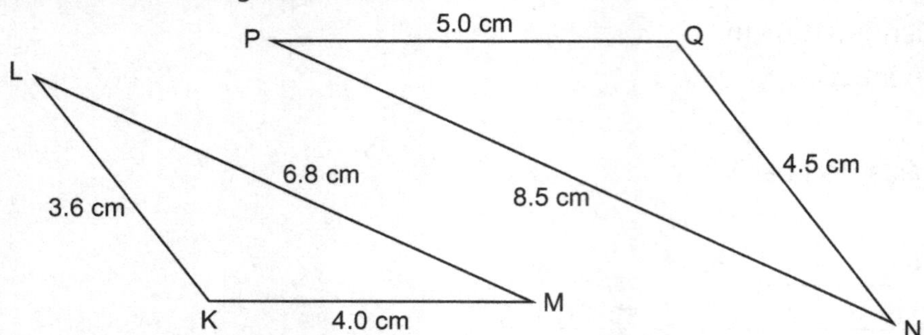

Find out if matching sides are proportional.

In △KLM, order the sides from shortest to longest: ____________________

In △NPQ, order the sides from shortest to longest: ____________________

Find the scale factors of matching sides.

$\frac{\text{length of\_\_\_\_}}{\text{length of\_\_\_\_}} = \frac{\_\_\_\_\_\_}{\_\_\_\_\_\_} = \_\_\_\_$

$\frac{\text{length of\_\_\_\_}}{\text{length of\_\_\_\_}} = \frac{\_\_\_\_\_\_}{\_\_\_\_\_\_} = \_\_\_\_$

$\frac{\text{length of\_\_\_\_}}{\text{length of\_\_\_\_}} = \frac{\_\_\_\_\_\_}{\_\_\_\_\_\_} = \_\_\_\_$

All scale factors are __________. So, the triangles __________.

The two shorter sides meet at vertices: ____ and ____

The longest and shortest sides meet at vertices: ____ and ____

The two longer sides meet at vertices: ____ and ____

So, △KLM ~ △______

7. At a certain time of day, a street light and a stop sign cast shadows.
Find the height of the street light.

Matching angles are ________.

So, △RST ~ △________

△________ is an enlargement of △________.

Use sides ____________ and ____________ to find the scale factor.

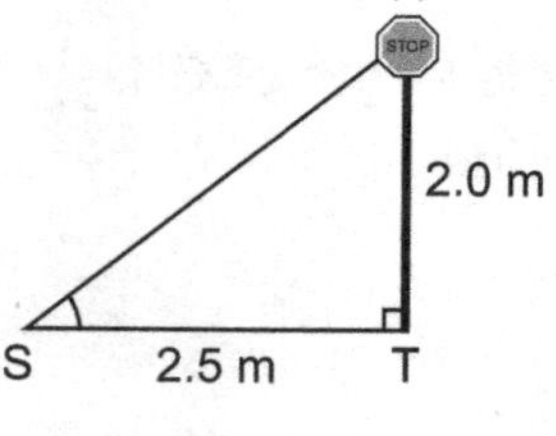

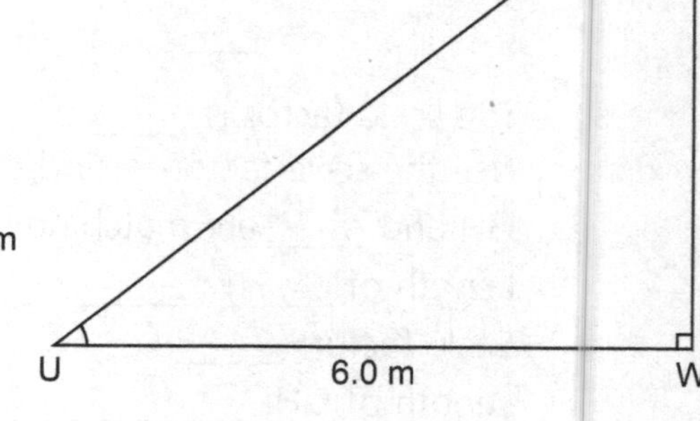

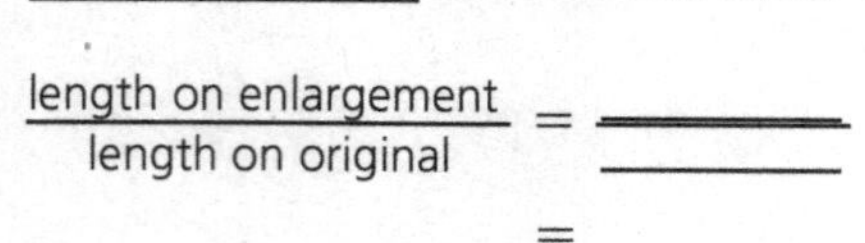

$\frac{\text{length on enlargement}}{\text{length on original}} = \frac{\_\_\_\_\_\_}{\_\_\_\_\_\_}$

$= \_\_\_\_$

The scale factor is ____.

Use the scale factor to find the height of the street light, VW.

VW and ____ are matching sides.

Length of ____ : ______ Scale factor: ____

Length of VW: ____________________

So, the height of the street light is ______.

7.5 Skill Builder

Lines of Symmetry in Quadrilaterals

A **line of symmetry** divides a shape into 2 matching, or **congruent** parts.
If we fold a shape along its line of symmetry, the parts match exactly.

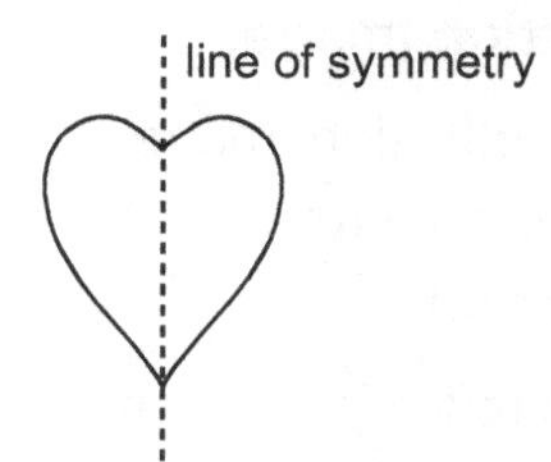

This trapezoid has 1 line of symmetry.

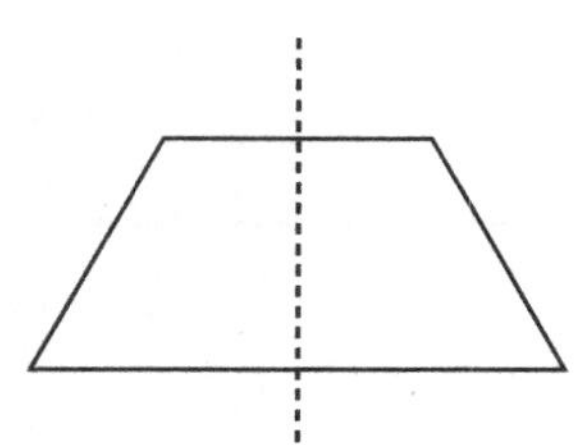

This rectangle has 2 lines of symmetry.

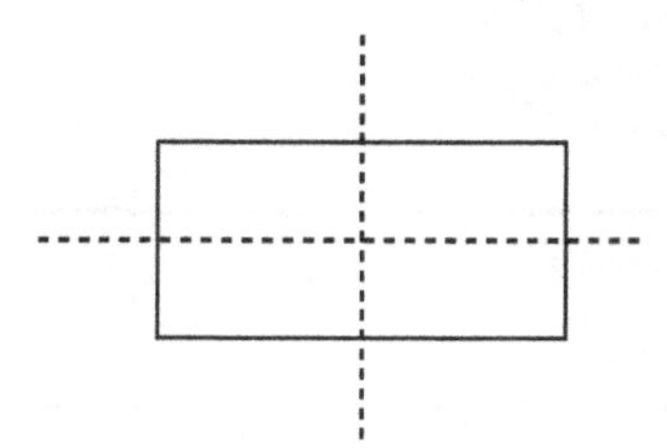

Check

1. How many lines of symmetry does each shape have? Draw in the lines.

a)

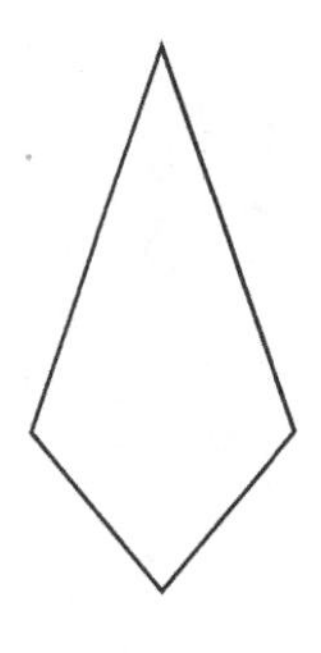

Number of lines of symmetry: ____

b)

Number of lines of symmetry: ____

c)

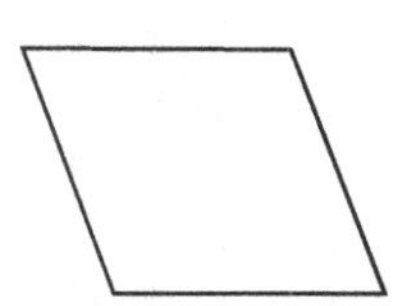

Number of lines of symmetry: ____

d)

Number of lines of symmetry: ____

Reflections

When a shape is reflected in a mirror, we see a **reflection image.**

A point and its reflection image are the same distance from a **line of reflection.**

A shape and its reflection image face opposite ways.

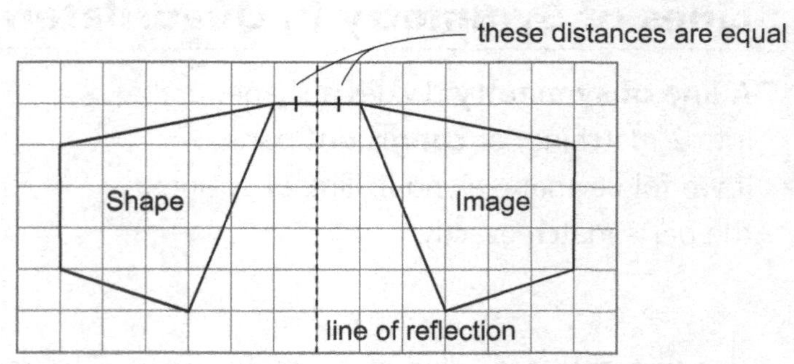

We can use a Mira to help us reflect a shape.

Check

1. Do these pictures show reflections?
If your answer is Yes, draw the line of reflection.

a)

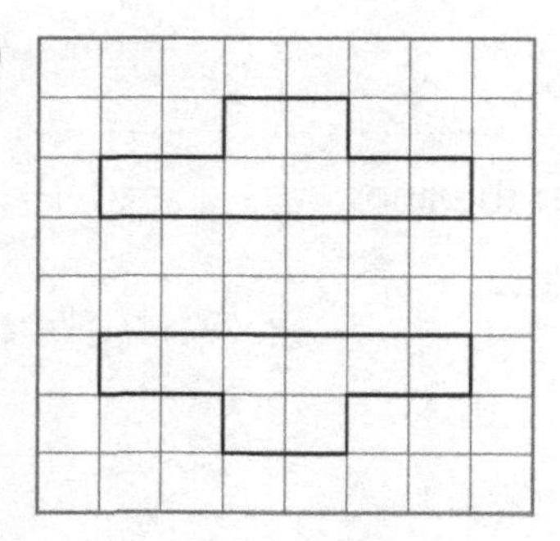

______

b)

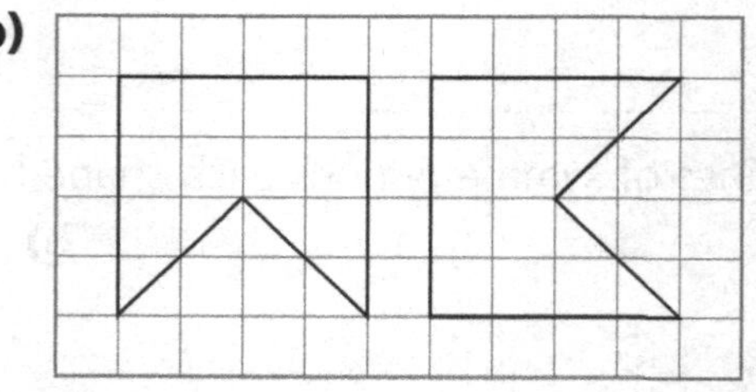

______

2. Draw each reflection image.

a)

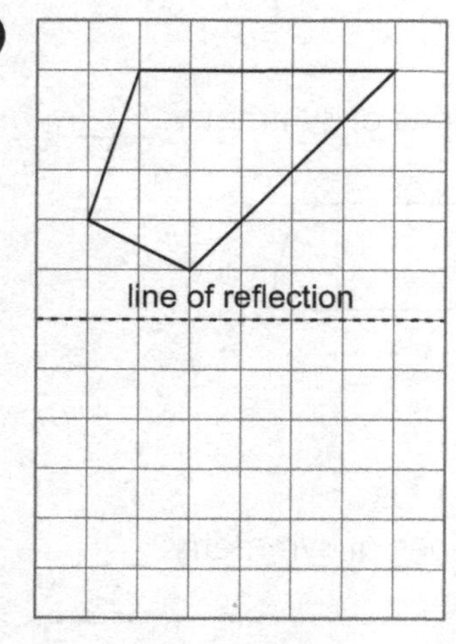

b)

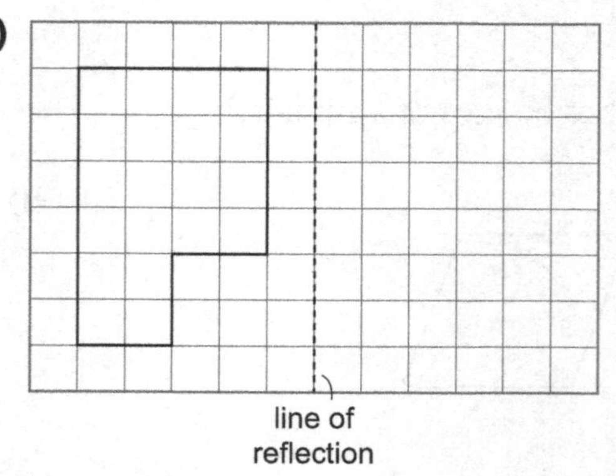

7.5 Reflections and Line Symmetry

FOCUS **Draw and classify shapes with line symmetry.**

When congruent copies of a polygon are used to cover a flat surface with no overlaps or gaps, a **tessellation** is created. Some tessellations have line symmetry.

Congruent polygons match exactly but may have different orientations.

Example 1 Identifying Lines of Symmetry in Tessellations

Identify the lines of symmetry in this tessellation.

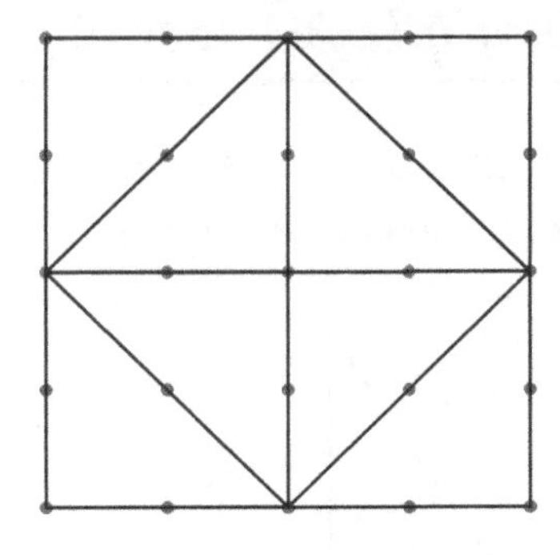

Solution

A line of symmetry must pass through the centre of the design. Use a Mira to check for vertical, horizontal, and diagonal lines of symmetry.

This tessellation has 4 lines of symmetry. The pattern on one side of each line is a mirror image of the pattern on the other side of the line.

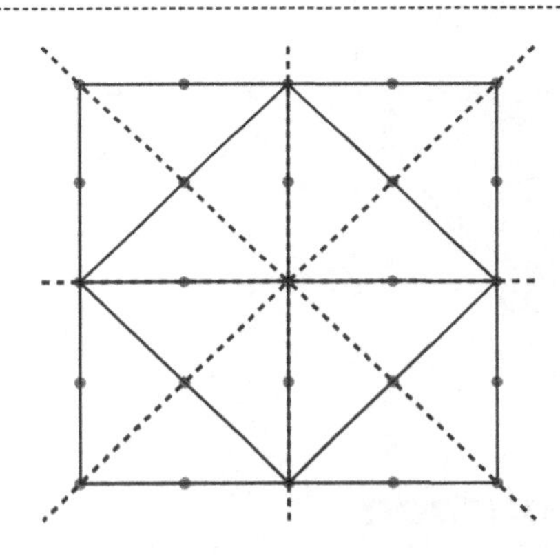

Check

1. Draw the lines of symmetry in each tessellation.

a) Use a Mira.
Is there a vertical line of symmetry? ______
Is there a horizontal line of symmetry? ______
Are there any diagonal lines of symmetry? ______
Draw the lines of symmetry.

Remember that a line of symmetry must pass through the centre of the design.

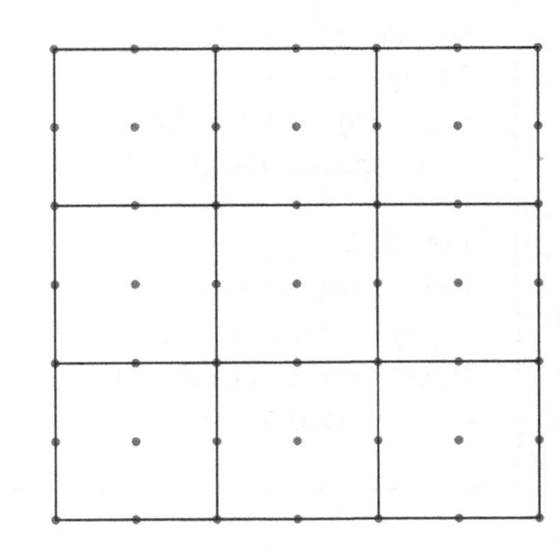

b) Is there a vertical line of symmetry? ______
Is there a horizontal line of symmetry? ______
Are there any diagonal lines of symmetry? ______
Draw the lines of symmetry.

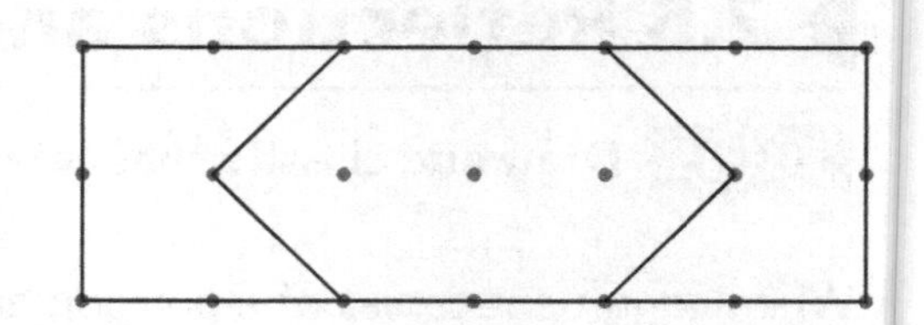

Two shapes may be related by a line of reflection.

Example 2 Identifying Reflected Shapes

Which triangle is a reflection of the shaded triangle?
Draw the line of reflection.

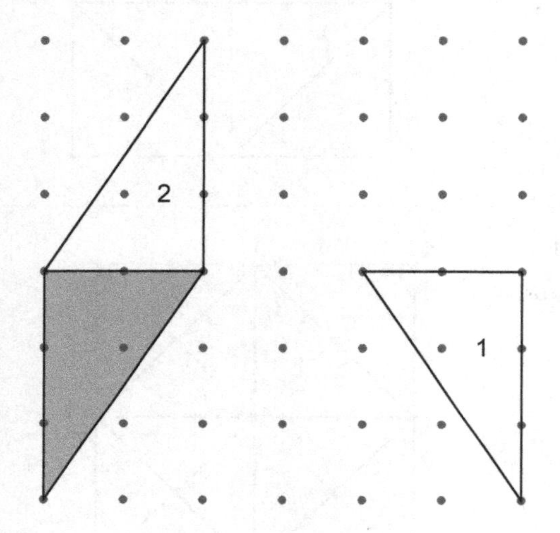

Solution

Use a Mira to check.
Triangle 1:
The triangle is to the right of the shaded triangle.
So, try a vertical line of reflection.
The triangle is the reflection image
of the shaded triangle in Line A.

Triangle 2:
The triangle is above the shaded triangle.
So, try a horizontal line of reflection.
The triangle is not a reflection image
of the shaded triangle.

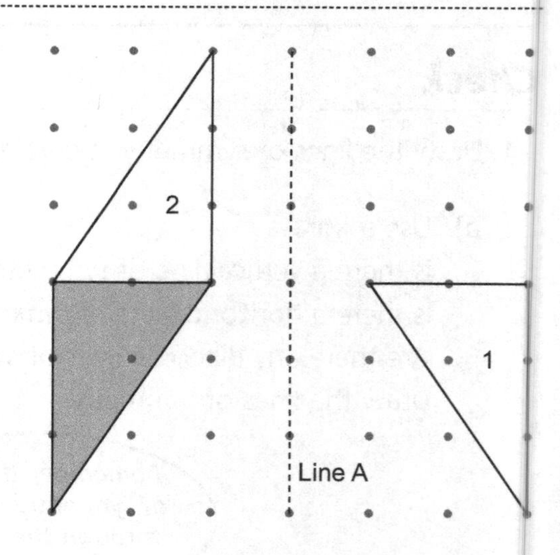

Check

1. Which polygon is a reflection of the shaded polygon?
Draw the line of reflection.

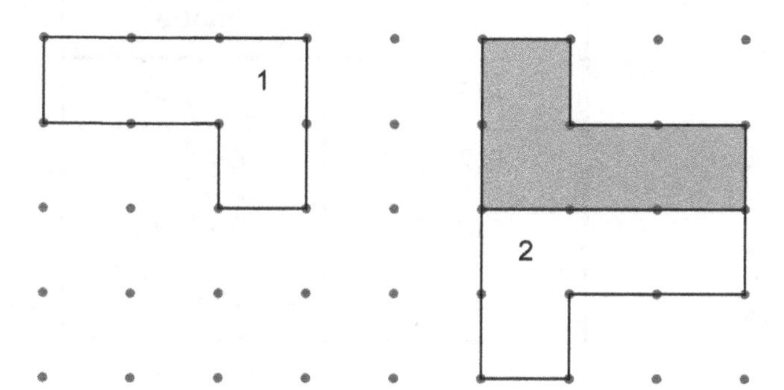

Use a Mira to check.
Polygon 1:
The polygon is to the __________ of the shaded polygon.
So, try a ____________ line of reflection.
The polygon __________ a reflection image
of the shaded polygon.
If the polygon is a reflection image, draw the line of reflection.

Polygon 2:
The polygon is __________ the shaded polygon.
So, try a ______________ line of reflection.
The polygon __________ a reflection image
of the shaded polygon.
If the polygon is a reflection image, draw the line of reflection.

Example 3 Completing a Shape Given its Line of Symmetry

Reflect quadrilateral ABCD in the line of reflection to make a larger shape.

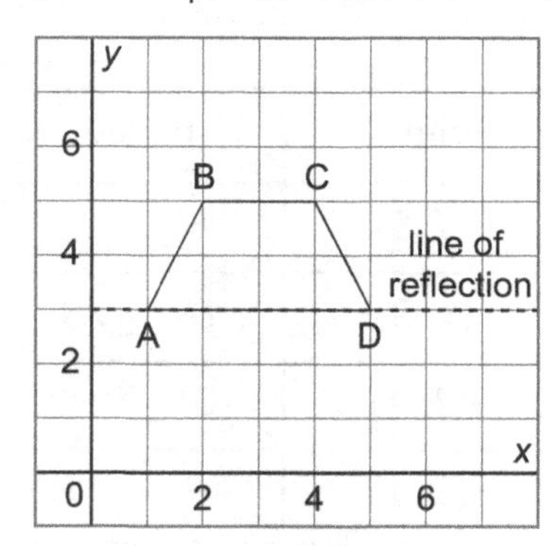

Solution

A point and its image must be the same distance from the line of reflection.

Point A: on the line of reflection
Reflection image: Point A reflects onto itself.

Point B: 2 squares above line of reflection
Reflection image: Point B′ is 2 squares below line of reflection.

Point C: 2 squares above line of reflection
Reflection image: Point C′ is 2 squares below line of reflection.

Point D: on the line of reflection
Reflection image: Point D reflects onto itself.

Point	Image
A(1, 3)	A(1, 3)
B(2, 5)	B′(2, 1)
C(4, 5)	C′(4, 1)
D(5, 3)	D(5, 3)

Plot the points. Join the points in order to complete the larger shape.

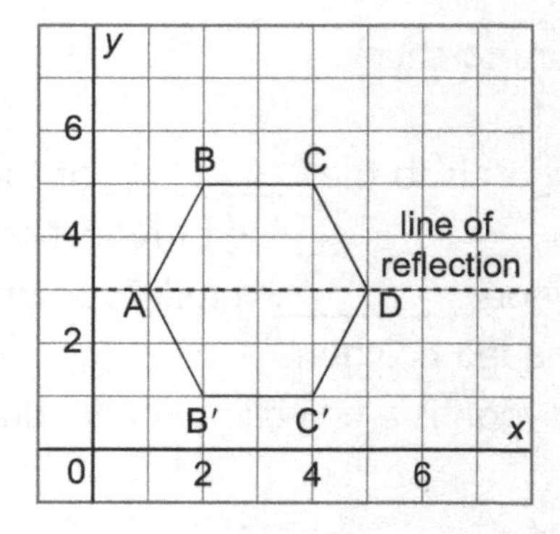

Point B′ is the image of point B.
We say: "B prime"

Check

1. Reflect quadrilateral EFGH in the line of reflection to make a larger shape.

Point E: on the line of reflection
Reflection image: ________________________

Point F: 2 squares left of line of reflection
Reflection image: ________________________
____

Point G: ____________________________
Reflection image: ________________________
____

Point H: ____________________________
Reflection image: ________________________

Plot the points.
Join the points to complete the larger shape.

Point	Image
E(3, 5)	E(___, 5)
F(1, 3)	F′(___, 3)
G(2, 1)	G′(___, 1)
H(3, 1)	H(___, 1)

Practice

1. Draw the lines of symmetry in each tessellation.

a)

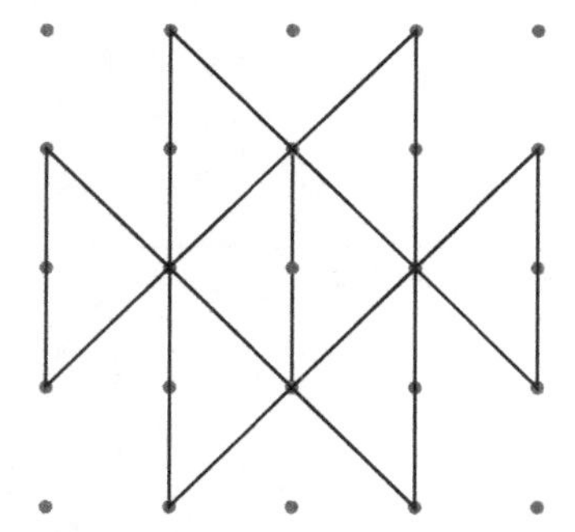

b)

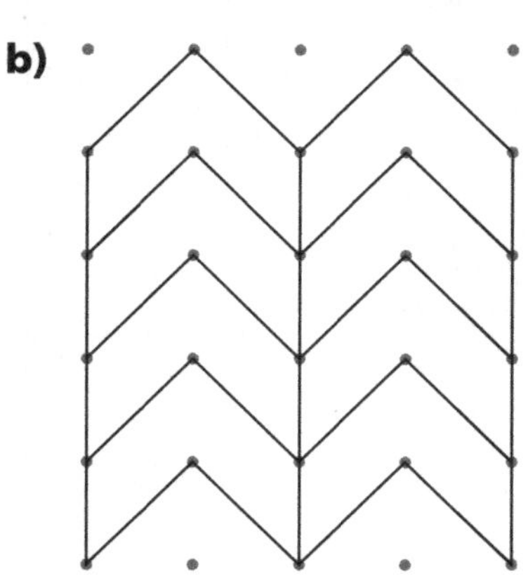

2. Which hexagons are reflections of the shaded hexagon?
Draw the line of reflection each time.

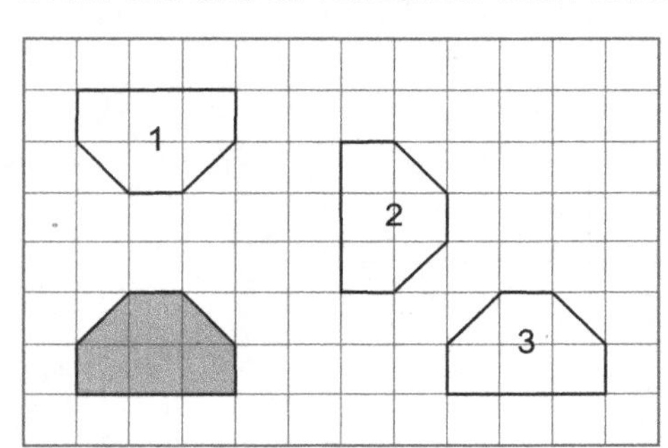

Hexagon 1:
The hexagon is ____________ the shaded hexagon.
So, try a ______________ line of reflection.
The hexagon ____________ a reflection image of the shaded hexagon.
If the polygon is a reflection image, draw the line of reflection, Line A.

Hexagon 2:
The hexagon is ____________ and to the ____________ of the shaded polygon.
So, try a ______________ line of reflection.
The hexagon ____________ a reflection image of the shaded hexagon.
If the polygon is a reflection image, draw the line of reflection, Line B.

Hexagon 3:
The hexagon is to the ___________ of the shaded hexagon.
So, try a ___________ line of reflection.
The hexagon ____________ a reflection image of the shaded hexagon.
If the polygon is a reflection image, draw the line of reflection, Line C.

3. Reflect each shape in the line of reflection to make a larger shape.

a)

Point	Image
A(0, 5)	A(__, __)
B(2, 5)	B(__, __)
C(3, 3)	C′(__, __)
D(2, 1)	D′(__, __)

b)

Point	Image
____	____
____	____
____	____
____	____
____	____
____	____

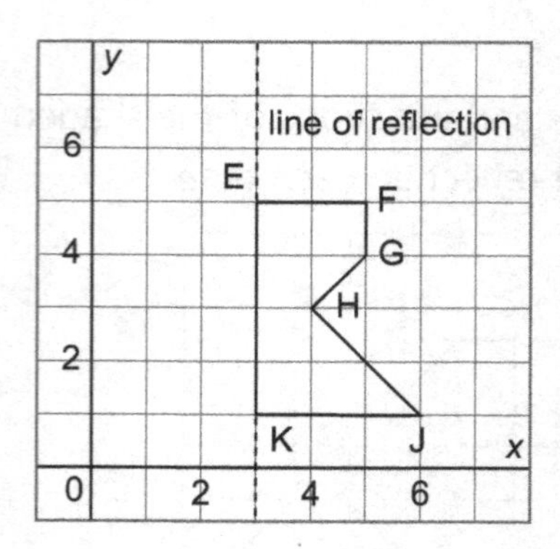

c)

Point	Image
____	____
____	____
____	____

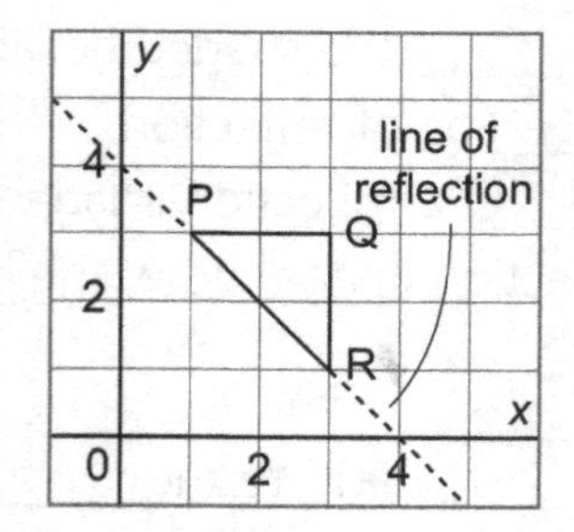

7.6 Skill Builder

Rotations

A **rotation** may be clockwise or counterclockwise.
Some common rotations are 90°, 180°, and 270°.

A complete turn measures 360°.

This shape was rotated 90° clockwise about point R.

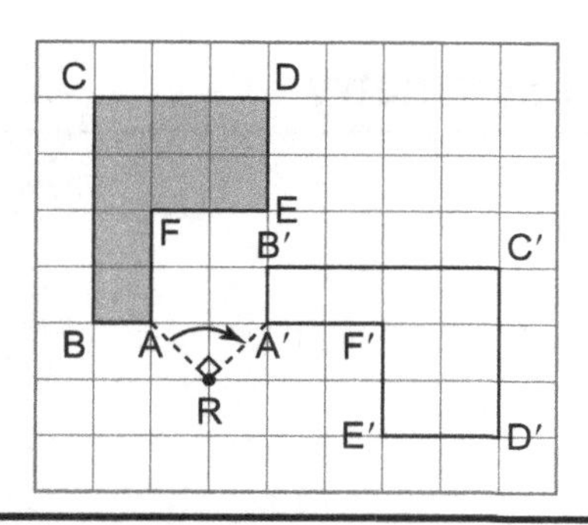

∠ARA′ = 90°, ∠BRB′ = 90°, and so on.
Each angle is the angle of rotation.
We can use a protractor to check.

Check

1. For each picture, write the angle of rotation.

a)

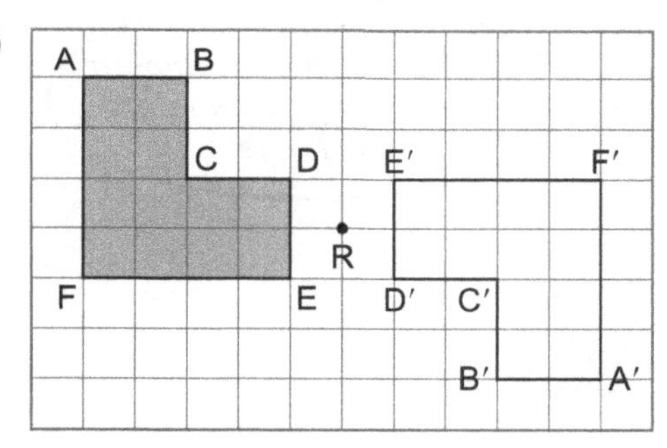

Angle of rotation: __________

b)

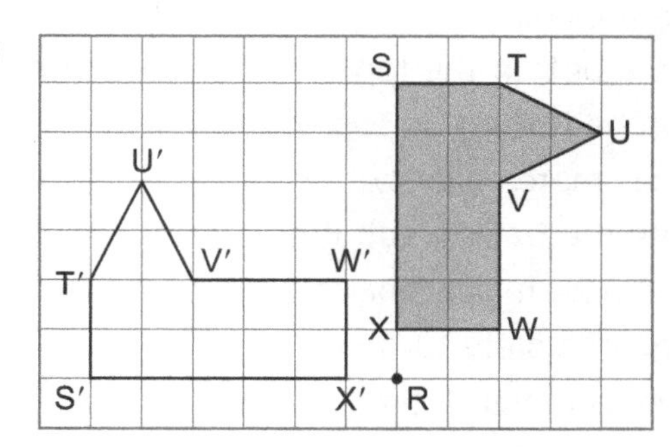

Angle of rotation: __________________

2. Draw the image after each rotation about point R.

a) 90° clockwise

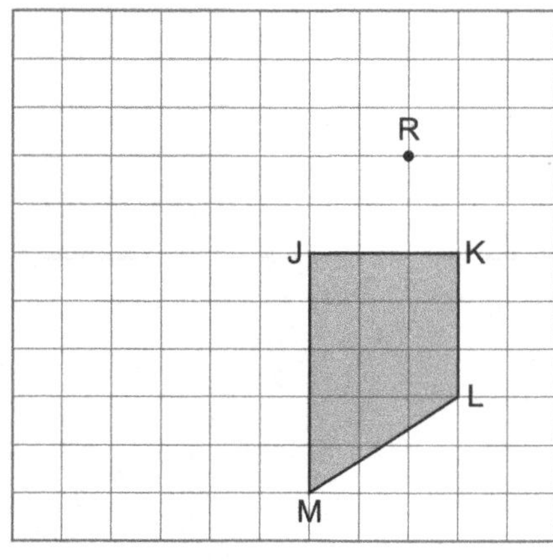

b) 180°

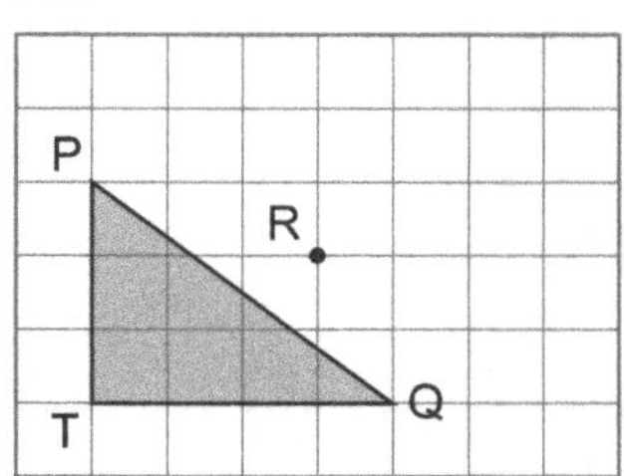

We can use tracing paper to help us rotate a shape.

7.6 Rotations and Rotational Symmetry

FOCUS **Draw and classify shapes with rotational symmetry.**

A shape has **rotational symmetry** when it can be turned less than 360° about its centre to match itself exactly.
The number of matches in a complete turn is the **order of rotation.**

Example 1 Determining the Order of Rotational Symmetry

Find the order of rotational symmetry for this star.

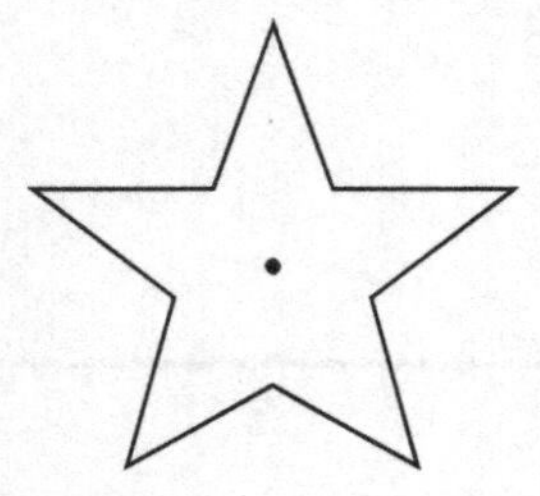

Solution

Trace the star.
Draw a dot on the top vertex of each star.
Place the tracing on top of the star
so they match exactly.
Rotate the tracing about its centre
to see how many times the stars match
in one complete turn.
The stars match 5 times.
So, the star has rotational symmetry of order 5.

You have made a complete turn when the two dots match again.

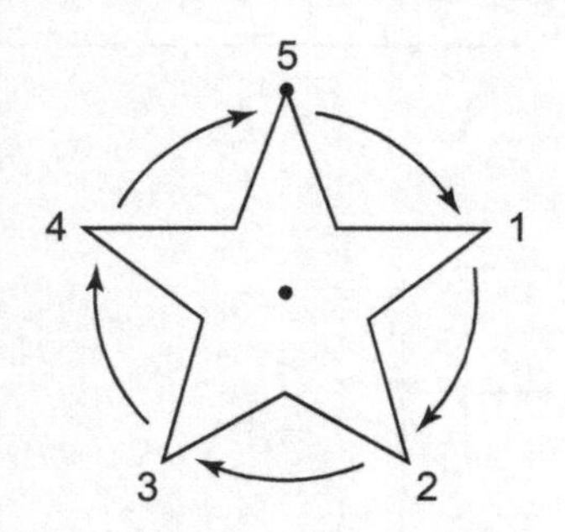

Check

1. Find the order of rotational symmetry for each shape.
   Use tracing paper to help.

   a)

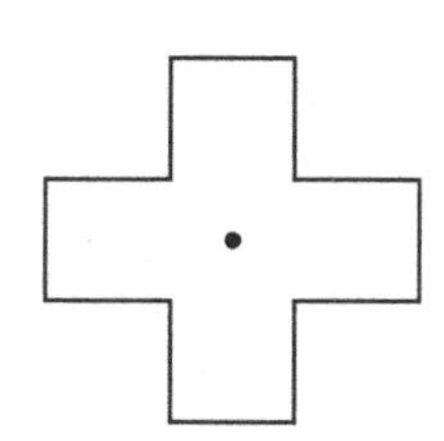

   The shape and its tracing match _____ times.
   So, the shape has rotational symmetry of order _____.

   b)

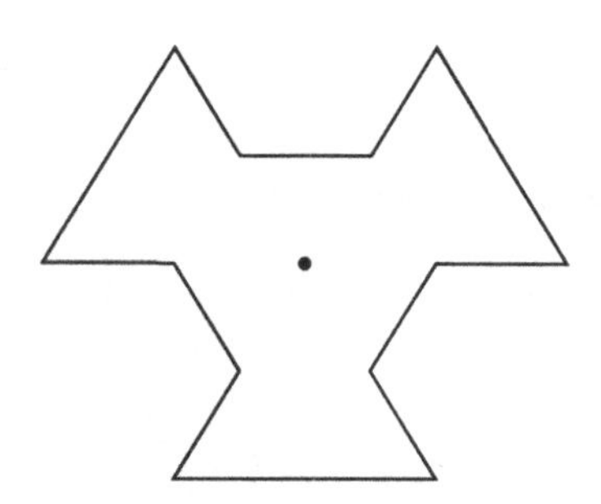

   The shape and its tracing match _____ times.
   So, the shape has rotational symmetry of order _____.

The smallest angle you need to turn for two shapes to match is the **angle of rotation.**

$$\text{The angle of rotation symmetry} = \frac{360^\circ}{\text{the order of rotation}}$$

Example 2 Determining the Angle of Rotation Symmetry

Find the angle of rotation symmetry for this shape.

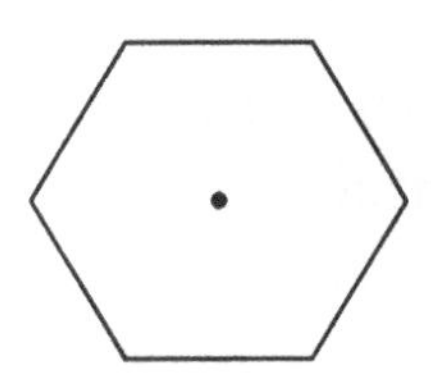

Solution

In one complete turn, the shape and its tracing match 6 times.
So, the order of rotation is 6.

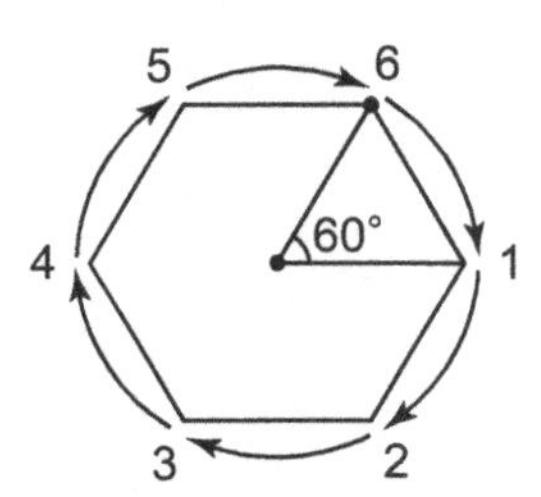

The shapes will match when the tracing is rotated by a multiple of 60°.

The angle of rotation symmetry is:

$$\frac{360^\circ}{\text{the order of rotation}} = \frac{360^\circ}{6}$$
$$= 60^\circ$$

The angle of rotation symmetry is 60°.

Check

1. Find the angle of rotation symmetry for each shape.

a)

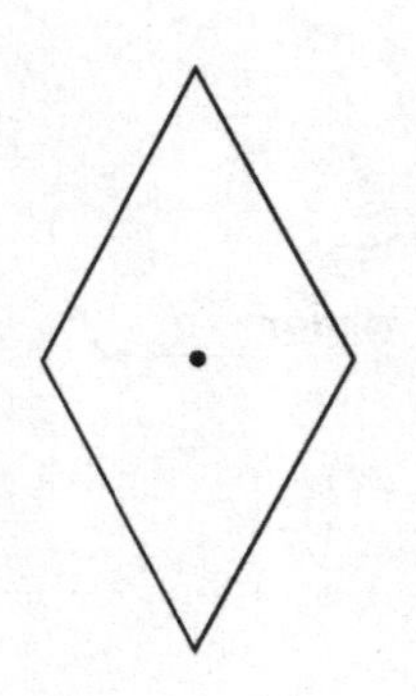

The shape and its tracing match ______ times.
So, the order of rotation is ______.
Angle of rotation symmetry is:

$$\frac{360^\circ}{\text{the order of rotation}} = \frac{360^\circ}{\_\_\_\_} = \_\_\_\_$$

The angle of rotation symmetry is ________.

b)

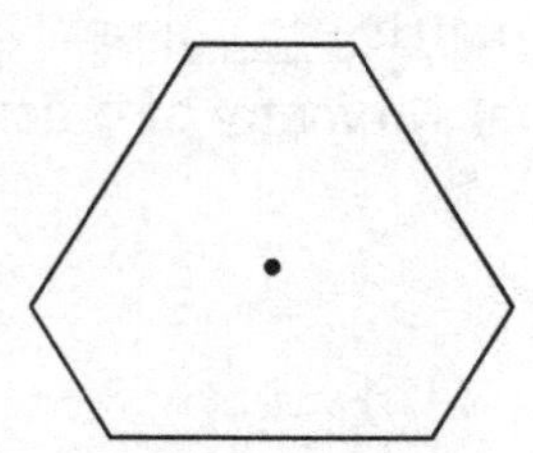

The shape and its tracing match ______ times.
So, the order of rotation is ______.
Angle of rotation symmetry is:

$$\frac{360^\circ}{\text{the order of rotation}} = \frac{360^\circ}{\_\_\_\_} = \_\_\_\_$$

The angle of rotation symmetry is ________.

Shapes that need a complete turn to match again do not have rotational symmetry.

We use isometric dot paper to draw images after rotations that are multiples of 60°.

We can use what we know about isometric dot paper to help us rotate a shape.

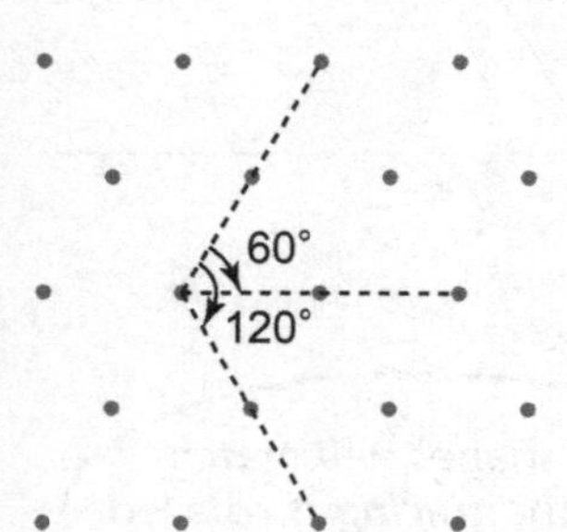

Example 3 Drawing Rotation Images

Rotate parallelogram ABCD 60° clockwise about vertex C.
Draw and label the rotation image.

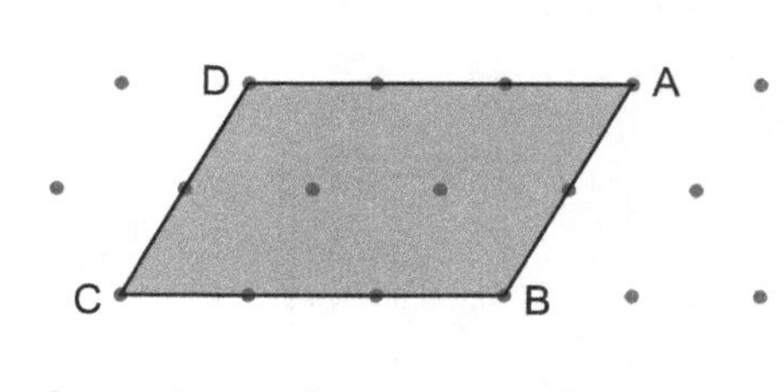

Solution

Trace the shape.
Label the vertices on the tracing.
Rotate the tracing 60° clockwise about vertex C.
Draw and label the rotation image.
The centre of rotation, C, does not move.
So, it is not labelled C′.

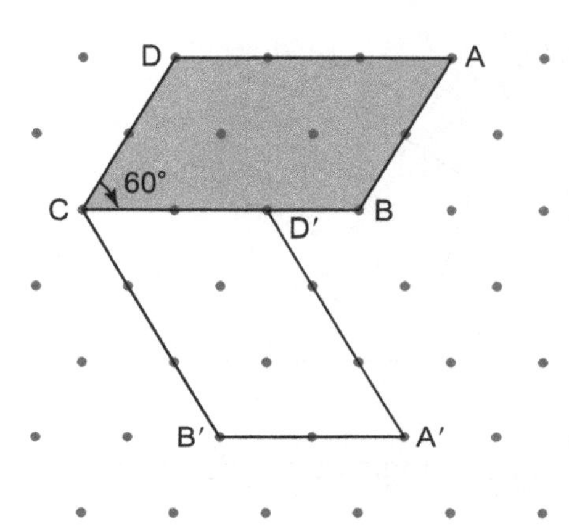

Check

1. Draw and label the image after each rotation.

a) 60° counterclockwise about vertex G

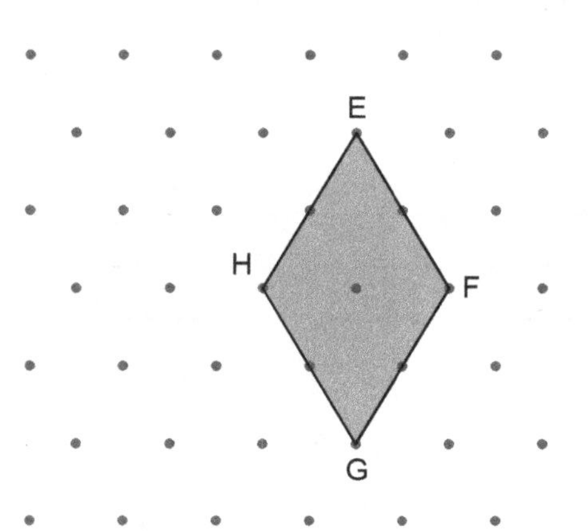

b) 120° clockwise about vertex S

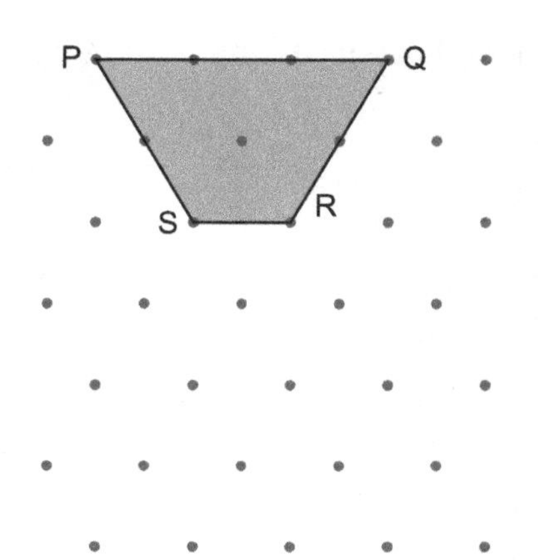

Practice

1. Find the order of rotational symmetry for each shape.

a)

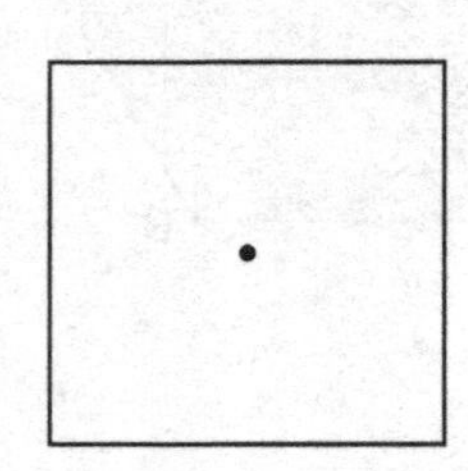

The shape and its image match _____ times.
So, the shape has rotational symmetry of order _____.

b)

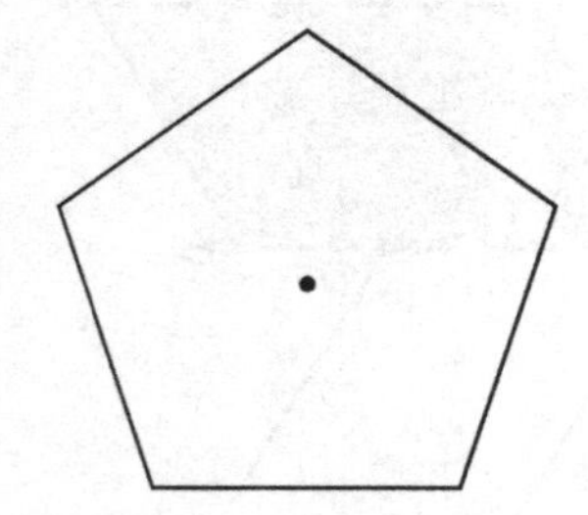

The shape and its image match _____ times.
So, the shape has rotational symmetry of order _____.

2. Find the angle of rotation symmetry for each shape in question 1.

a) The order of rotation is _____.
Angle of rotation symmetry is:

$$\frac{360°}{\text{the order of rotation}} = \frac{360°}{\_\_\_\_}$$
$$= \_\_\_\_\_$$

The angle of rotation symmetry is _______.

b) The order of rotation is _____.
Angle of rotation symmetry is:

$$\frac{360°}{\text{the order of rotation}} = \frac{360°}{\_\_\_\_}$$
$$= \_\_\_\_\_$$

The angle of rotation symmetry is _______.

3. Does this shape have rotational symmetry?

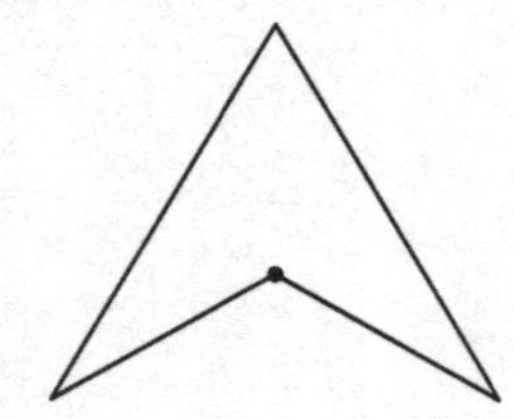

____

____

____

4. The angle of rotation symmetry for a shape is 36°.
What is the shape's order of rotation?

The angle of rotation symmetry is: $\frac{360°}{\text{the order of rotation}}$

So, $36° = \frac{360°}{\text{order of rotation}}$

Think: Which number divides into 360 exactly 36 times?

I know 360 ÷ ____ = 36

So, the order of rotation is ____.

5. Draw the image after each rotation.

a) 90° counterclockwise about vertex A

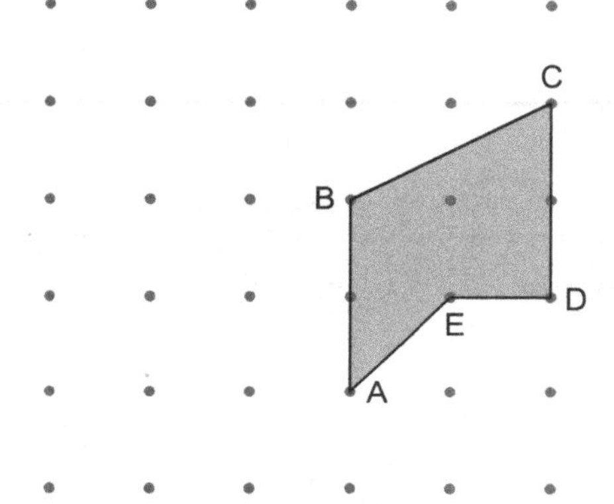

b) 180° about vertex J

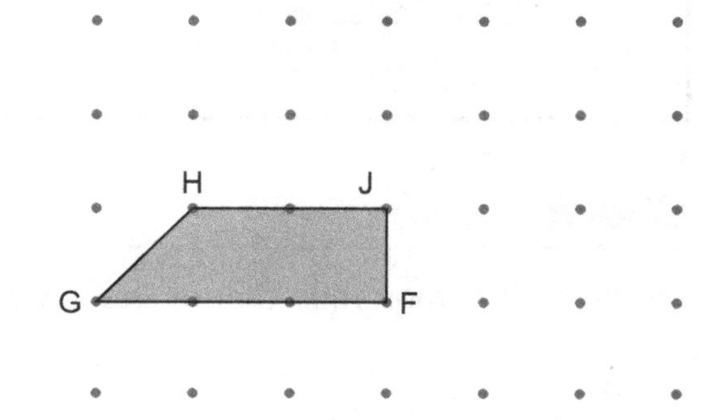

c) 60° clockwise about vertex N

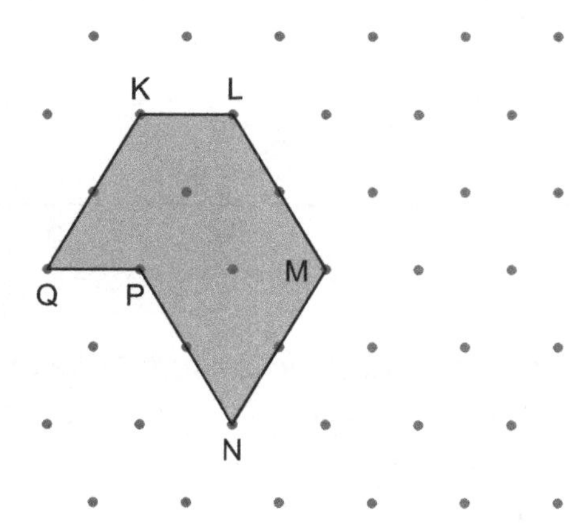

d) 120° counterclockwise about vertex T

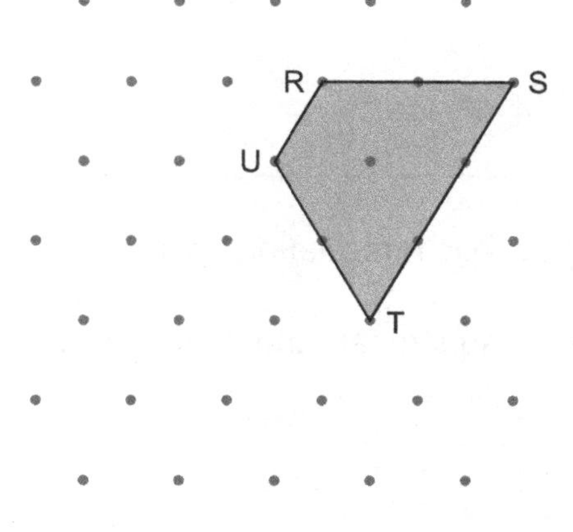

7.7 Skill Builder

Translations

A **translation** moves a shape along a straight line.
A shape and its translation image face the same way.

This shape was translated 2 squares right and 3 squares up.

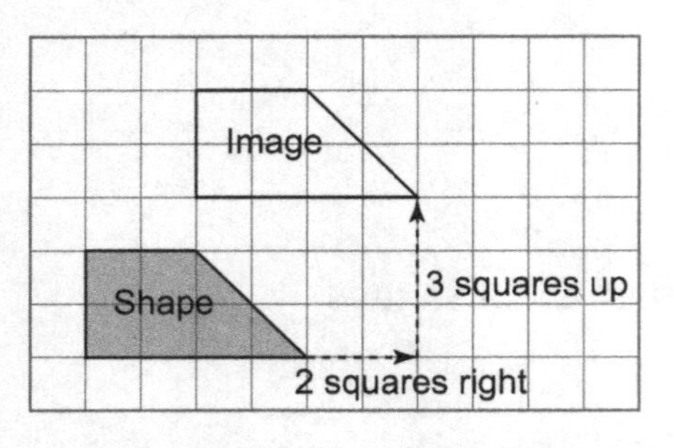

We say how many squares left or right before we say how many up or down.

Check

1. Write the translation that moves each shape to its image.

a)

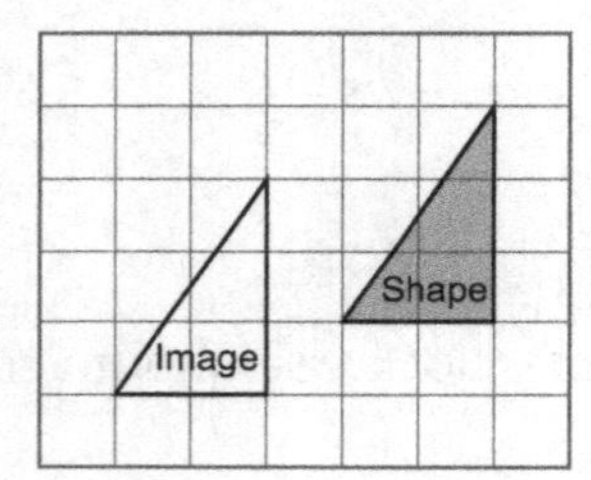

____

b)

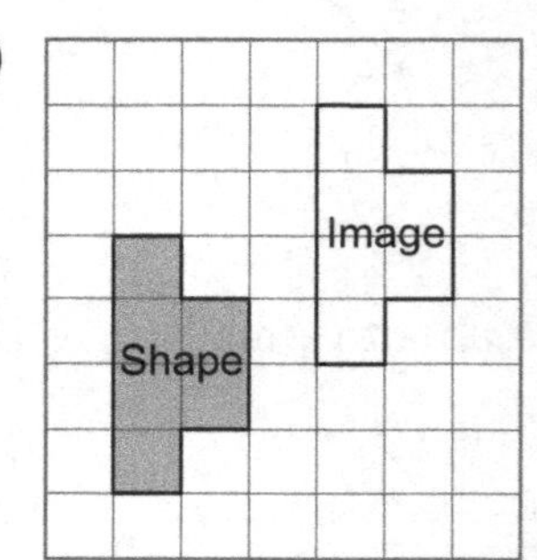

____

2. Draw each translation image.

a) 1 square left and 3 squares up

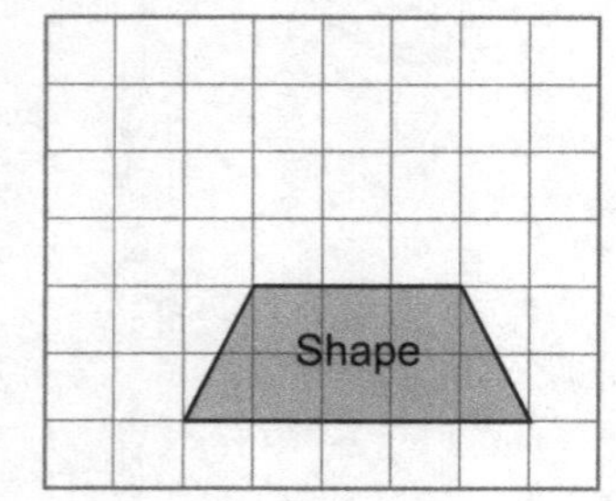

b) 3 squares right and 2 squares down

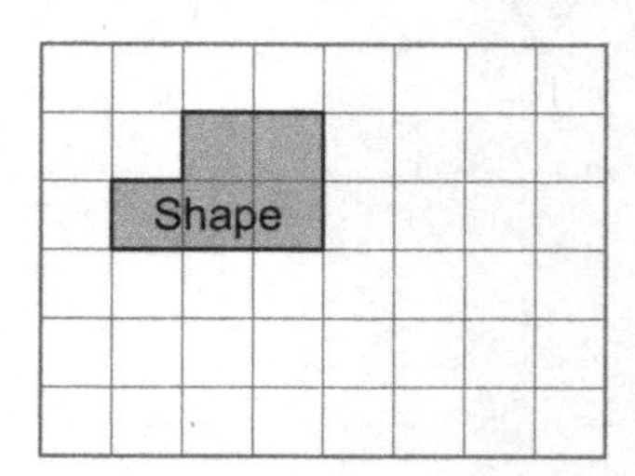

7.7 Identifying Types of Symmetry on the Cartesian Plane

FOCUS **Identify and classify line and rotational symmetry.**

A diagram of a shape and its transformation image may have:

- line symmetry
- rotational symmetry
- both line symmetry and rotational symmetry
- no symmetry

Example 1 Determining whether Shapes Are Related by Symmetry

Are rectangles ABCD and EFGH related by symmetry?

Solution

Check for line symmetry:
Rectangle ABCD is to the left of rectangle EFGH.
So, try a vertical line of reflection.
When I place a Mira on Line A,
the rectangle and its image match.
So, the rectangles are related by line symmetry.

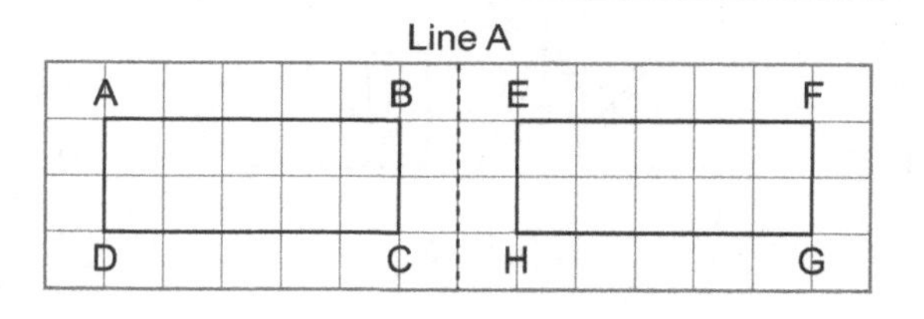

Matching points are the same distance from the line of reflection.

Check for rotational symmetry:
The rectangles do not touch.
So, try a point of rotation off the rectangles.
Try different points to see if the rectangles ever match. When I rotate rectangle ABCD 180° about point R, the rectangles match.
So, the rectangles are related by rotational symmetry.

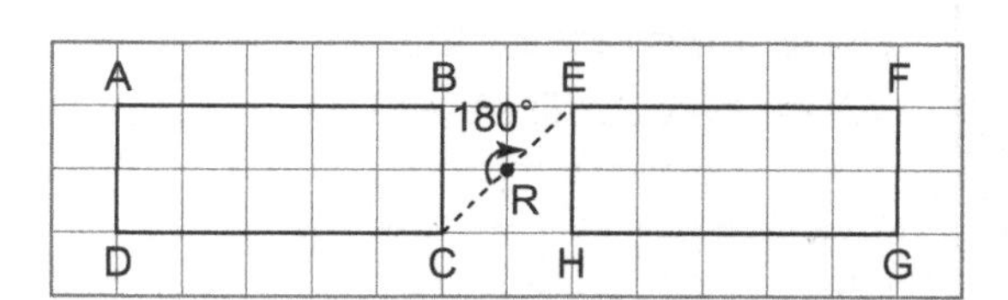

Check

1. For each diagram, find out if the polygons are related by symmetry.

a) Do the polygons face opposite ways? ________
One polygon is above the other,
so try a ____________ line of reflection.
Use a Mira to find the line of reflection.
Are the polygons related by a reflection? ________
If they are, draw the line of reflection.

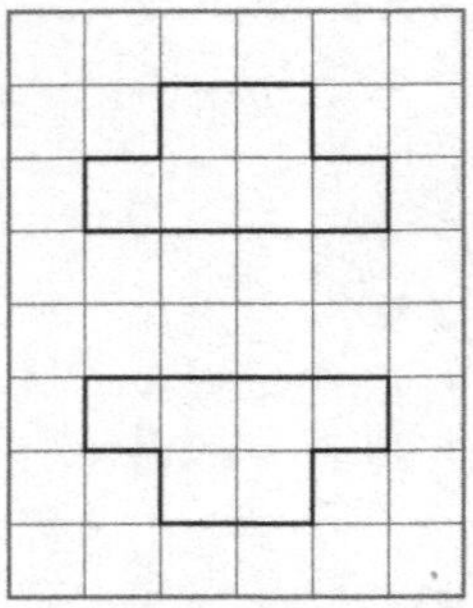

Do the polygons touch? ________
So, try a point of rotation ________ the polygons.
Try different points of rotation.
Do the polygons ever match? ________
Are the polygons related by a rotation? ________
If they are, label the point of rotation.

b) Do the polygons face different ways? ________
Do the polygons face opposite ways? ________
So, are the polygons related by a reflection? ________

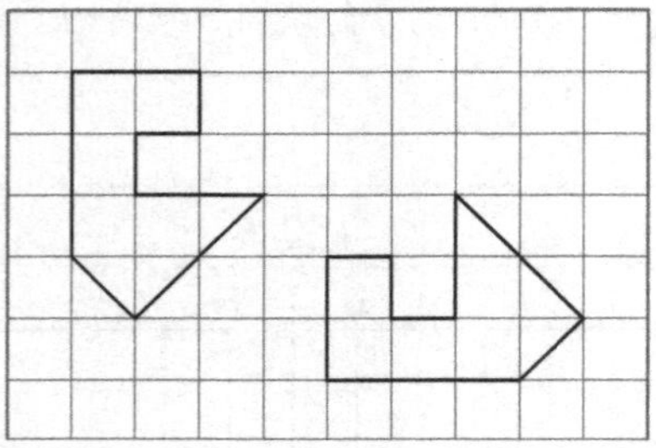

Do the polygons touch? ________
So, try a point of rotation ________ the polygons.
Try different points of rotation.
Do the polygons ever match? ________
Are the polygons related by a rotation? ________

Example 2 Identifying Symmetry in a Shape and Its Transformation Image

Draw the image of this parallelogram after a translation of 2 squares down and 1 square right.
Write the coordinates of each vertex and its image. Describe any symmetry that results.

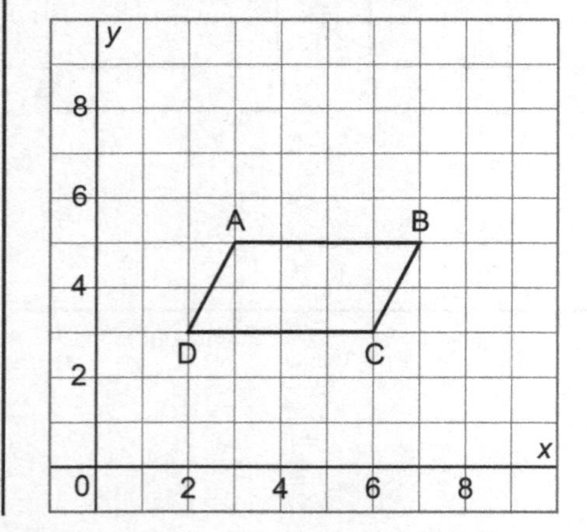

Solution

Translate parallelogram ABCD 2 squares down
and 1 square right.
Draw and label the translation image.
Write the coordinates of each vertex and its image.

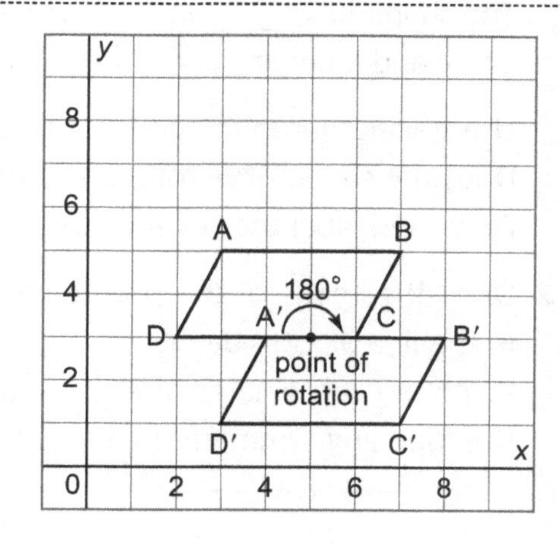

Point	Image
A(3, 5)	A′(4, 3)
B(7, 5)	B′(8, 3)
C(6, 3)	C′(7, 1)
D(2, 3)	D′(3, 1)

Use a Mira to check for line symmetry.
There is no line on which I can place a Mira
so one parallelogram matches the other.
So, the shape does not have line symmetry.

Use tracing paper to check for rotational symmetry.
The shape and its tracing match after a rotation
of 180° about (5, 3).
So, the shape has rotational symmetry.

Check

1. Draw the image of this polygon after
a translation of 2 squares down.
Write the coordinates of each vertex and its image.
Describe any symmetry that results.

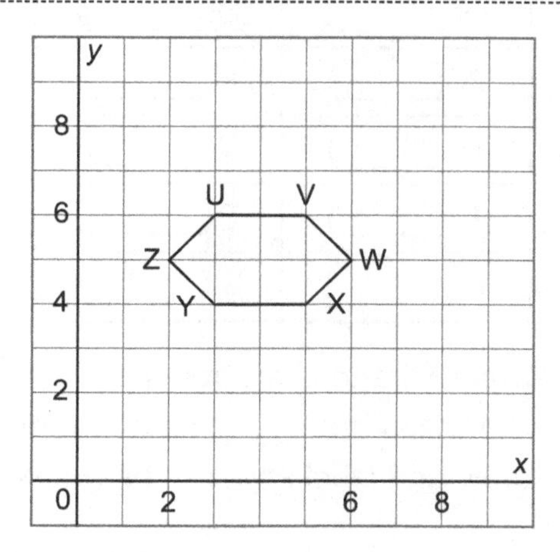

Translate the polygon 2 squares down.
Draw and label the translation image.

Point	Image
U(3, 6)	Y(3, 4)
V(5, 6)	X(5, 4)
W(__, __)	W′(__, __)
X(__, __)	X′(__, __)
Y(__, __)	Y′(__, __)
Z(__, __)	Z′(__, __)

Use a Mira to check for line symmetry.
The shape has _____ lines of symmetry:
Draw and label any lines of symmetry you found.

Use tracing paper to check for rotational symmetry.
Does the shape have rotational symmetry? _______
Draw and label the point of rotation.

2. Draw the image of this polygon after a reflection in the line along side QR.
Write the coordinates of each vertex and its image.
Describe any symmetry that results.

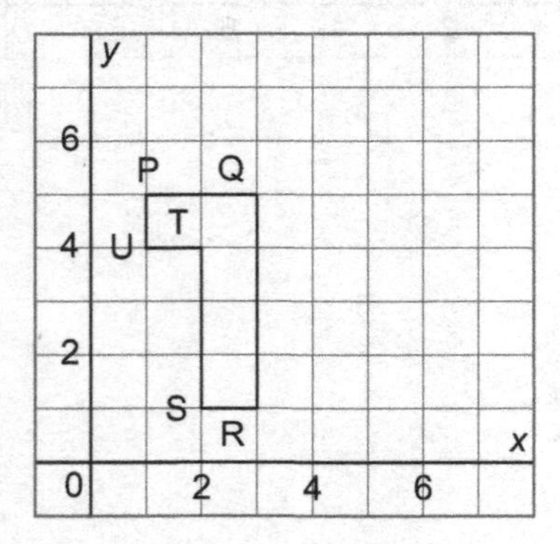

Reflect the polygon.
Draw and label the reflection image.

Point	Image
P(___, ___)	P′(___, ___)
Q(___, ___)	Q(___, ___)
R(___, ___)	R(___, ___)
S(___, ___)	S′(___, ___)
T(___, ___)	T′(___, ___)
U(___, ___)	U′(___, ___)

Use a Mira to check for line symmetry.
The shape has _____ line of symmetry:
Draw and label any lines of symmetry you found.

Use tracing paper to check for rotational symmetry.
Is there a point about which you can turn the tracing so it matches the shape? _______
Does the shape have rotational symmetry? _______

Practice

1. Which of these polygons are related by line symmetry?

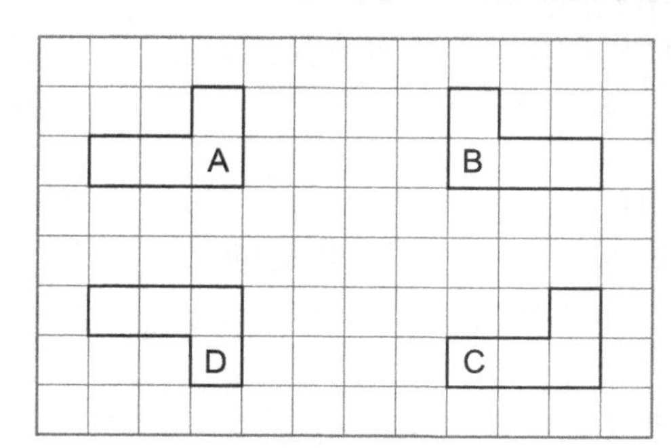

Which pairs of polygons face opposite ways?

____

Draw in the line of reflection for each pair of polygons.

Which polygons are related by line symmetry?

____

2. Which of these polygons are related by rotational symmetry about point R?

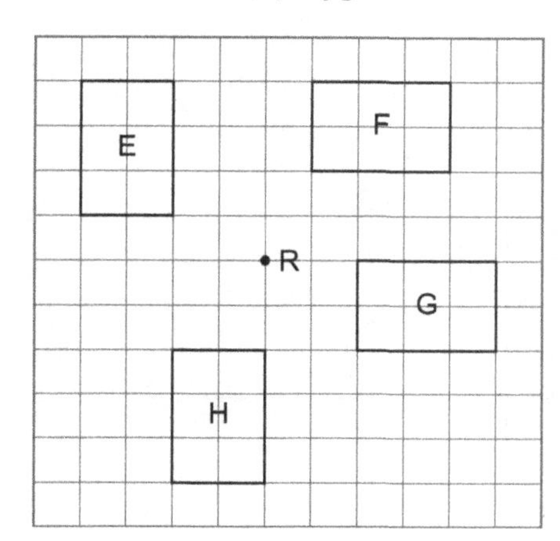

Trace rectangle E.
Rotate the tracing about point R.
Which rectangle does it match? ______________
Trace rectangle G.
Rotate the tracing about point R.
Which rectangle does it match? ______________
Which rectangles are related by rotational symmetry?

____

3. For each diagram, find out if the triangles are related by symmetry.
Use tracing paper and a Mira to help.

a)

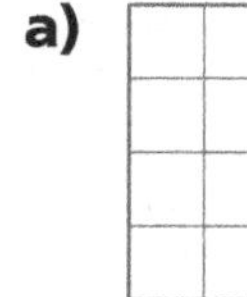

L
N
P
M
N

Do the triangles face opposite ways? ________
So, are the triangles related by a reflection? ________

Do the triangles touch? ________
So, try a point of rotation ________ the triangles.
Which vertex is common to both triangles?

____

Try different rotations about this vertex.
When do the triangles match? ______________

____

Are the triangles related by a rotation? ________
If they are, label the point of rotation.

b)

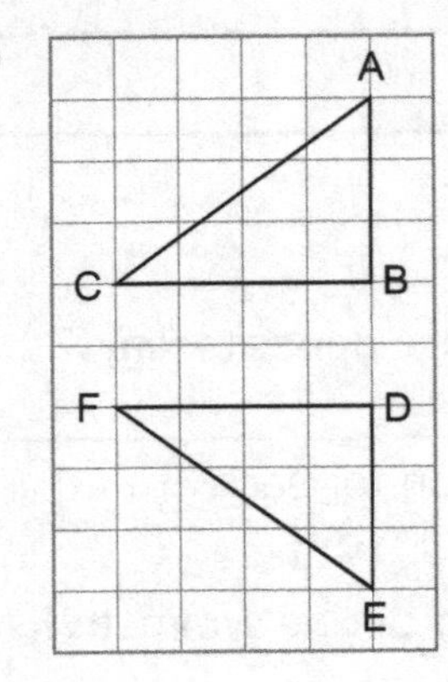

Do the triangles face opposite ways? ______
One triangle is above the other,
so try a __________ line of reflection.
Use a Mira to find the line of reflection.
Are the triangles related by a reflection? ______
If they are, draw the line of reflection.

Do the triangles touch? ______
So, try a point of rotation ______ the triangles.
Try different points of rotation.
Do the triangles ever match? ______
Are the triangles related by a rotation? ______
If they are, label the point of rotation.

4. Draw the image of this polygon after a rotation of 180° about point A.
Write the coordinates of each vertex and its image.
Describe any symmetry that results.

Rotate the polygon.
Draw and label the rotation image.

Point	Image
P(___, ___)	P′(___, ___)
Q(___, ___)	Q′(___, ___)
R(___, ___)	S(___, ___)
S(___, ___)	R(___, ___)
T(___, ___)	T′(___, ___)

Use a Mira to check for line symmetry.

____

____

____

Use tracing paper to check for rotational symmetry.
Does the shape have rotational symmetry? ______
If it does, label the point of rotation.

Unit 7 Puzzle

Mystery Logo!

A friend designed a logo for Hal's new gift-wrapping business.
Follow these instructions to create the logo on the coordinate grid below.

Instructions:

1. a) Plot and label the points H(1, 7), A(3, 5), L(1, 3).
Join the points in order to form a triangle. Shade the triangle.

b) Rotate △HAL 90° counterclockwise about H. Shade the triangle.

c) Rotate △HAL 90° clockwise about L. Shade the triangle.

d) Reflect △HAL in the vertical line through A. Shade the triangle.

2. Reflect the shape from part 1 in Line A.
Shade to match the shape in part 1.

3. Plot the points (5, 6), (7, 6), (7, 4), (5, 4).
Join the points in order to form a square. Shade the square a different colour.

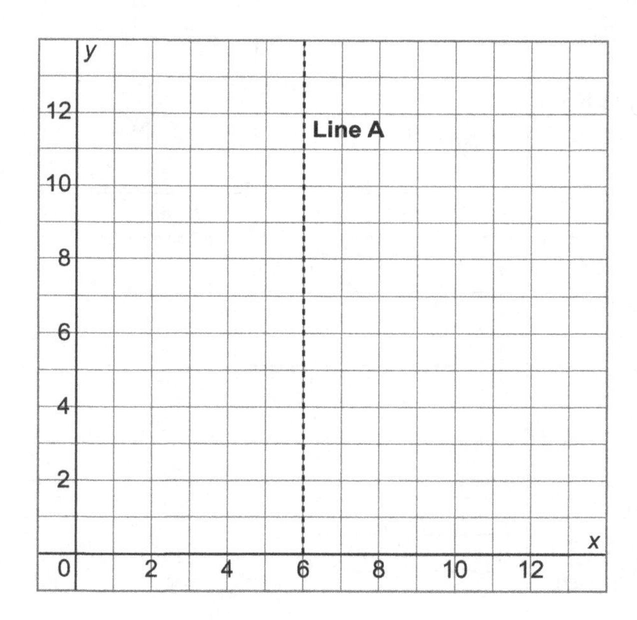

Does the logo have any symmetry?

____

____

____

____

Unit 7 Study Guide

Skill	Description	Example
Find the scale factor of a scale diagram.	Scale factor = $\frac{\text{length on scale diagram}}{\text{length on original diagram}}$ An enlargement has a scale factor > 1. A reduction has scale a factor < 1.	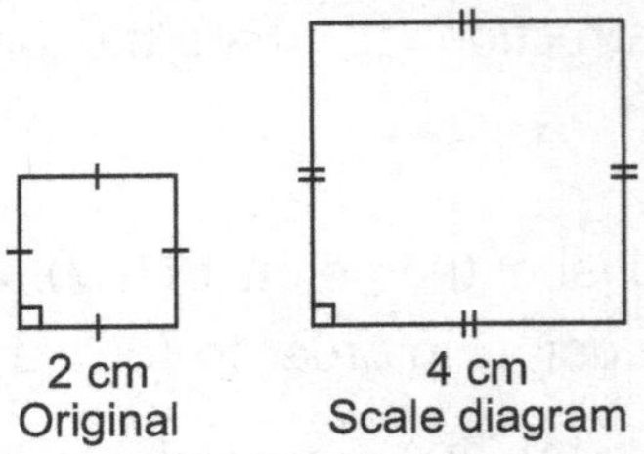 2 cm Original 4 cm Scale diagram Scale factor: $\frac{\text{length on scale diagram}}{\text{length on original diagram}} = \frac{4}{2} = 2$
Find out if two polygons are similar.	In two similar polygons: – matching angles are equal *and* – all pairs of matching sides have the same scale factor.	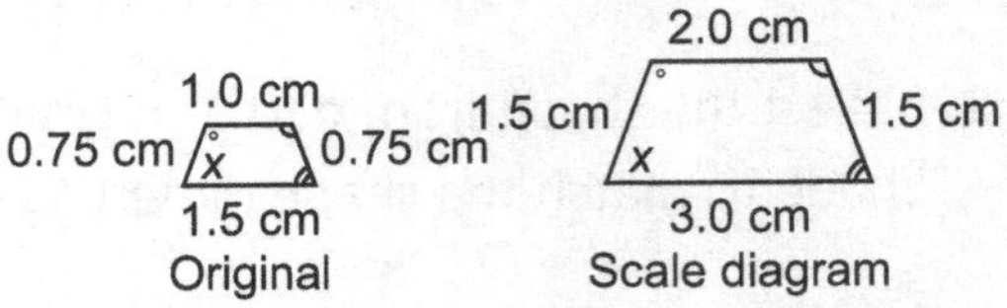 1.0 cm, 0.75 cm, x, 0.75 cm, 1.5 cm Original 2.0 cm, 1.5 cm, x, 1.5 cm, 3.0 cm Scale diagram
Find out if two triangles are similar.	In two similar triangles: – matching angles are equal *or* – all pairs of matching sides have the same scale factor.	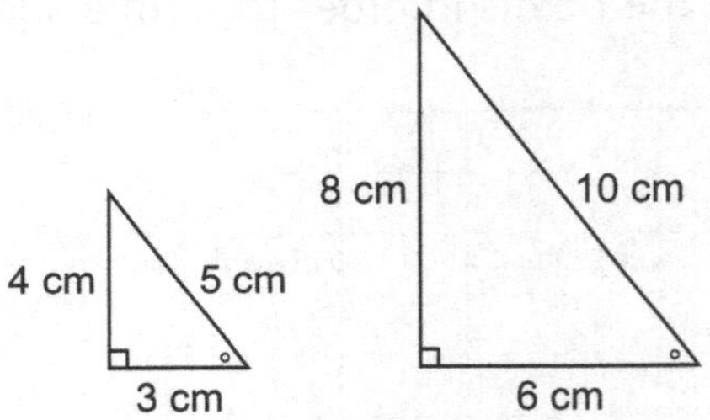 4 cm, 5 cm, 3 cm 8 cm, 10 cm, 6 cm
Identify lines of symmetry.	A line of symmetry divides a shape into 2 congruent parts. When one part is reflected in the line of symmetry, it matches the other part exactly.	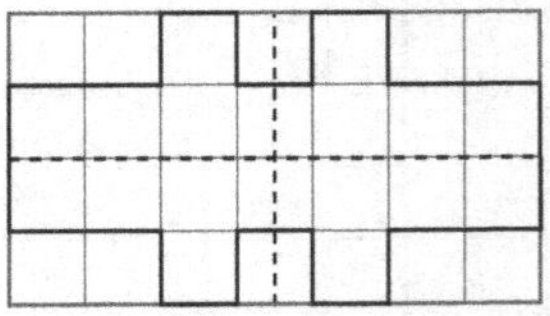
Find out if a shape has rotational symmetry.	A shape has rotational symmetry when it can be turned less than 360° about its centre to match itself exactly.	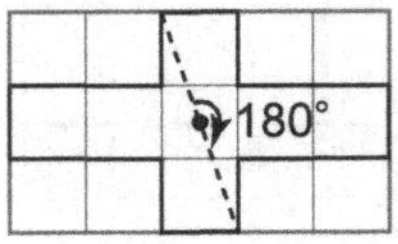 180°
Find the order of rotation and the angle of rotation symmetry for a polygon.	The number of times a shape matches itself in one complete turn is the order of rotation. The angle of rotation symmetry is: $\frac{360°}{\text{the order of rotation}}$	A square has order of rotation 4. 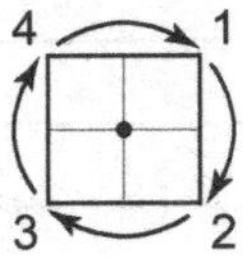 4 1 3 2 So, its angle of rotation symmetry is: $\frac{360°}{4} = 90°$

Unit 7 Review

7.1 **1.** A photo of a baby giraffe is to be enlarged for a newspaper.
The actual photo measures 4 cm by 6 cm.

Find the dimensions of the enlargement with a scale factor of $\frac{7}{2}$.

Write the scale factor as a decimal: $\frac{7}{2}$ = ________

Length of original photo: ________
Length of enlargement: ________ × ________ = ________

Width of original photo: ________
Width of enlargement: ________ × ________ = ________

The enlargement has dimensions ______________________.

7.2 **2.** Find the scale factor for this reduction.

Original

Reduction

Length of original line segment: ________ cm
Length of reduction: ________ cm

$$\text{Scale factor} = \frac{\text{length on reduction}}{\text{length on original}}$$

$$= \frac{\_\_\_\_\_\_}{\_\_\_\_\_\_}$$

$= \_\_\_\_\_\_$

The scale factor is ______.

7.3 **3.** Are these parallelograms similar?

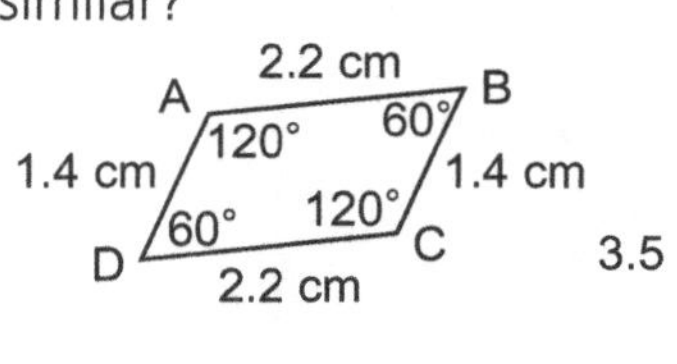

E
5.5 cm
F
120°
60°
3.5 cm
3.5 cm
120°
G
60°
H
5.5 cm

Check matching angles.

∠A = _____ = ______   ∠B = _____ = _____   __________________   __________________

All matching angles ______ equal.

Check matching sides.
The matching sides are: ______ and ______, and ______ and ______. Find the scale factors.

$$\frac{\text{length of } \_\_\_\_}{\text{length of } \_\_\_\_} = \frac{\_\_\_\_\_\_}{\_\_\_\_\_\_}$$

$= \_\_\_\_\_\_$

$$\frac{\text{length of } \_\_\_\_}{\text{length of } \_\_\_\_} = \frac{\_\_\_\_\_\_}{\_\_\_\_\_\_}$$

$= \_\_\_\_\_\_$

The scale factors ______ equal. So, the parallelograms ______ similar.

7.4 **4.** Are these two triangles similar?

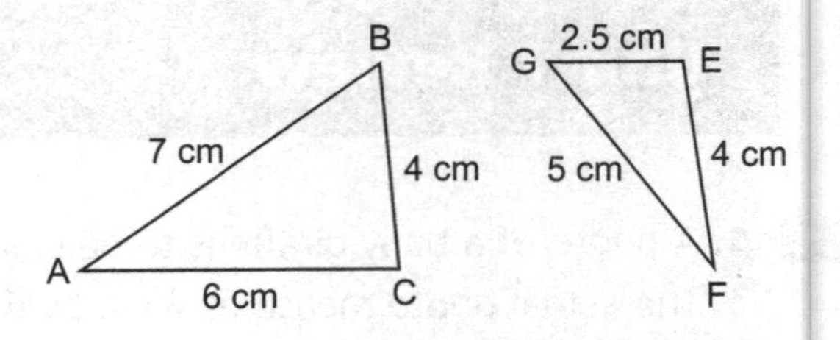

In $\triangle ABC$, order the sides from shortest to longest:

____

In $\triangle EFG$, order the sides from shortest to longest:

____

Find the scale factors of matching sides.

$\frac{\text{length of } \_\_\_\_}{\text{length of } \_\_\_\_} = \frac{\_\_\_\_\_\_}{\_\_\_\_\_\_} = \_\_\_\_\_\_$

$\frac{\text{length of } \_\_\_\_}{\text{length of } \_\_\_\_} = \frac{\_\_\_\_\_\_}{\_\_\_\_\_\_} = \_\_\_\_\_\_$

$\frac{\text{length of } \_\_\_\_}{\text{length of } \_\_\_\_} = \frac{\_\_\_\_\_\_}{\_\_\_\_\_\_} = \_\_\_\_\_\_$

All scale factors are ____________. So, the triangles ________________.

5. Triangle EFG is similar to $\triangle JKL$.
Find the length of JK.

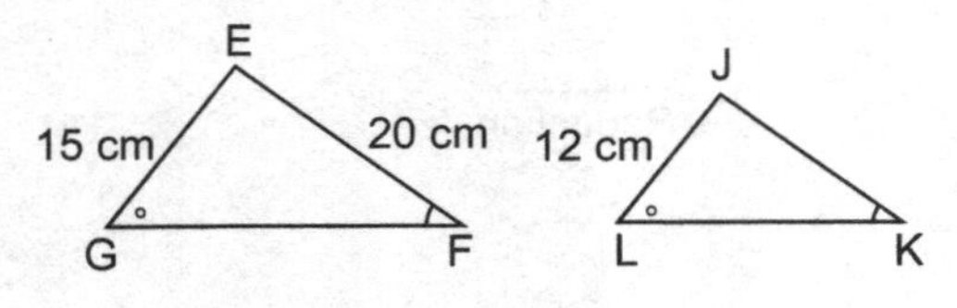

________ is a reduction of ________.

Choose a pair of matching sides whose lengths are both known:

____

$\text{Scale factor} = \frac{\text{length on reduction}}{\text{length on original}}$

$= \frac{\_\_\_\_\_\_}{\_\_\_\_\_\_}$

$= \_\_\_\_\_\_$

The scale factor is ________.

Use the scale factor to find the length of JK.
JK and EF are matching sides.
Length of EF: ________
Scale factor: ________
Length of JK: ________________________
So, JK has length ________.

7.5 **6.** Draw the lines of symmetry in each tessellation.

a)

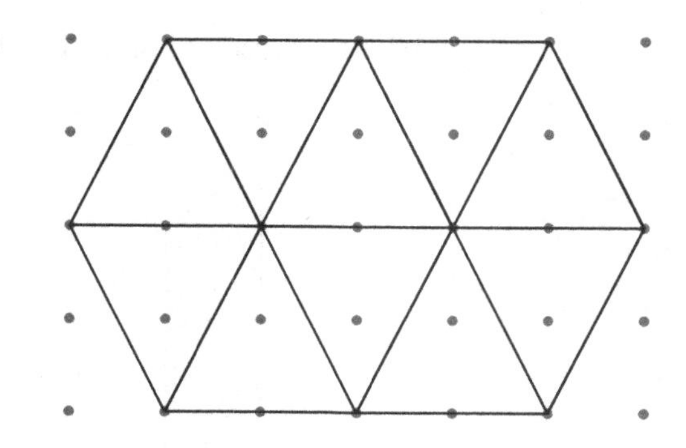

b)

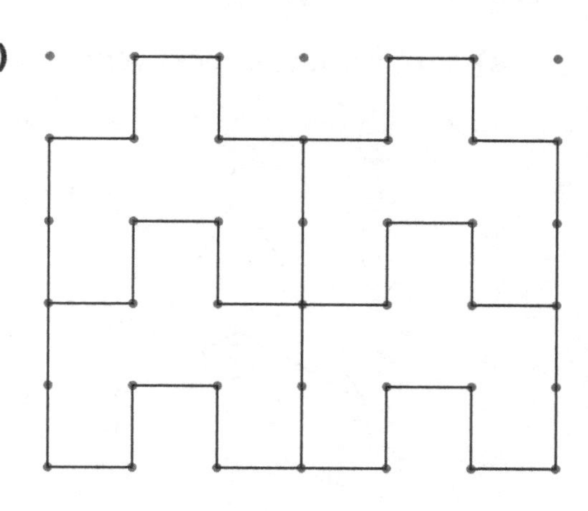

7. Reflect the shape in the line of reflection to make a larger shape.

Point	Image
P(___, ___)	________
Q(___, ___)	________
R(___, ___)	________
S(___, ___)	________
T(___, ___)	________
U(___, ___)	________

y
line of reflection
6
P Q
R
4 S
2 T
U
x
0 2 4 6

7.6 **8.** Find the order of rotational symmetry and the angle of rotation symmetry for this shape.

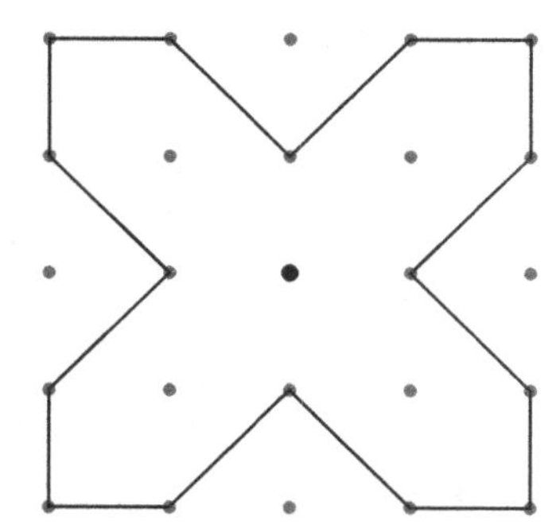

The shape and its image match ____ times.
So, the shape has rotational symmetry of order ____.
Angle of rotation symmetry is:

$$\frac{360°}{\text{the order of rotation}} = \frac{360°}{\_\_\_\_\_\_}$$

$$= \_\_\_\_\_\_$$

9. Draw the image after each rotation.

a) 120° clockwise about vertex B

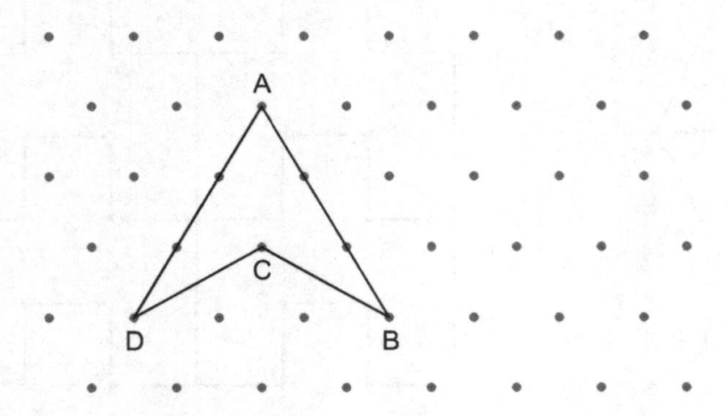

b) 180° about vertex L

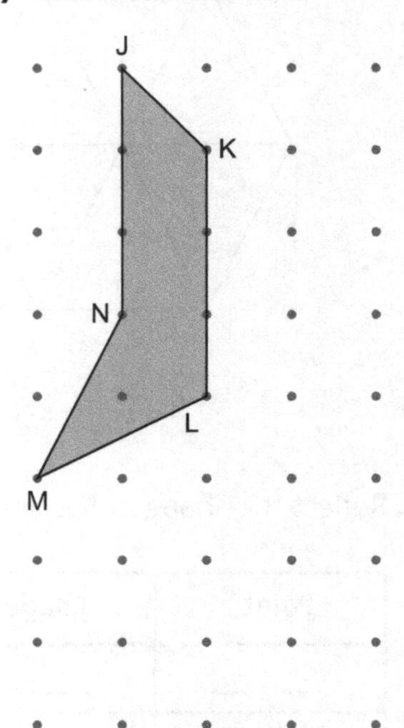

7.7 **10.** Find out if the polygons are related by symmetry.
Use tracing paper and a Mira to help.

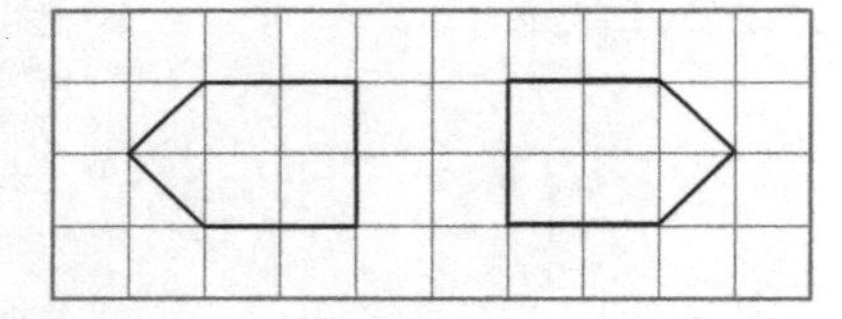

Do the polygons face opposite ways? ________
So, are the polygons related by a reflection? ________
Draw and label the line of reflection.

Do the polygons touch? ________
So, try a point of rotation ________ the polygons.
Are the polygons related by a rotation? ________
If they are, label the point of rotation.

11. a) Reflect the polygon in the vertical line through 3 on the *x*-axis.
Draw and label the image.

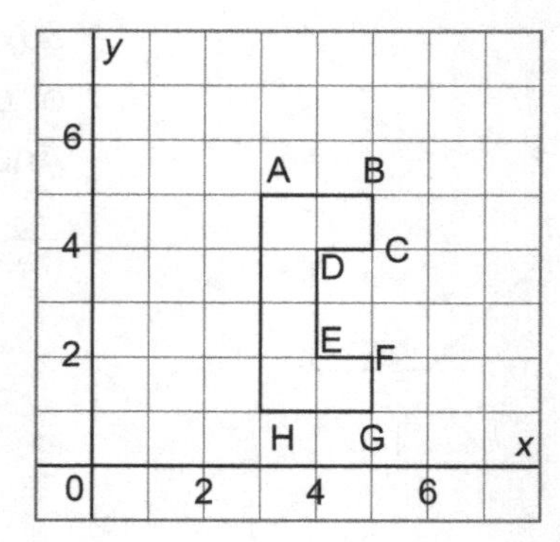

b) Describe the symmetry in the shape that results.

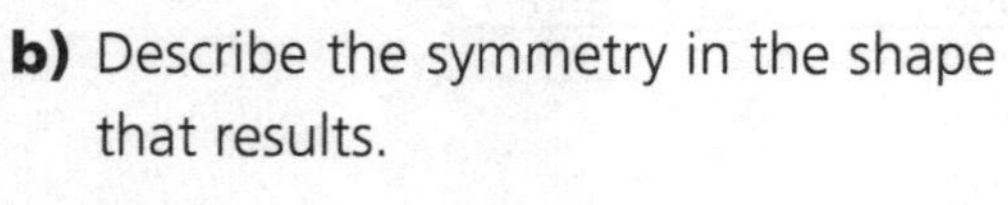

The shape has _____ lines of symmetry:
Draw and label any lines of symmetry you found.
Does the shape have rotational symmetry? ________
If it does, label the point of rotation.

UNIT 8

Circle Geometry

What You'll Learn

How to

- Solve problems involving tangents to a circle
- Solve problems involving chords of a circle
- Solve problems involving the measures of angles in a circle

Why Is It Important?

Circle properties are used by

- artists, when they create designs and logos

Key Words

radius (radii)
right angle
tangent
point of tangency
diameter
right triangle
isosceles triangle
chord
perpendicular bisector
central angle
inscribed angle
arc
subtended
semicircle

8.1 Skill Builder

Solving for Unknown Measures in Triangles

Here are 2 ways to find unknown measures in triangles.

Angle Sum Property

In any triangle:

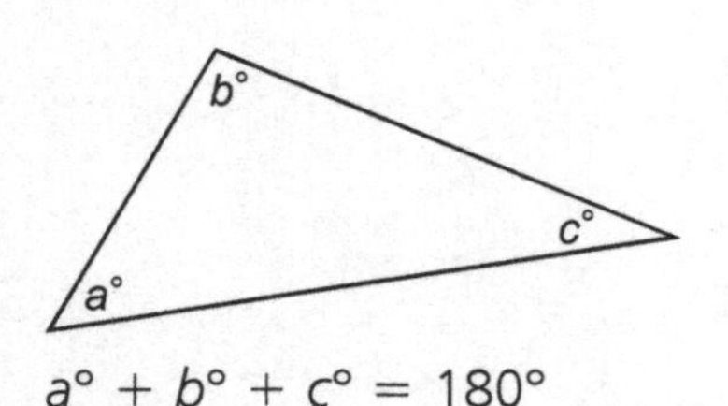

$a° + b° + c° = 180°$

Pythagorean Theorem

In any right △PQR:

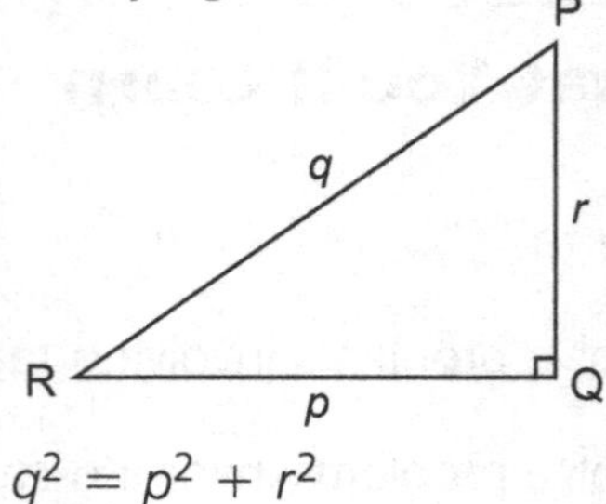

$q^2 = p^2 + r^2$

Here is how to find the unknown measures in right △PQR.

In △PQR, the angles add up to 180°.
To find $x°$, start at 180° and subtract the known measures.

$x° = 180° - 90° - 60°$
$= 30°$

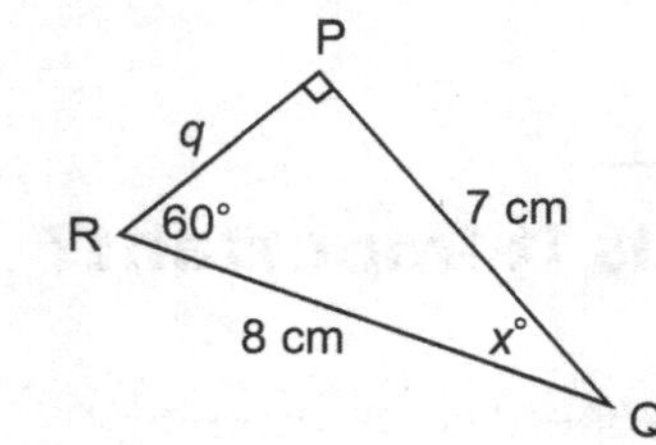

By the Pythagorean Theorem:

$QR^2 = PR^2 + PQ^2$
$8^2 = q^2 + 7^2$
So: $q^2 = 8^2 - 7^2$
$q = \sqrt{8^2 - 7^2}$
$\doteq 3.87$

So, $x°$ is 30° and q is about 4 cm.

Answer to the same degree of accuracy as the question uses.

Check

1. Find each unknown measure.

a)

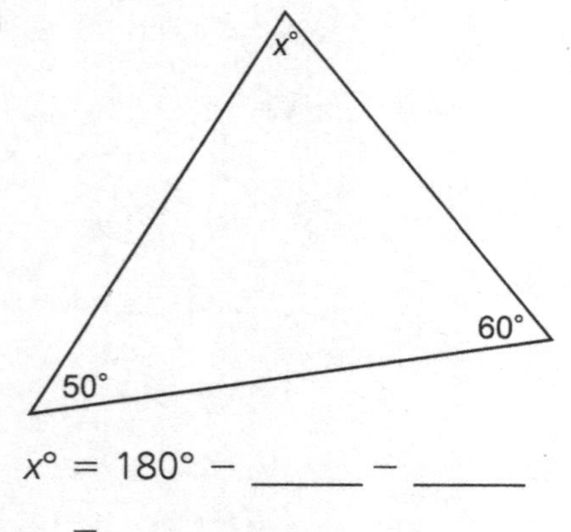

$x° = 180° -$ _____ $-$ _____
$=$ _____

b)

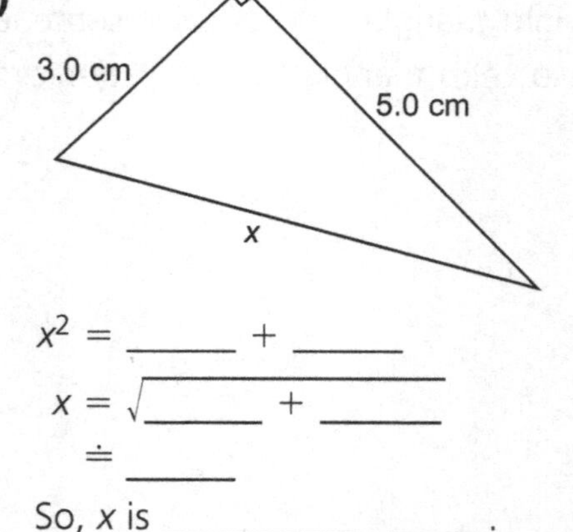

$x^2 =$ _____ $+$ _____
$x = \sqrt{\_\_\_\_\_ + \_\_\_\_\_}$
$\doteq$ _____

So, x is _______________.

8.1 Properties of Tangents to a Circle

FOCUS **Use the relationship between tangents and radii to solve problems.**

A **tangent** touches a circle at exactly one point.

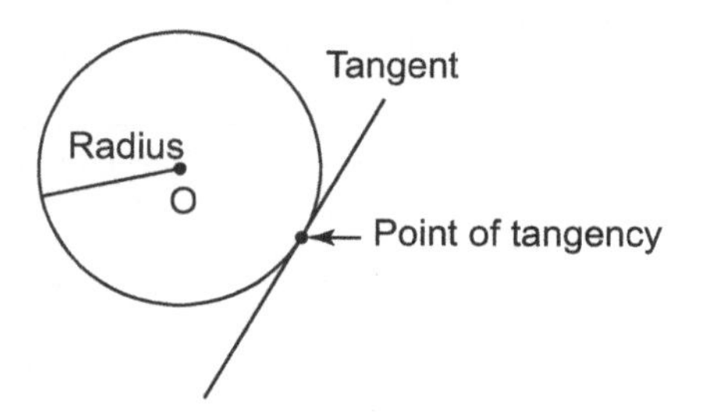

Tangent-Radius Property

A tangent to a circle is perpendicular to the radius drawn to the point of tangency.
So, $OP \perp AB$, $\angle OPA = 90°$ and $\angle OPB = 90°$

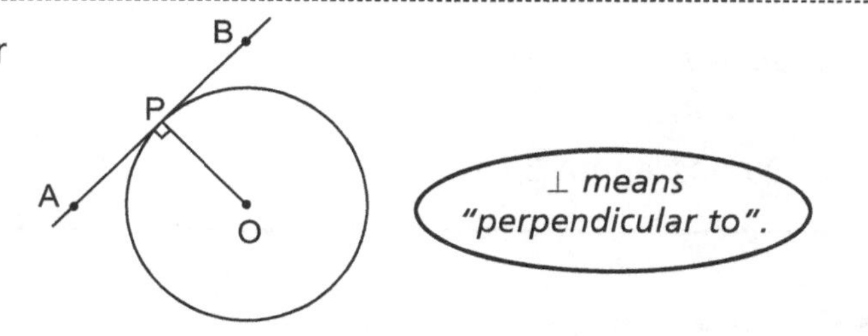

Example 1 Finding the Measure of an Angle in a Triangle

BP is tangent to the circle at P.
O is the centre of the circle.
Find the measure of $x°$.

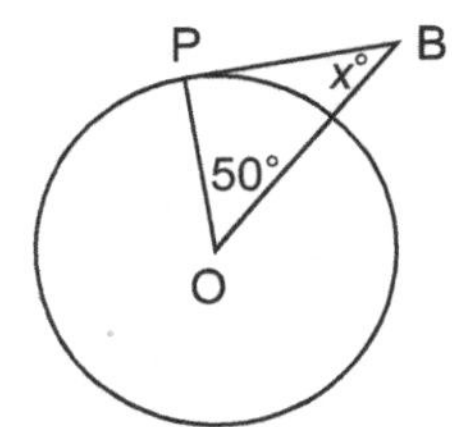

Solution

By the tangent-radius property: $\angle OPB = 90°$
Since the sum of the angles in $\triangle OPB$ is 180°:
$x° = 180° - 90° - 50°$
$= 40°$
So, $x°$ is 40°.

Check

1. Find the value of $x°$.

$\angle$______ $= 90°$
$x° = 180° -$ _____ $-$ _____
$=$ _____

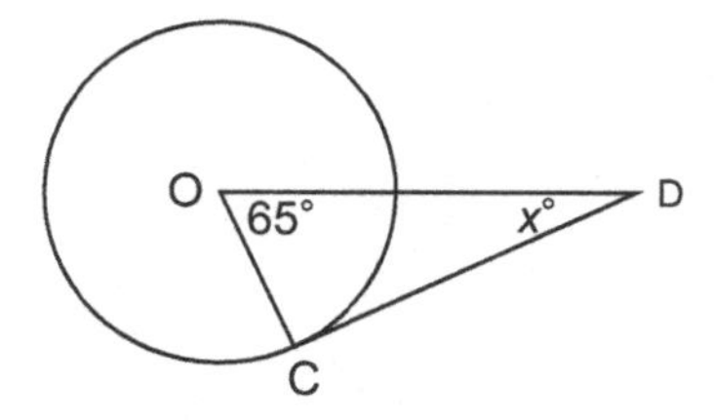

Example 2 Using the Pythagorean Theorem in a Circle

MB is a tangent to the circle at B. O is the centre.
Find the length of radius OB.

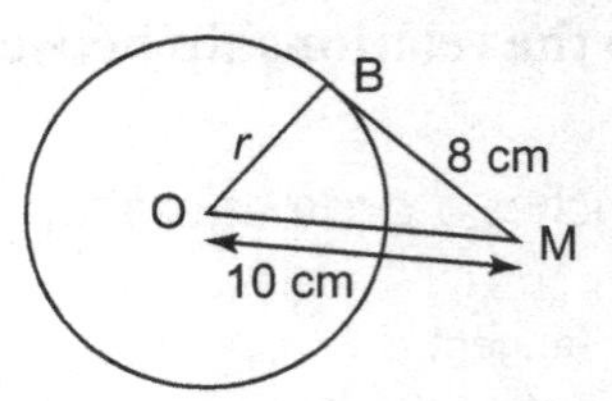

Solution

By the tangent-radius property: $\angle OBM = 90°$
By the Pythagorean Theorem in right $\triangle MOB$:

$$OM^2 = OB^2 + BM^2$$
$$10^2 = r^2 + 8^2$$
$$100 = r^2 + 64$$
$$100 - 64 = r^2$$
$$36 = r^2$$
$$\sqrt{36} = r$$
$$r = 6$$

Radius OB has length 6 cm.

Check

1. ST is a tangent to the circle at S. O is the centre.
Find the length of radius OS.
Answer to the nearest millimetre.

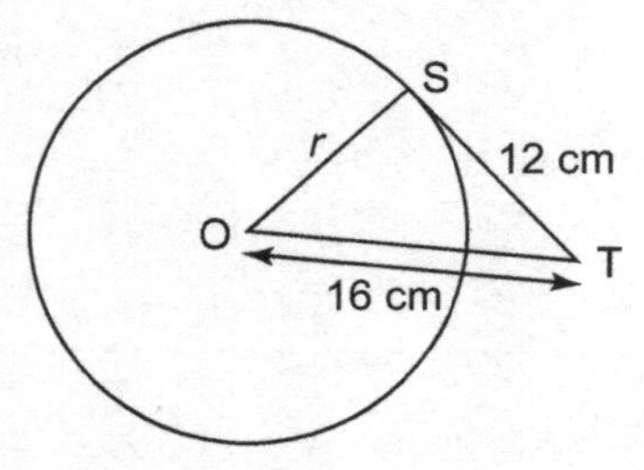

$\angle OST =$ _____ By the tangent-radius property

$OT^2 =$ _____ + _____ By the Pythagorean Theorem

_____ $= r^2 +$ _____

_____ $= r^2 +$ _____

_____ − _____ $= r^2$

_____ $= r^2$

$\sqrt{\_\_\_\_\_} = r$

$r \doteq$ _____

OS is about _____ cm long.

Practice

In each question, O is the centre of the circle.

1. From the diagram, identify:

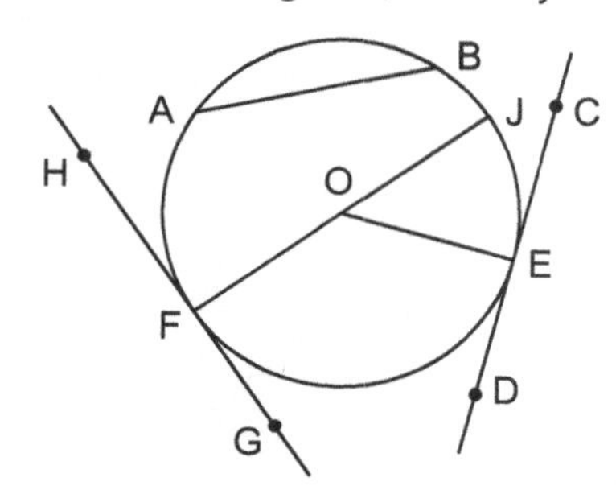

a) 3 radii _______, _______, _______

b) 2 tangents _______, _______

c) 2 points of tangency _______, _______

d) 4 right angles ∠_______, ∠_______, ∠_______, ∠_______

2. What is the measure of each angle?

a)

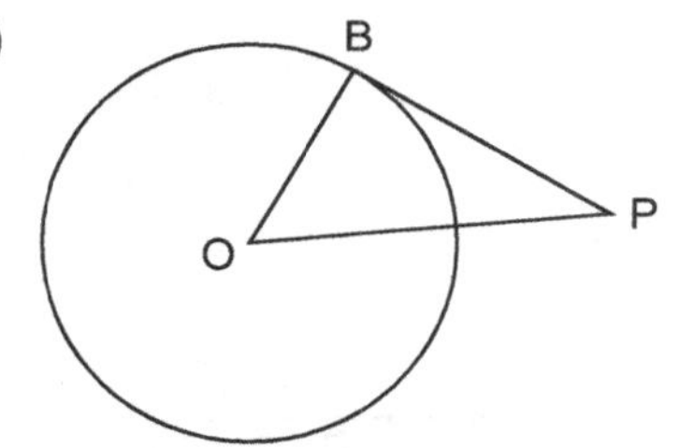

∠OBP = _____

b)

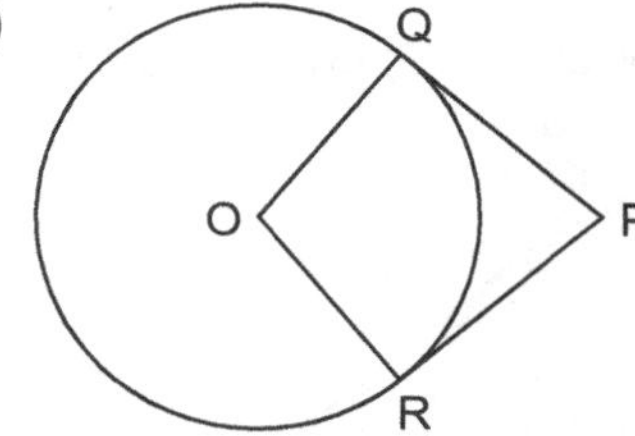

∠PQO = _____ ∠PRO = _____

3. Find each value of $x°$.

a)

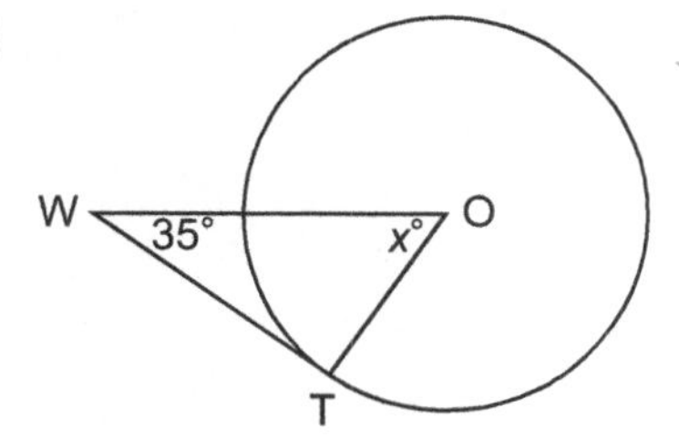

∠OTW = ______

$x°$ = 180° − ______ − ______

= ______

b)

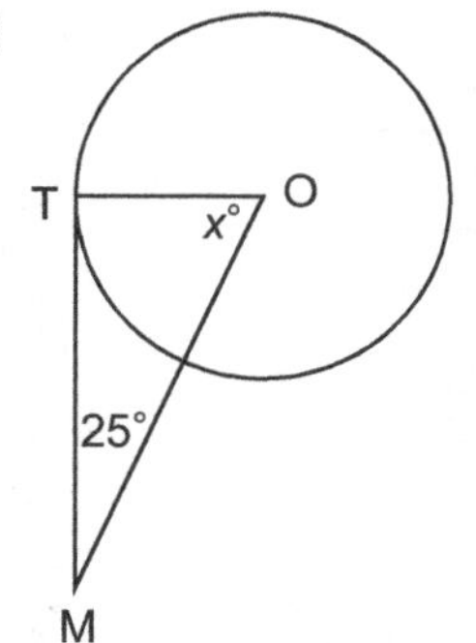

____

____

= ______

4. Find each value of x.

Answer to the nearest tenth of a unit.

a)

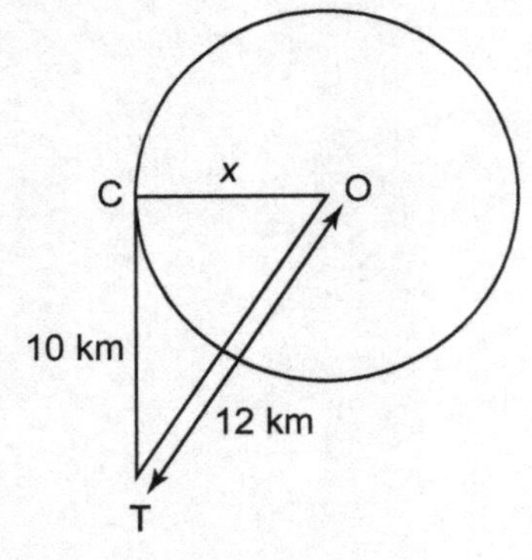

$\angle OCT = 90°$ By the tangent-radius property

$\_\_\_\_ = x^2 + \_\_\_\_$ By the Pythagorean Theorem in $\triangle OCT$

$\_\_\_\_ = x^2 + \_\_\_\_$

$\_\_\_\_ - \_\_\_\_ = x^2$

$\_\_\_\_ = x^2$

$\sqrt{\_\_\_\_} = x$

$x \doteq \_\_\_\_$

So, OC is about ____ km.

b)

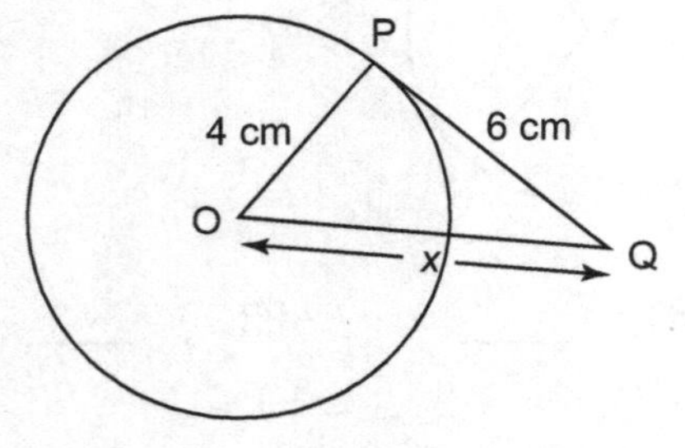

$\angle OPQ = \_\_\_\_$, and:

$x^2 = \_\_\_\_ + \_\_\_\_$

$x^2 = \_\_\_\_ + \_\_\_\_$

$x^2 = \_\_\_\_$

$x = \sqrt{\_\_\_\_}$

$x \doteq \_\_\_\_$

So, OQ is about ____ cm.

c)

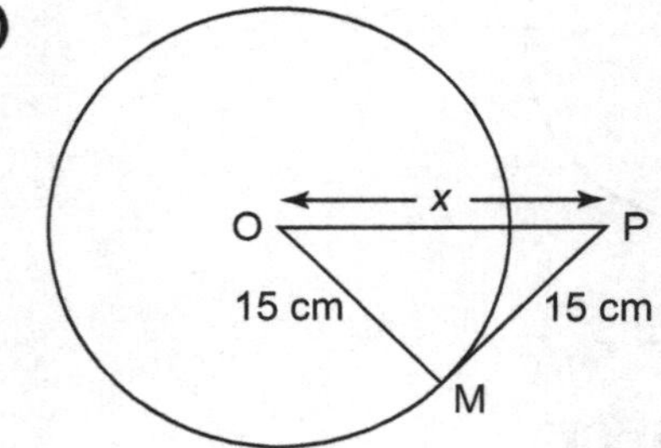

$x^2 = \_\_\_\_ + \_\_\_\_$

$x^2 = \_\_\_\_ + \_\_\_\_$

$x^2 = \_\_\_\_$

$x = \sqrt{\_\_\_\_}$

$x \doteq \_\_\_\_$

So, OP is about ____ cm.

8.2 Properties of Chords in a Circle

FOCUS **Use chords and related radii to solve problems.**

A **chord** of a circle joins 2 points on the circle.

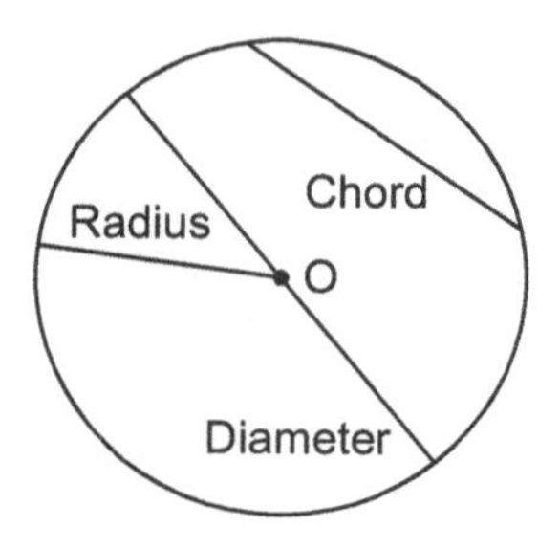

Chord Properties

In any circle with centre O and chord AB:

- If OC bisects AB, then OC ⊥ AB.
- If OC ⊥ AB, then AC = CB.
- The perpendicular bisector of AB goes through the centre O.

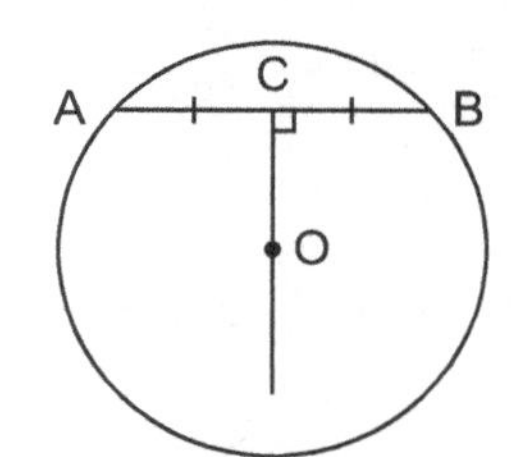

Example 1 Finding the Measure of Angles in a Triangle

Find $x°$, $y°$, and $z°$.

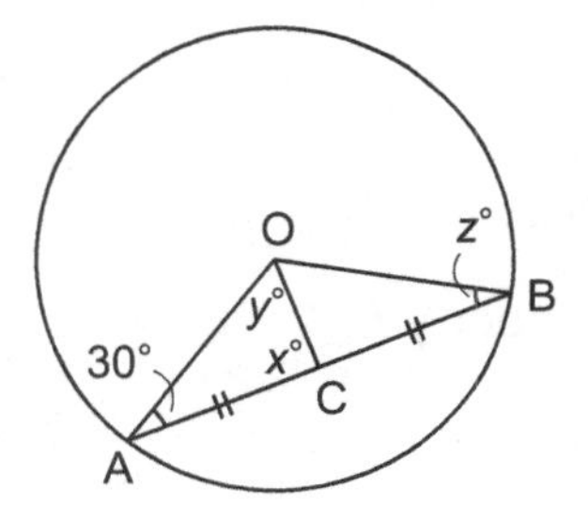

Solution

OC bisects chord AB, so OC ⊥ AB
Therefore, $x° = 90°$

By the angle sum property in △OAC:
$y° = 180° - 90° - 30°$
$= 60°$
Since radii are equal, OA = OB, and △OAB is isosceles.
∠OBA = ∠OAB

So, $z° = 30°$

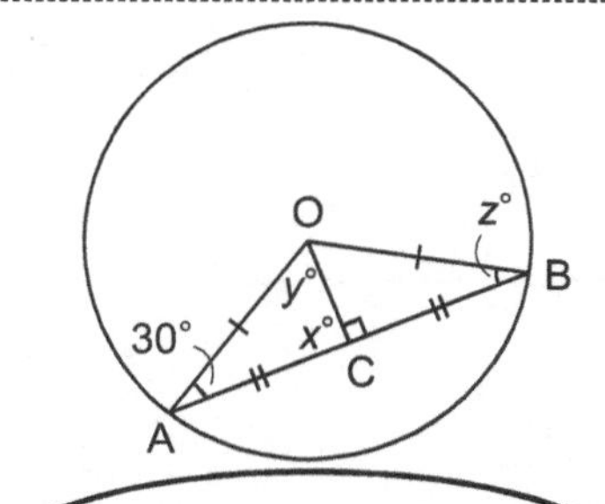

In an isosceles triangle, 2 base angles are equal.

Check

1. Find the values of $x°$ and $y°$.

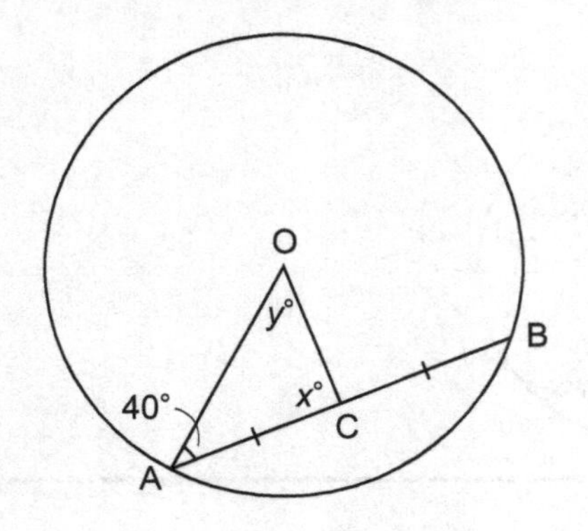

____ ⊥ ____ So, $x°$ = ____ By the chord properties

$y°$ = ____ − ____ − ____ By the angle sum property

= ____

2. Find the values of $x°$, $y°$, and $z°$.

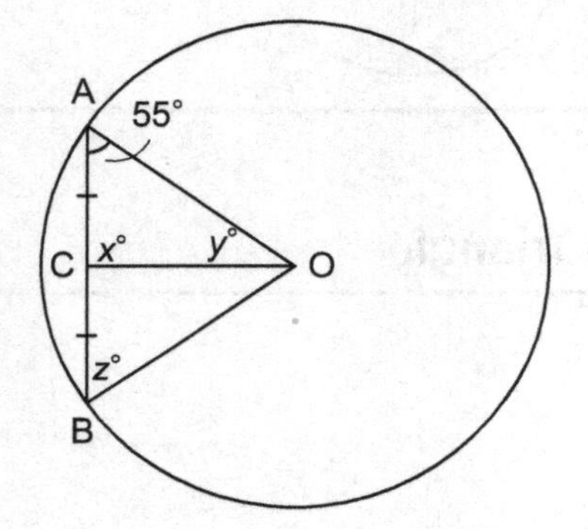

____ ⊥ ____ So, $x°$ = ____ By the chord properties

$y°$ = ____ − ____ − ____ By the angle sum property

= ____

Since OA = ____,

△____ is isosceles and ∠____ = ∠____

So, $z°$ = ____

Example 2 Using the Pythagorean Theorem in a Circle

O is the centre of the circle.
Find the length of chord AB.

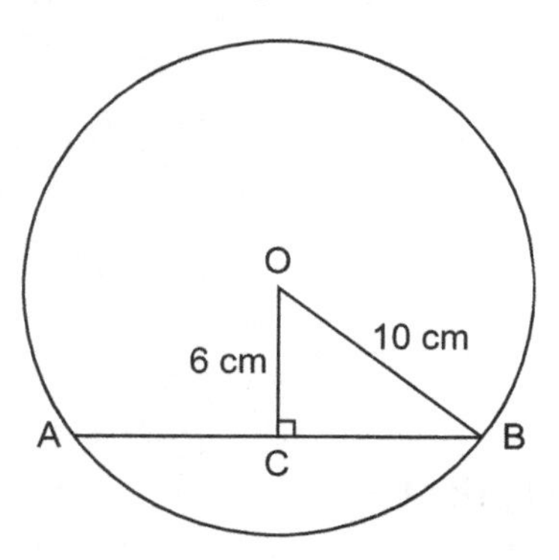

Solution

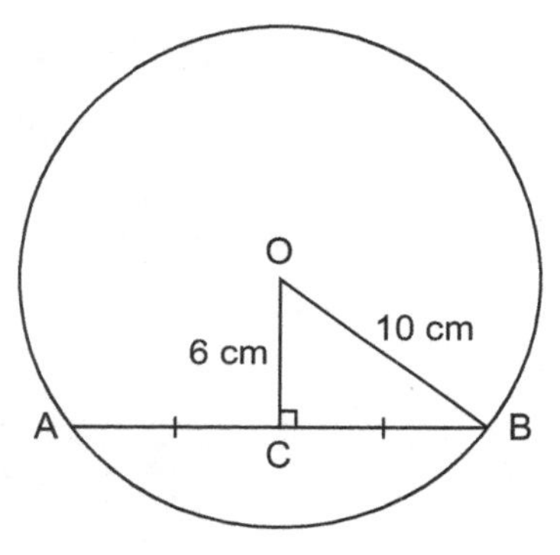

$$
\begin{aligned}
10^2 &= 6^2 + BC^2 \\
100 &= 36 + BC^2 \\
100 - 36 &= BC^2 \\
64 &= BC^2 \\
BC &= \sqrt{64} \\
&= 8
\end{aligned}
$$

By the Pythagorean Theorem in right $\triangle OCB$

So, BC = 8 cm

Since $OC \perp AB$, OC bisects AB. By the chord properties

So, AC = BC = 8 cm

The length of chord AB is: 2×8 cm = 16 cm

Check

1. Find the values of *a* and *b*.

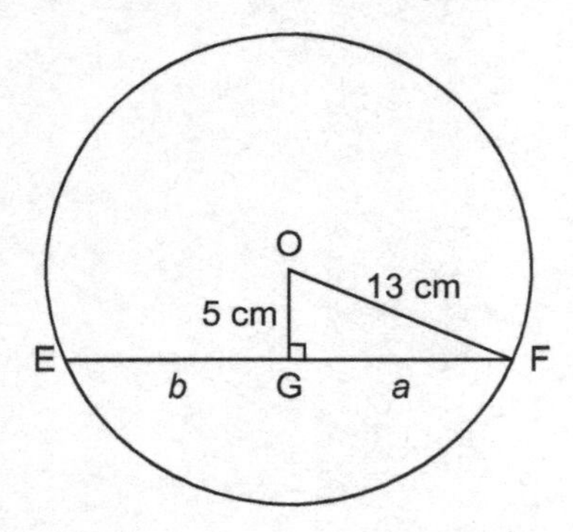

____ = ____ + a^2 By the Pythagorean Theorem in right $\triangle$OFG

____

____

____

____

____

So, a = ____ cm

____ = ____ By the chord properties

So, b = ____ cm

Practice

In each diagram, O is the centre of the circle.

1. Name all radii, chords, and diameters.

a)

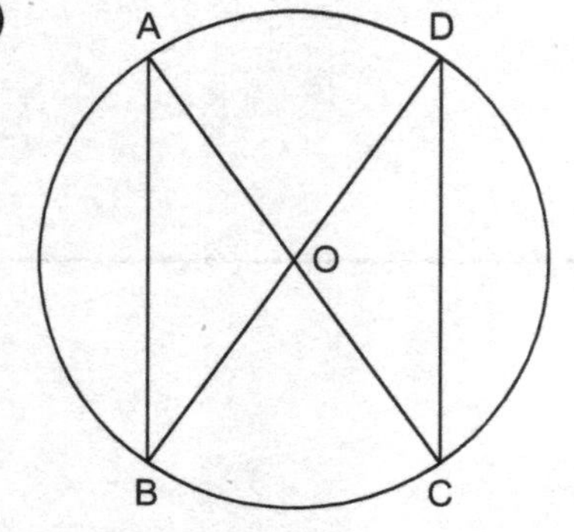

Radii: ________________

Chords: ________________

Diameters: ________________

b)

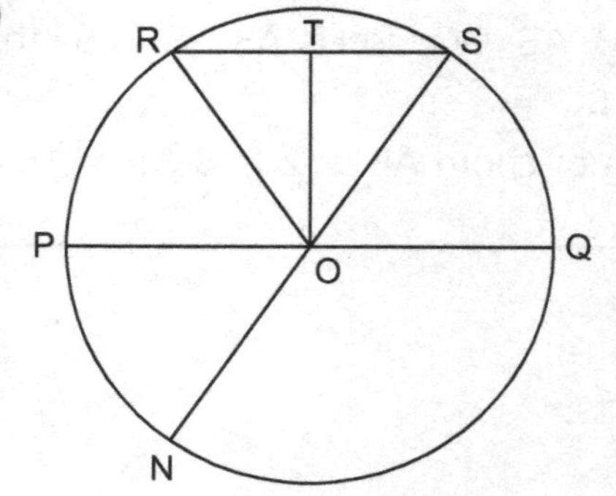

Radii: ________________

Chords: ________________

Diameters: ________________

2. On each diagram, mark line segments with equal lengths.
Then find each value of *a*.

a)

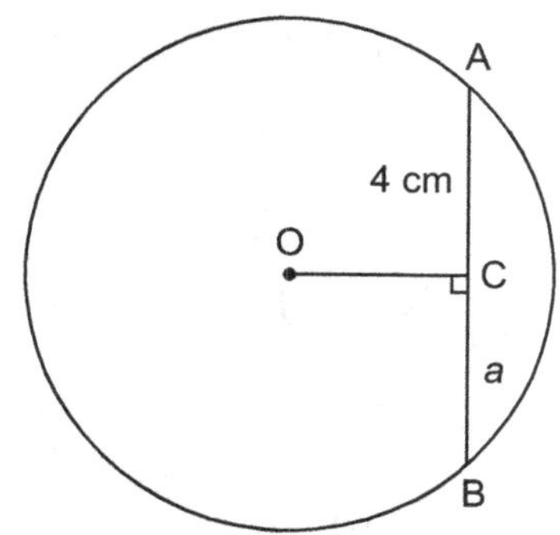

AC = CB = ______ cm
So, a = ______ cm

b)

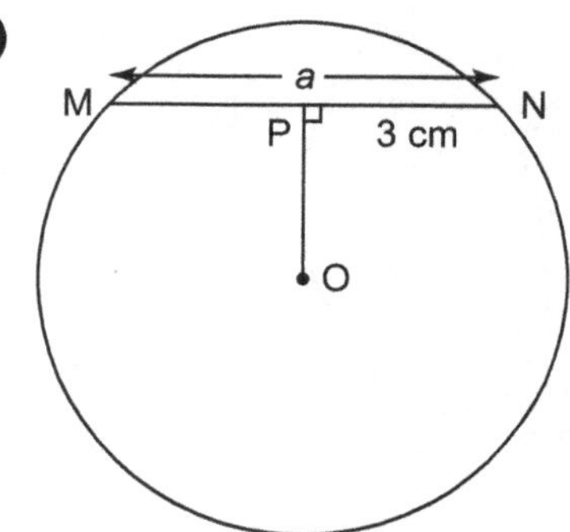

MN = 2 × ______
= 2 × ______ cm
= ______ cm
So, a = ______

c)

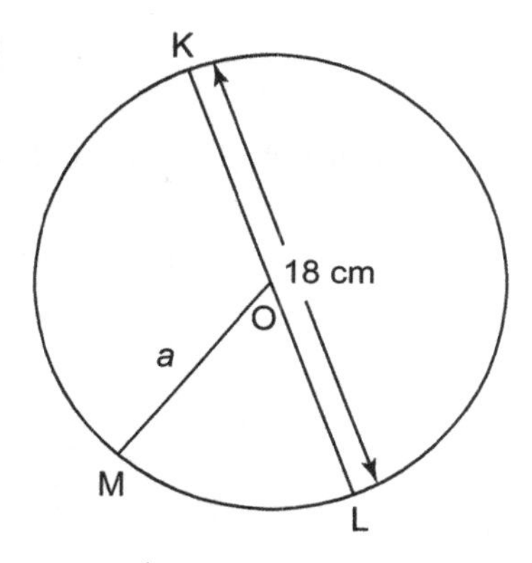

OL = $\frac{1}{2}$ × ______
= $\frac{1}{2}$ × ______ cm
= ______ cm
So, a = ______

d)

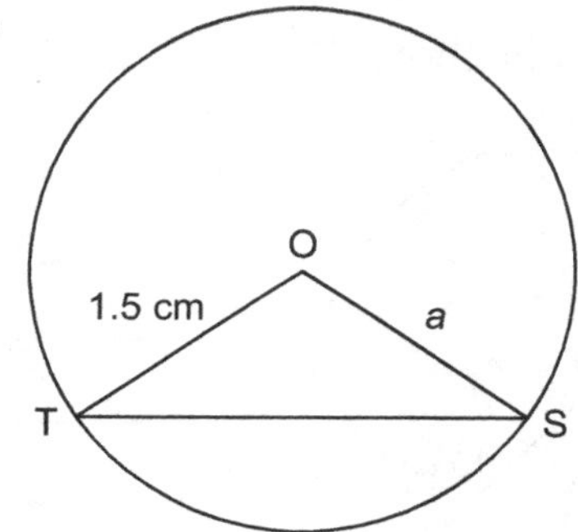

OS = ______ = ______ cm
So, a = ______

3. Find each value of $x°$ and $y°$.

a)

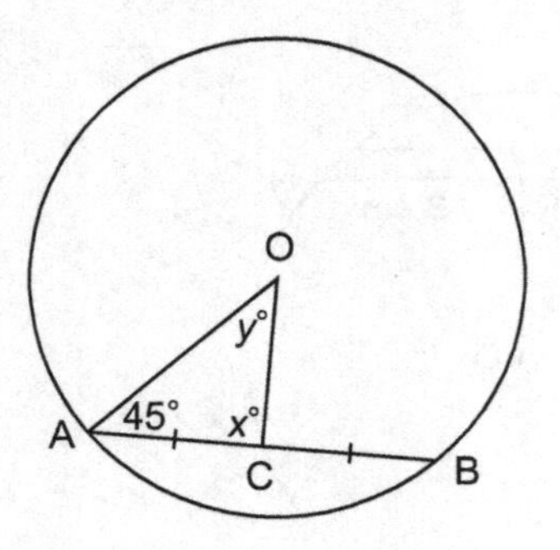

$x° =$ ______

$y° = 180° -$ ______ $-$ ______

$=$ ______

b)

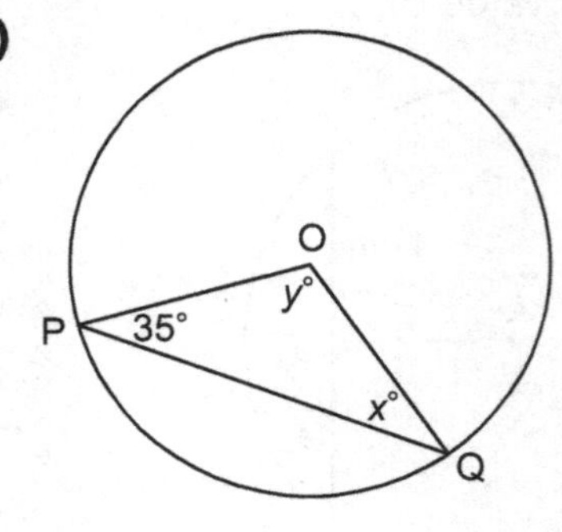

______ $=$ ______ $\triangle OPQ$ is ______________

$\angle$______ $= \angle$______

So, $x° =$ ______

$y° = 180° -$ ______ $-$ ______

$=$ ______

4. Find the length of chord BC.

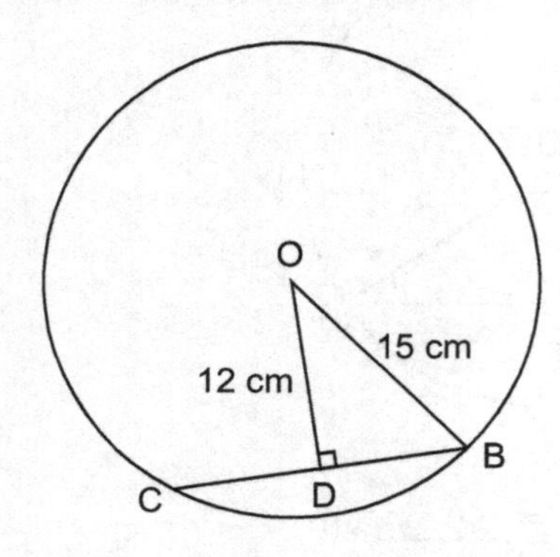

______ $=$ ______ $+ DB^2$ By the Pythagorean Theorem

______ $=$ ______ $+ DB^2$

____

____

____

____

So, DB = ______ cm

______ $=$ ______ $=$ ______ cm By the chord properties

So, chord BC has length: $2 \times$ ______ cm $=$ ______ cm

5. Find ON.

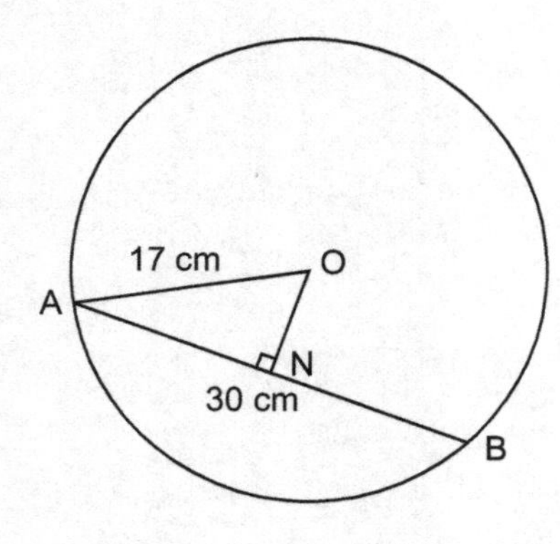

$AN = \frac{1}{2} \times$ ______

$= \frac{1}{2} \times$ ______ cm By the chord properties

$=$ ______ cm

______ $=$ ______ $+ ON^2$ By the Pythagorean Theorem

______ $=$ ______ $+ ON^2$

____

____

____

____

____

So, ON is ______ cm.

Can you ...

- Solve problems using tangent properties?
- Solve problems using chord properties?

8.1 In each diagram, O is the centre of the circle.
Assume that lines that appear to be tangent are tangent.

1. Name the angles that measure 90°.

a)

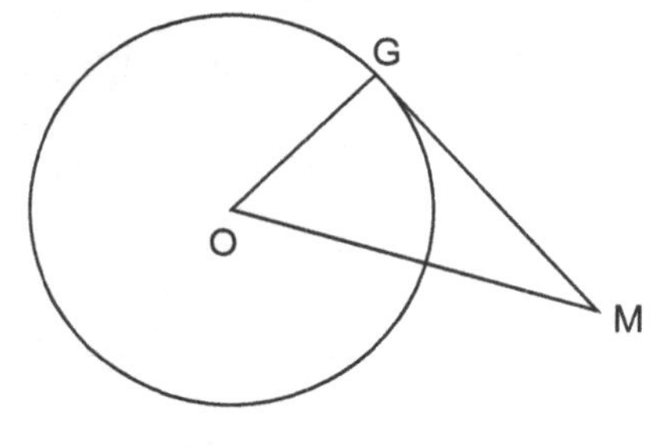

____

b)

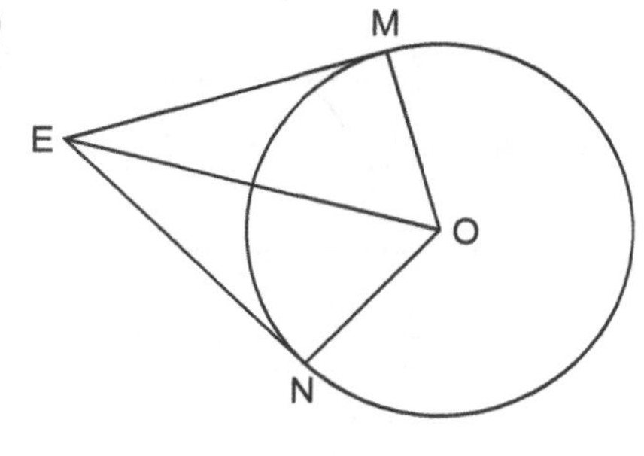

____

2. Find the unknown angle measures.

a)

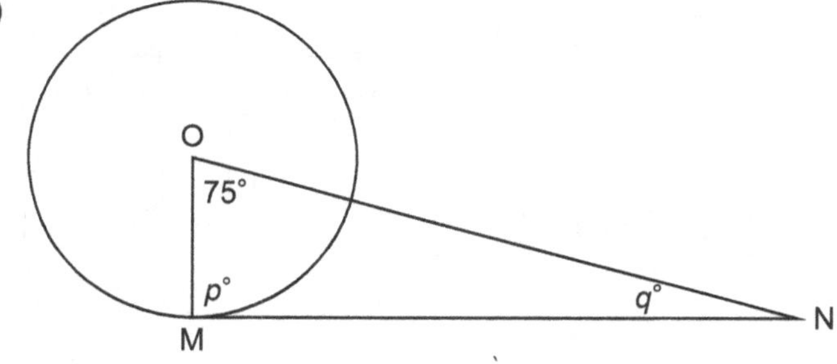

$p° =$ _____ Tangent-radius property

$q° = 180° -$ _____ $-$ _____ Angle sum property

$q° =$ _____

b)

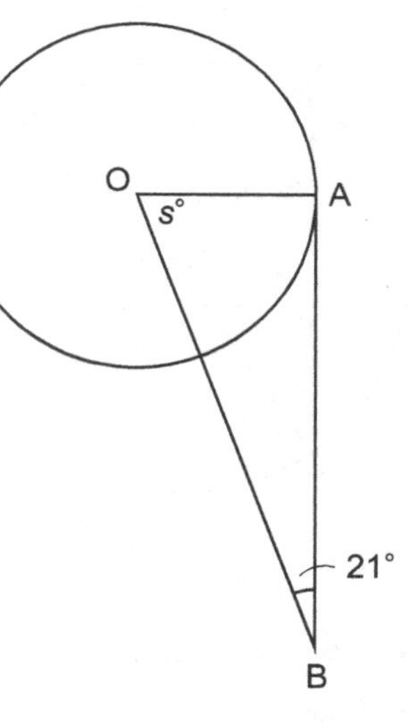

______ $= 90°$

$s° =$ _____ $-$ _____ $-$ _____

$s° =$ _____

3. Find the values of a and b to the nearest tenth.

a)

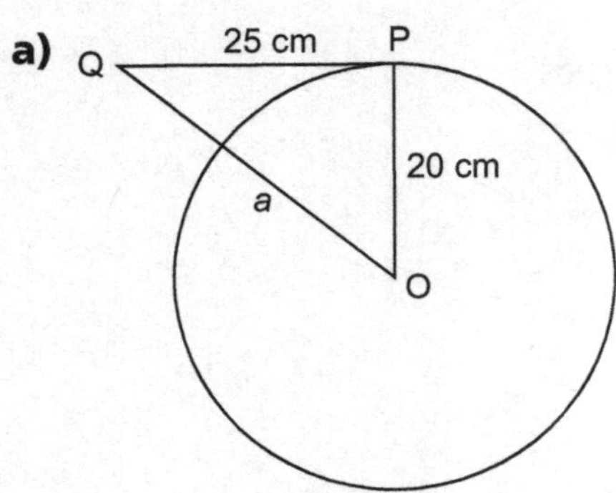

$\angle OPQ =$ ______ By the tangent-radius property

OQ is ____________ of $\triangle OPQ$.

$a^2 =$ ______ $+$ ______ By the Pythagorean Theorem

____

____

____

So, $a \doteq$ ______ cm

b)

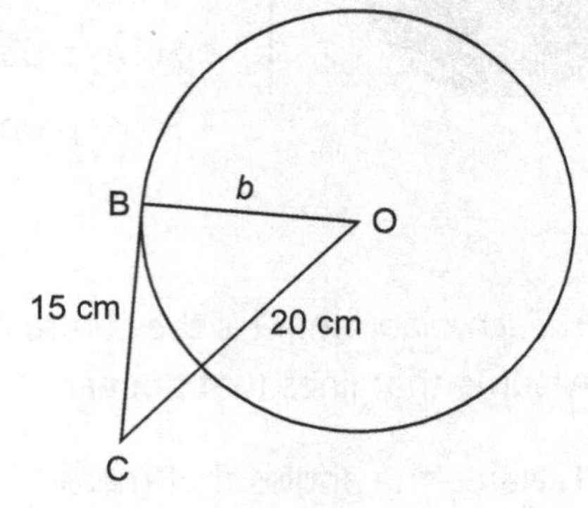

$\angle OBC =$ ______

OB is ______ of $\triangle OBC$.

______ $= b^2 +$ ______

____

____

____

____

So, $b \doteq$ ______ cm

8.2 **4.** Find the unknown measures.

a)

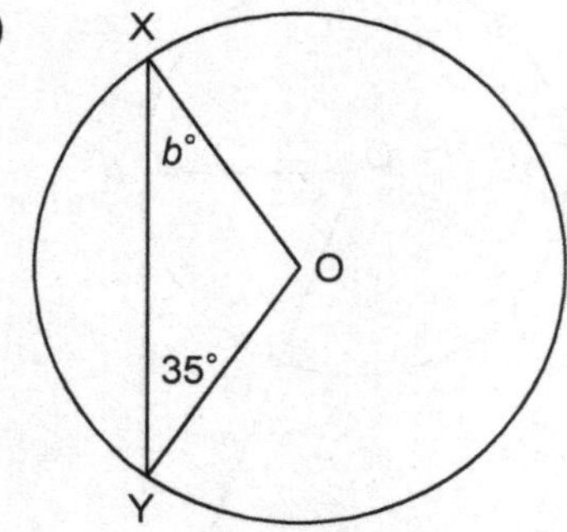

OX $=$ ______

$\triangle OXY$ is __________.

So, $b° =$ ______

b)

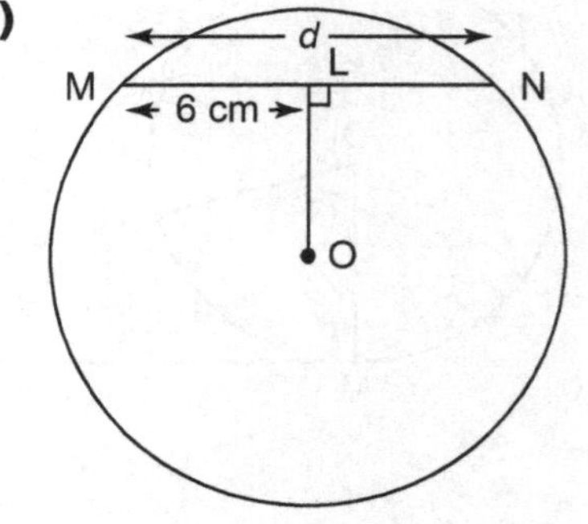

MN $= 2 \times$ ______

MN $=$ ______ $\times$ ______ cm

$=$ ______ cm

So, $d =$ ______ cm

5. Find each value of $x°$, $y°$, and $z°$.

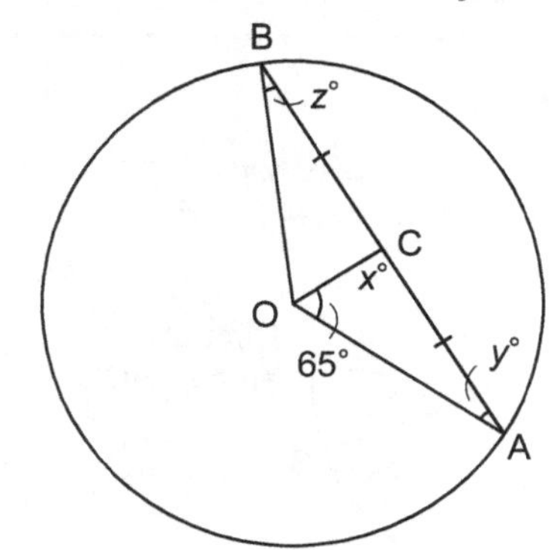

$x° =$ ______ By the chord properties

$y° =$ ______ $-$ ______ $-$ ______ By the angle sum property

$=$ ______

______ $=$ ______, so ______ is isosceles.

$\angle$______ $= \angle$______

So, $z° =$ ______

6. Find the length of OP.

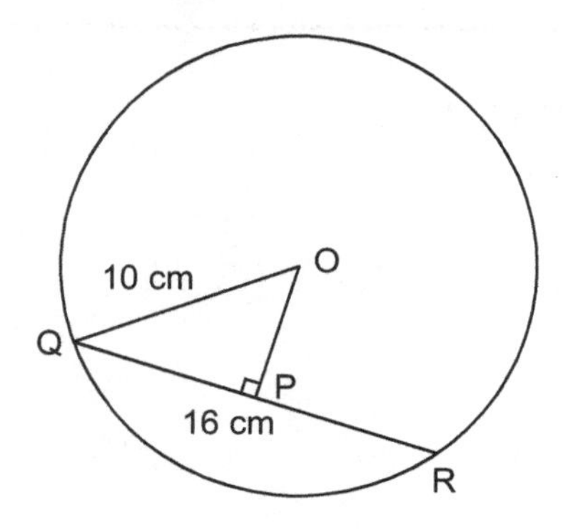

$QP = \frac{1}{2} \times QR$ By the chord properties

$= \frac{1}{2} \times$ ______ cm

$=$ ______ cm

$OQ^2 =$ ______ $+ OP^2$ By the Pythagorean Theorem

______ $=$ ______ $+ OP^2$

______

______

______

______

$=$ ______

So, the length of OP is ______ cm.

8.3 Properties of Angles in a Circle

FOCUS Use inscribed angles and central angles to solve problems.

In a circle:

- A **central angle** has its vertex at the centre.
- An **inscribed angle** has its vertex on the circle.

Both angles in the diagram are **subtended** by **arc** AB.

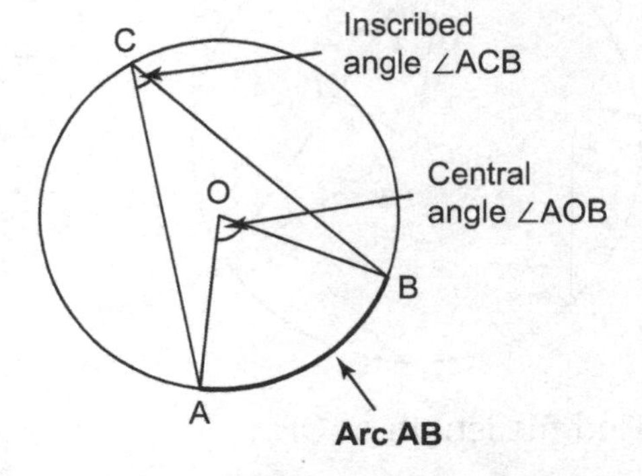

Central Angle and Inscribed Angle Property

The measure of a central angle is twice the measure of an inscribed angle subtended by the same arc.

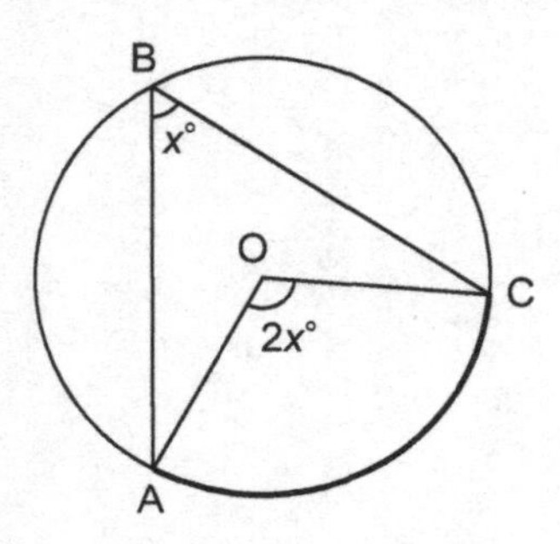

So, $\angle AOC = 2\angle ABC$, or

$\angle ABC = \frac{1}{2}\angle AOC$

Inscribed Angles Property

Inscribed angles subtended by the same arc are equal.
So, $\angle ACB = \angle ADB = \angle AEB$

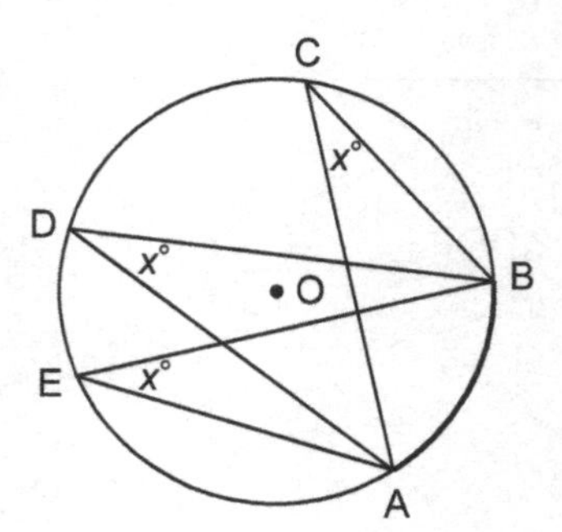

Example 1 Using Inscribed and Central Angles

Find the values of $x°$ and $y°$.

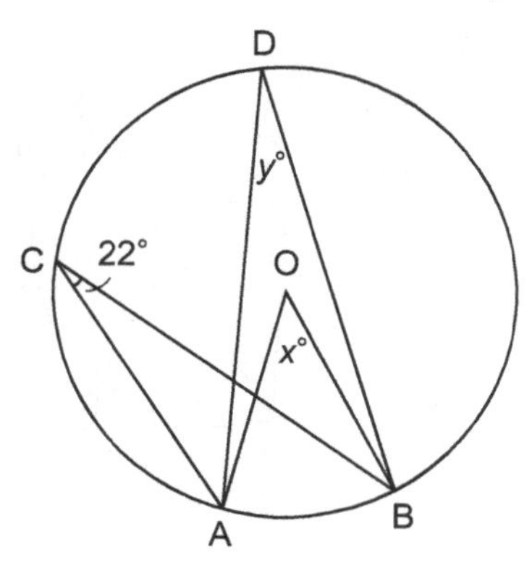

Solution

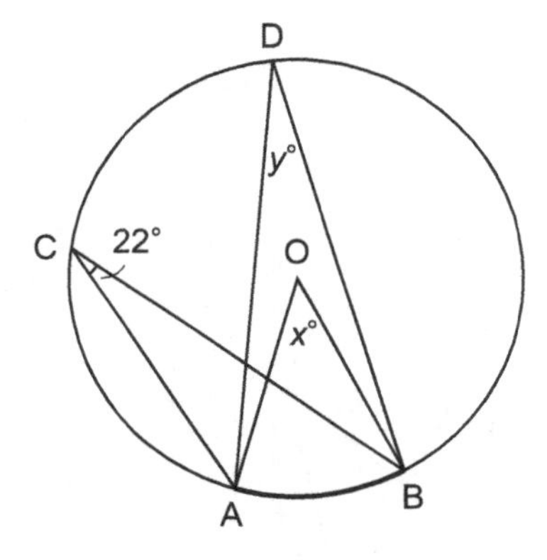

Central $\angle AOB$ and inscribed $\angle ACB$ are both subtended by arc AB.

So, $\angle AOB = 2\angle ACB$

$x° = 2 \times 22°$

$= 44°$

$\angle ACB$ and $\angle ADB$ are inscribed angles subtended by the same arc AB.

So, $\angle ADB = \angle ACB$

$y° = 22°$

Check

1. Find each value of $x°$.

a)

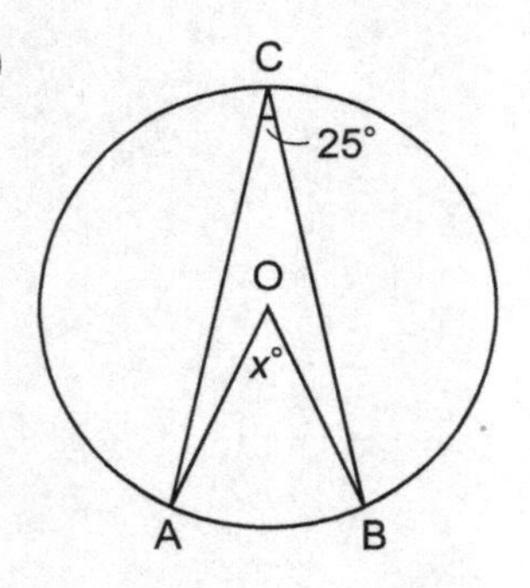

b)

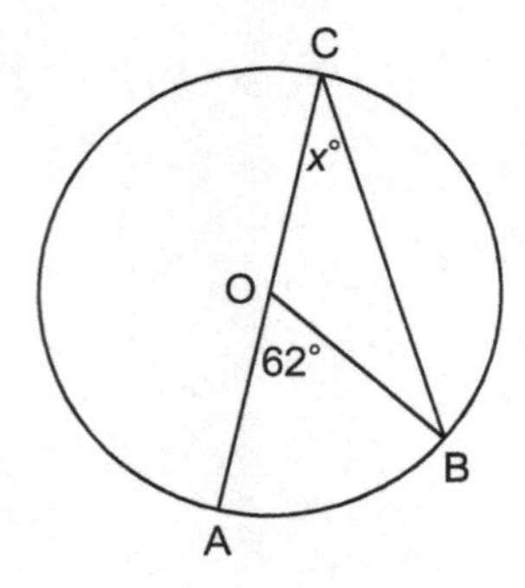

c)

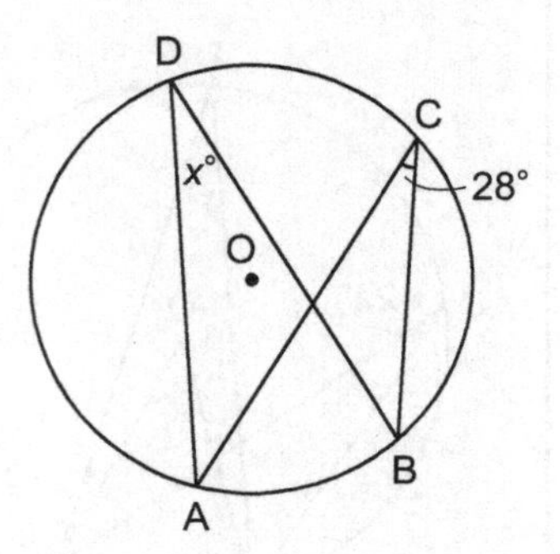

a) $\angle AOB = 2 \times \angle ACB$

$x° = 2 \times$ ______

$=$ ______

b) $\angle ACB = \frac{1}{2} \times$ ______

$x° = \frac{1}{2} \times$ ______

$=$ ______

c) $\angle ADB =$ ______

$x° =$ ______

2. Find the values of $x°$ and $y°$.

a)

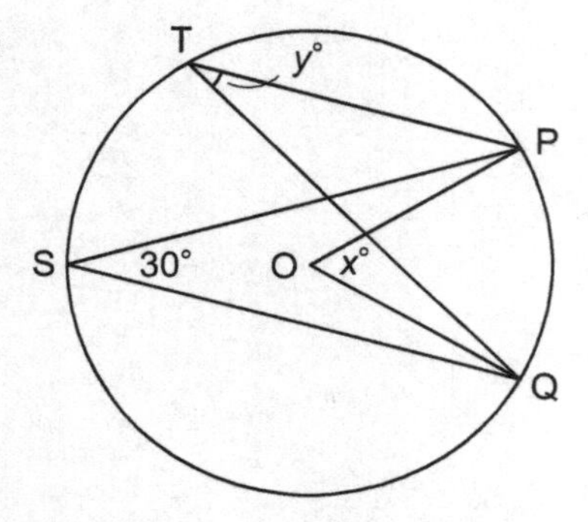

$\angle QOP = 2 \times \angle QSP$

$x° = 2 \times$ ______

$x° =$ ______

$\angle QTP =$ ______

$y° =$ ______

b)

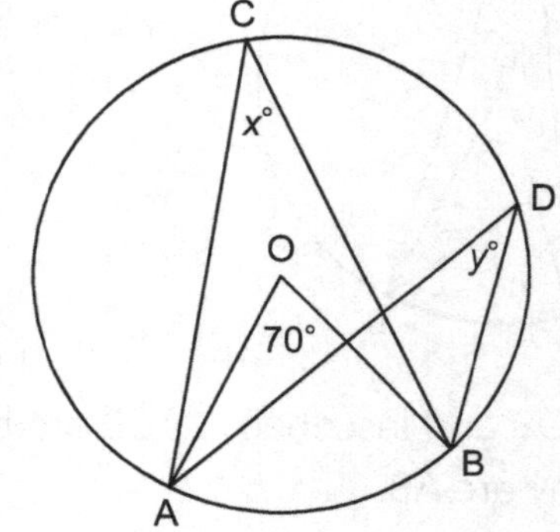

$\angle ACB = \frac{1}{2} \times$ ______

$x° = \frac{1}{2} \times$ ______

$x° =$ ______

$\angle ADB =$ ______

$y° =$ ______

Angles in a Semicircle Property

Inscribed angles subtended by a semicircle are right angles.

$\angle AFB = \angle AGB = \angle AHB = 90°$

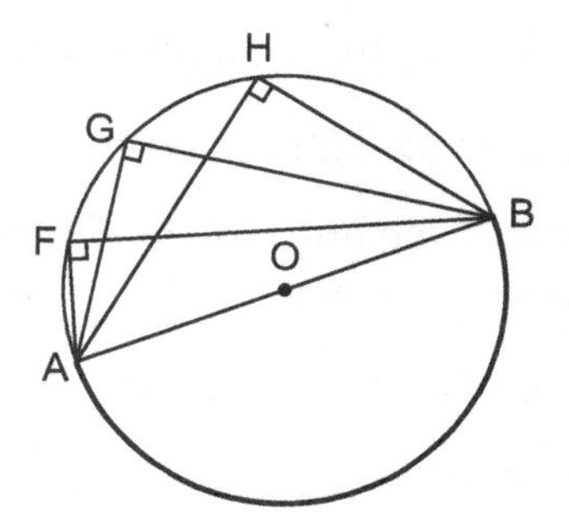

Example 2 Finding Angles in an Inscribed Triangle

Find $x°$ and $y°$.

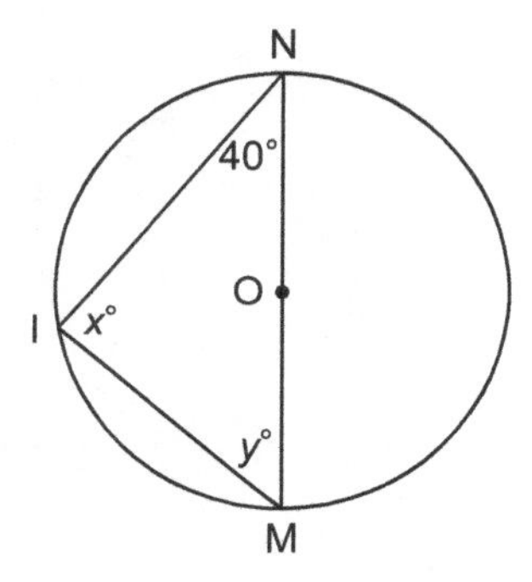

Solution

$\angle MIN$ is an inscribed angle subtended by a semicircle.

So, $x° = 90°$

$y° = 180° - 90° - 40°$ By the angle sum property in $\triangle MIN$

$= 50°$

Check

1. Find the values of $x°$ and $y°$.

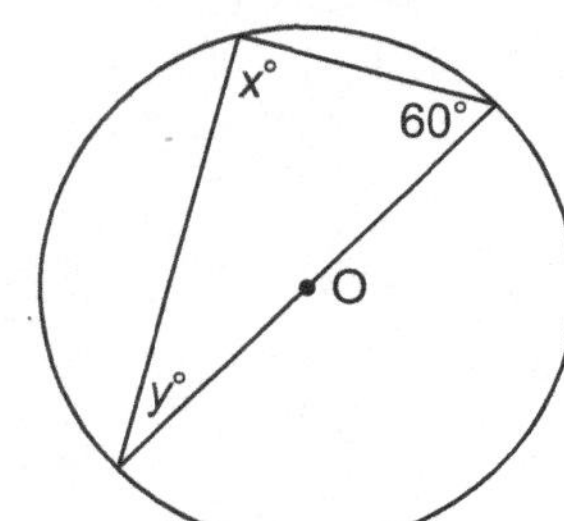

$x° =$ ______

$y° = 180° -$ ______ $-$ ______

$y° =$ ______

Practice

1. Name the following from the diagram.

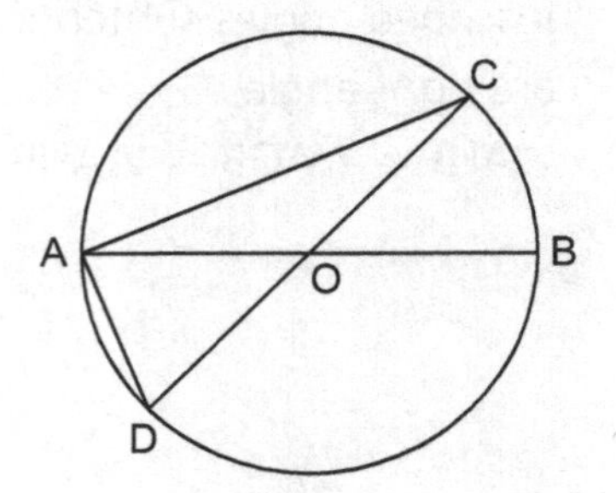

a) the central angle subtended by arc CB: ∠______

b) the central angle and inscribed angle subtended by arc AD: ∠______ and ∠______

c) the inscribed angle subtended by a semicircle: ∠______

d) the right angle: ∠______

2. In each circle, name a central angle and an inscribed angle subtended by the same arc. Shade the arc.

a)

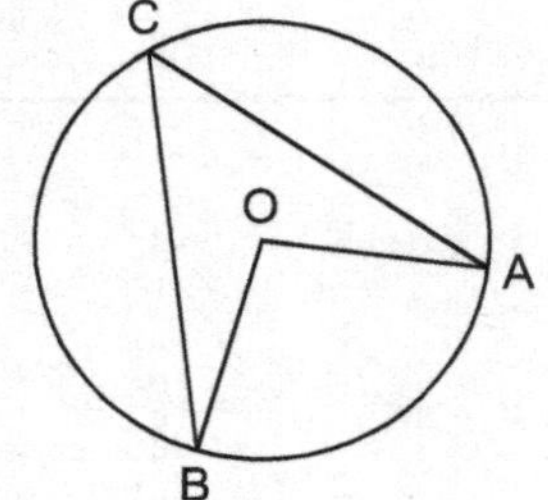

Central angle: ∠______
Inscribed angle: ∠______

b)

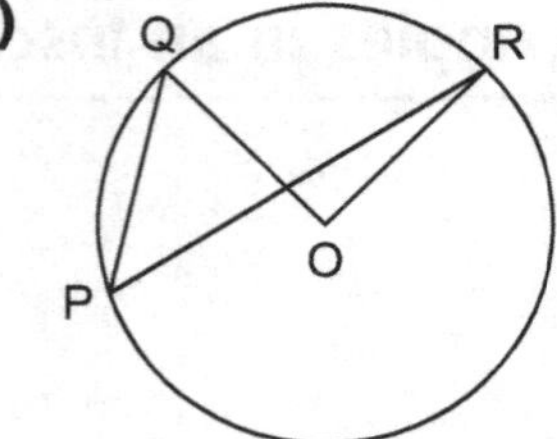

Central angle: ∠______
Inscribed angle: ∠______

3. Determine each indicated measure.

a)

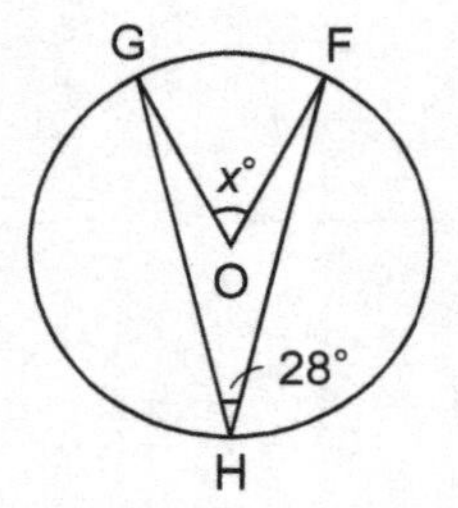

∠GOF = 2 × ∠GHF

$x° = 2 \times$ ______

= ______

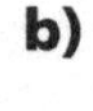

b)

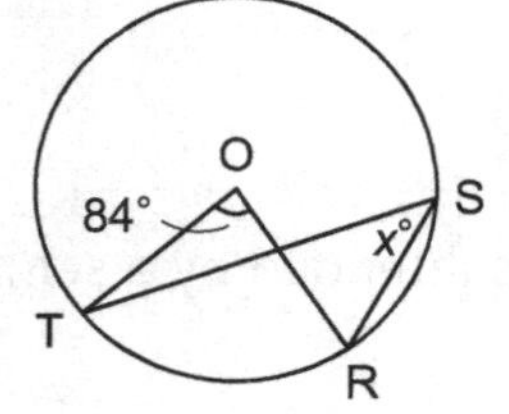

$∠TSR = \frac{1}{2} \times ∠$______

$x° = \frac{1}{2} \times$ ______

$x° =$ ______

c)

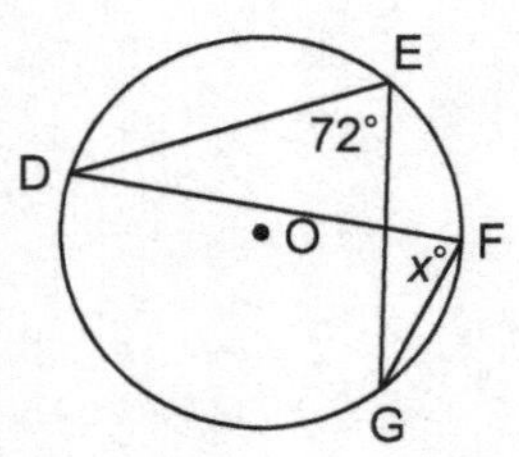

∠DEG = ______

$x° =$ ______

d)

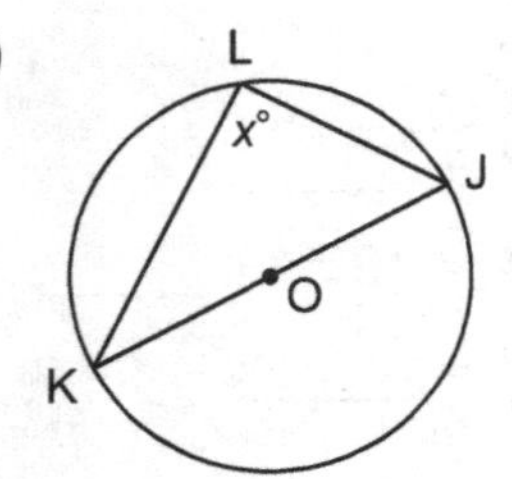

$x° =$ ______

4. Determine each value of $x°$ and $y°$.

a)

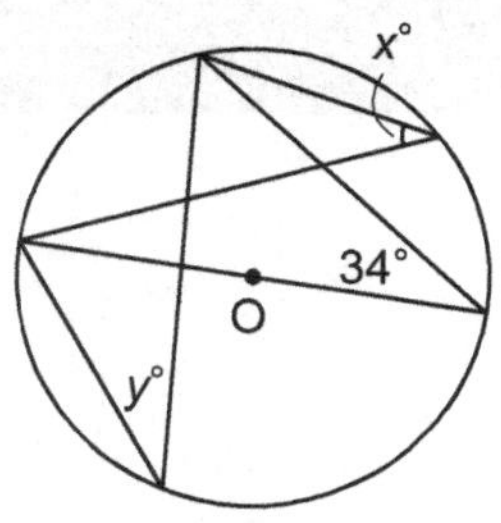

$x° =$ ______

$y° =$ ______

b)

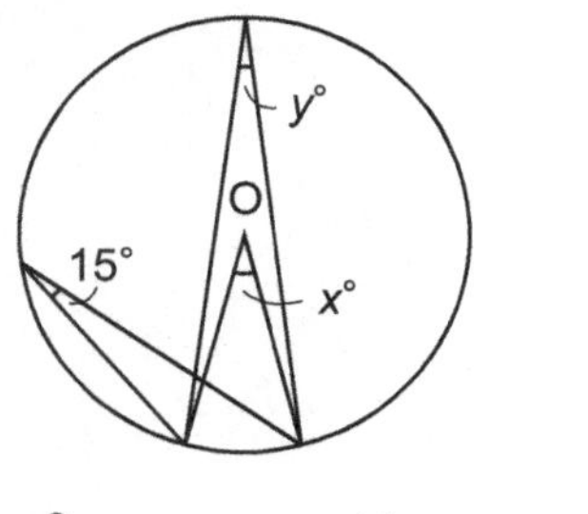

$x° =$ ______ $\times$ ______ $=$ ______

$y° =$ ______

5. Find the value of $x°$ and $y°$.

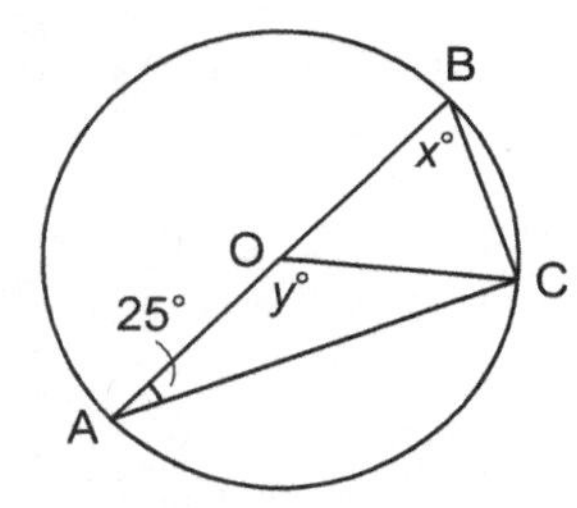

$\angle ACB =$ ______

$x° = 180° -$ ______ $-$ ______ By the angle sum property

$=$ ______

$y° =$ ______ $\times$ ______

$=$ ______

6. Find the value of $x°$, $y°$, and $z°$.

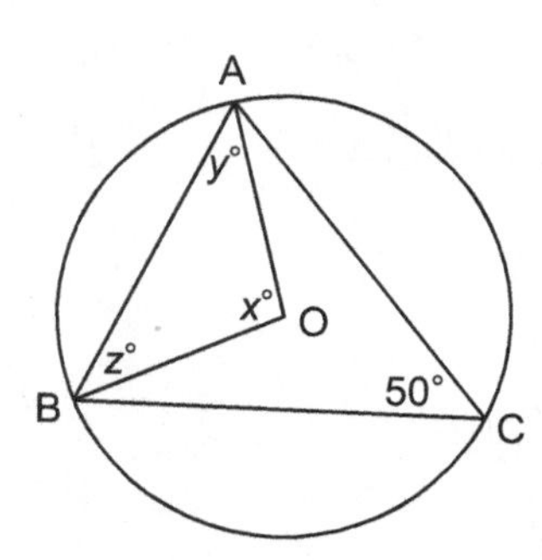

$\angle AOB = 2 \times$ ______

$x° = 2 \times$ ______ $=$ ______

In $\triangle OAB$, ______ $=$ ______

$\triangle OAB$ is ______________.

In $\triangle OAB$:

$y° = z°$

$y° + y° =$ ______ $-$ ______ By the angle sum property

$2y° =$ ______

$y° = \frac{\_\_\_}{2}$

So, $y° =$ ______ and $z° =$ ______

Unit 8 Puzzle

Circle Geometry Word Search

R E S E M I C I R C L E E K P
A L C T S E S Y E C P L N O S
L A I N F Y E O H L G O I M U
U U R E E R B O S N G N F M B
C Q C G T R R I A C T N P C T
I E L N A D E L S O E E A K E
D S E A C V A F F E D L L T N
N C E T C R E T M I C S E Z D
E L C E T E A L A U T T I S E
P L A N R N Z M D V C U J E D
R K E C G G E E G P U R B M M
E C R E Z T E E L G N A I R T
P A N M E R A D I U S O I C G
H C Q R T N I O P F R B G R Q
Y E L G N A D E B I R C S N I

Find these words in the puzzle above.
You can move in any direction to find the entire word.
A letter may be used in more than one word.

degrees
centre
diameter
tangent
circumference
arc
point of tangency

radius
inscribed angle
central angle
chord
point
isosceles
semicircle

perpendicular
bisect
circle
triangle
angle
equal
subtended

Unit 8 Study Guide

Skill	Description	Example
Recognize and apply tangent properties	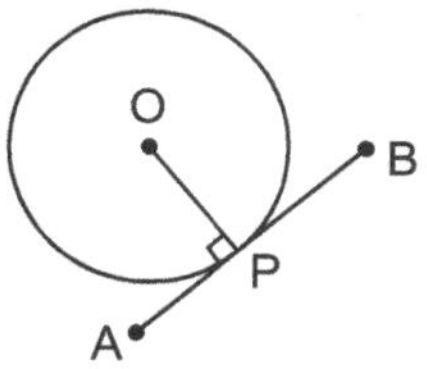 $\angle APO = \angle BPO = 90°$	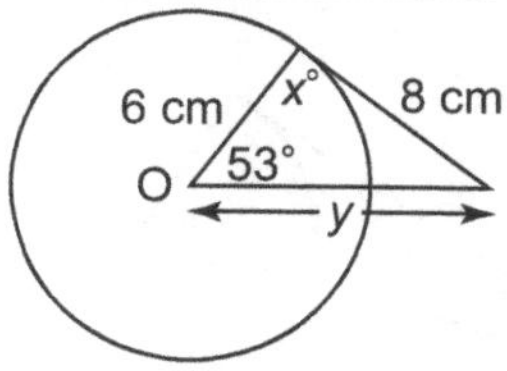 $x° = 90°$
Recognize and apply chord properties in circles	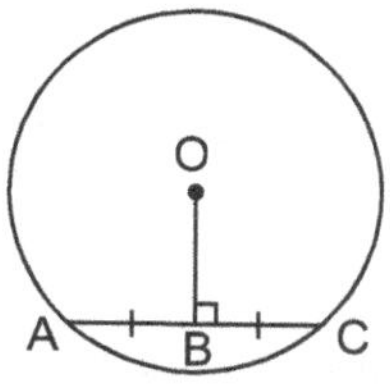 If $OB \perp AC$, then $AB = CB$. If $AB = CB$, then $OB \perp AC$.	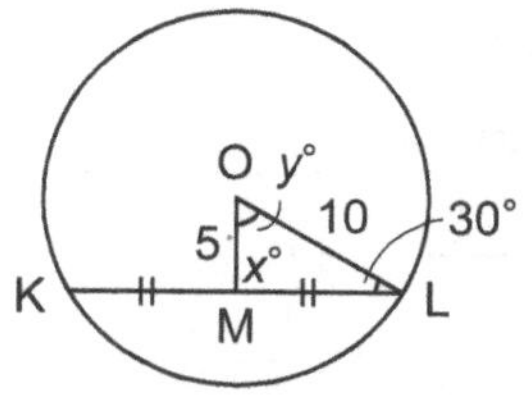 $x° = 90°$ and $y° = 60°$ $ML^2 = 10^2 - 5^2$
Recognize and apply angle properties in a circle	• Inscribed and central angles 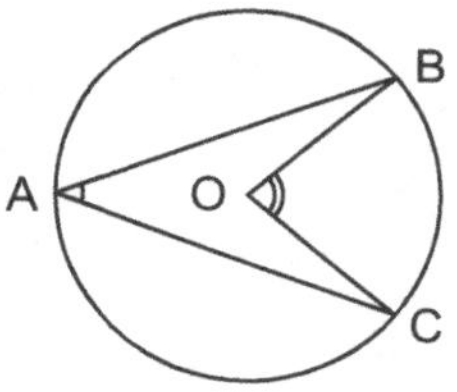 $\angle BOC = 2\angle BAC$, or $\angle BAC = \frac{1}{2}\angle BOC$ • Inscribed angles 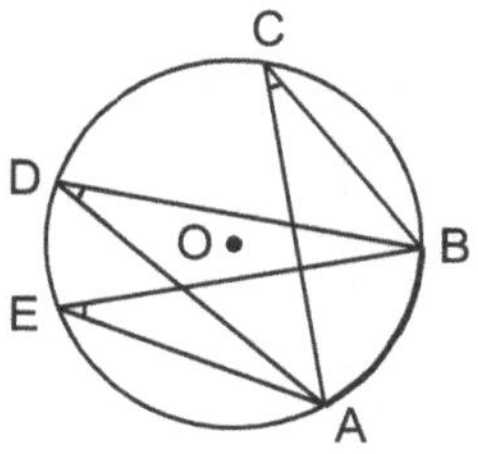 $\angle ACB = \angle ADB = \angle AEB$ • Angles on a semicircle 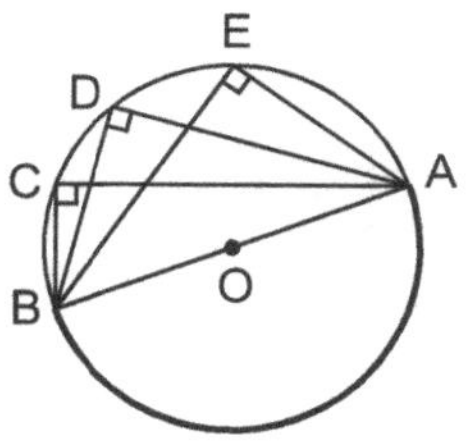 $\angle ACB = \angle ADB = \angle AEB = 90°$	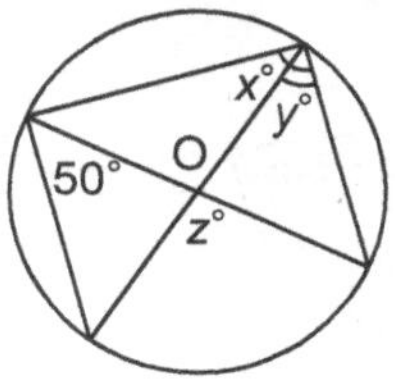 $x° = 90°$ $y° = 50°$ $z° = 100°$

Unit 8 Review

8.1 **1.** Find each value of $x°$ and $y°$. Segments RS and MN are tangents.

a)

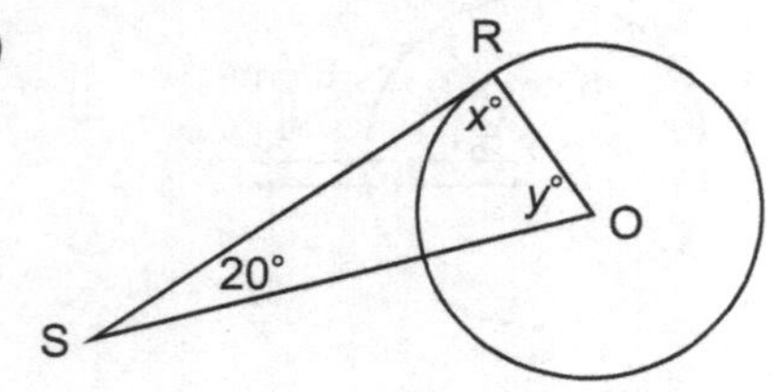

$x° =$ _____

$y° = 180° -$ _____ $-$ _____

$=$ _____

b)

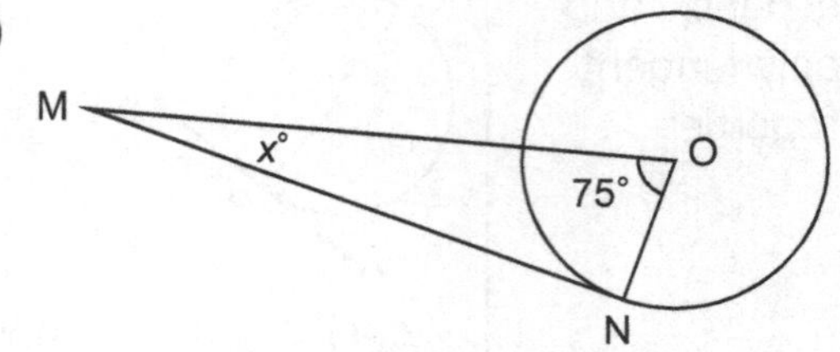

$\angle ONM =$ _____

$x° = 180° -$ _____ $-$ _____

$=$ _____

2. Find each value of x to the nearest tenth. Segments GH and ST are tangents.

a)

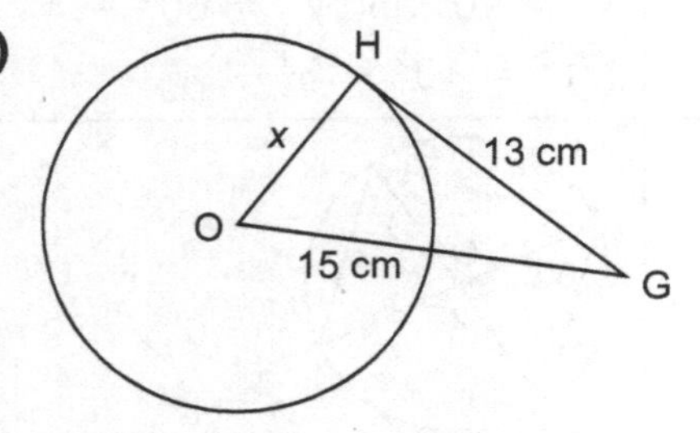

$\angle OHG =$ _____

_____ $= x^2 +$ _____

____

____

____

____

So, $x \doteq$ _______ cm

b)

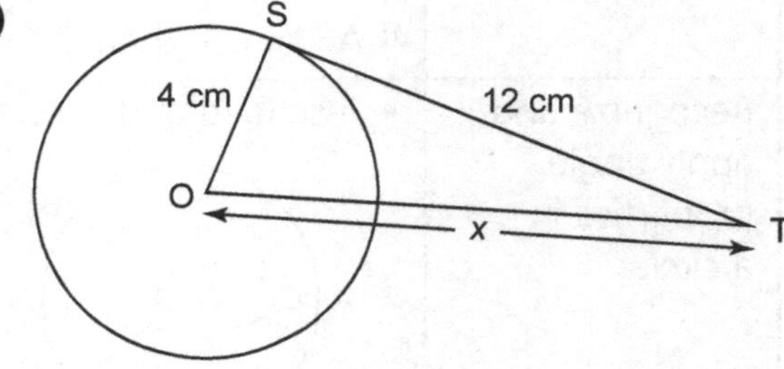

$\angle OST =$ _____

____

____

____

____

So, $x \doteq$ _______ cm

8.2 **3.** Find the values of $x°$ and $y°$.

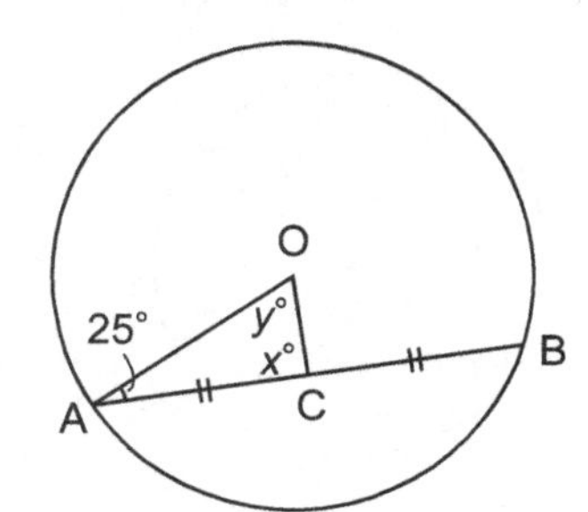

$x° =$ ______ By the chord properties

$y° =$ ______ − ______ − ______ By the angle sum property

____

$y° =$ ______

4. Find the values of $x°$, $y°$, and $z°$.

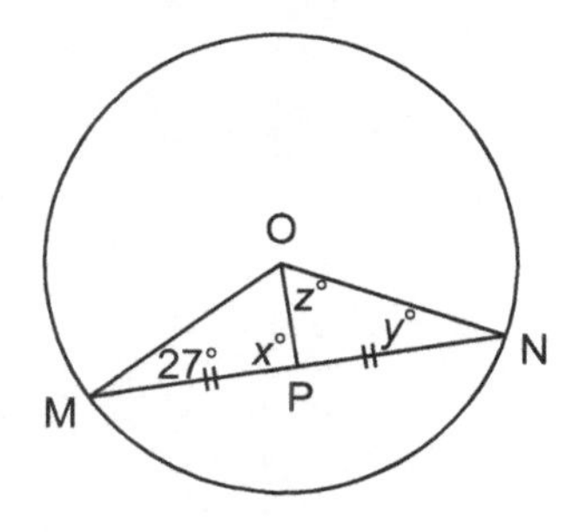

$x° =$ ______ By the ______________

OM = ON, so △______ is isosceles.

∠ONP = ∠OMP

So, $y° =$ ______

$z° =$ ______ − ______ − ______ ______________

$z° =$ ______

5. Find the length of the radius of the circle to the nearest tenth.

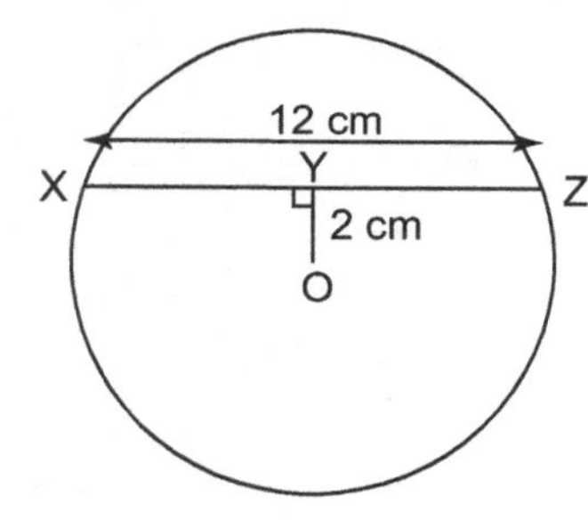

$XY = \frac{1}{2} \times$ ______

$= \frac{1}{2} \times$ ______ cm

$=$ ______ cm

Draw radius OX.

$OX^2 =$ ______ $+ XY^2$

$OX^2 =$ ______ + ______

____

____

____

$OX \doteq$ ______

The radius is about ______ cm.

8.3 **6.** Find each value of $x°$.

a)

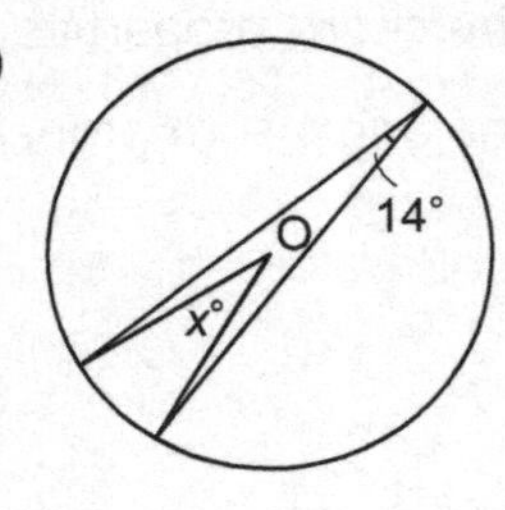

$x° = 2 \times$ ______

$x° =$ ______

b)

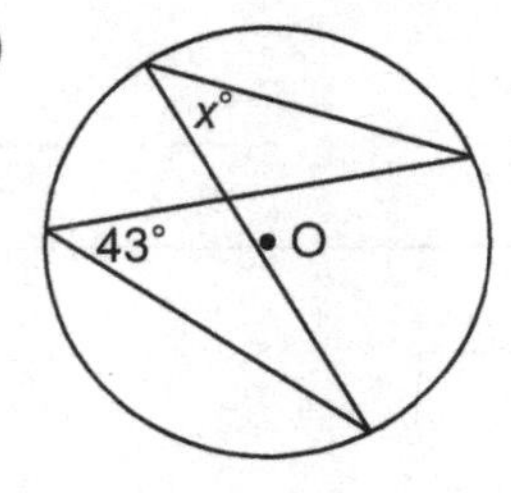

$x° =$ ______

c)

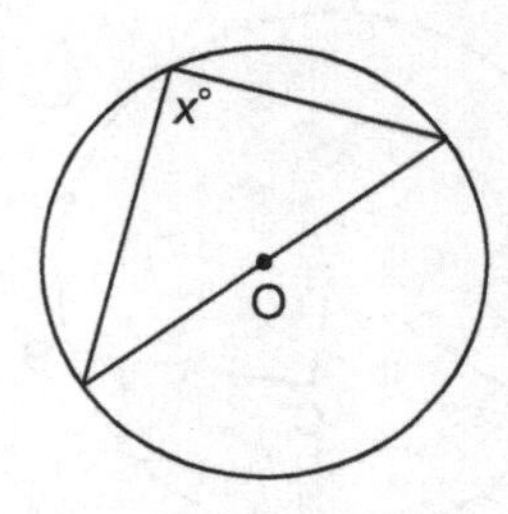

$x° =$ ______

7. Find each value of $x°$ and $y°$.

a)

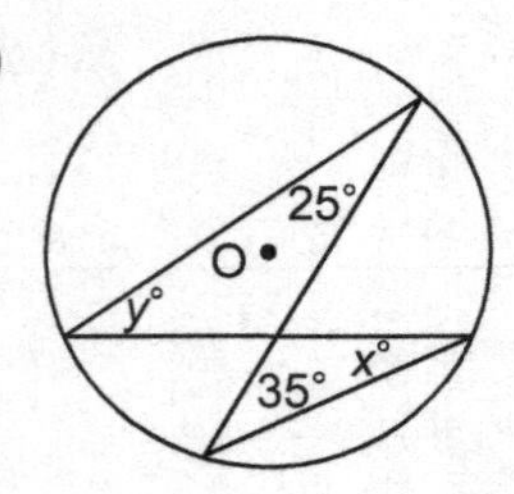

$x° =$ ______

$y° =$ ______

b)

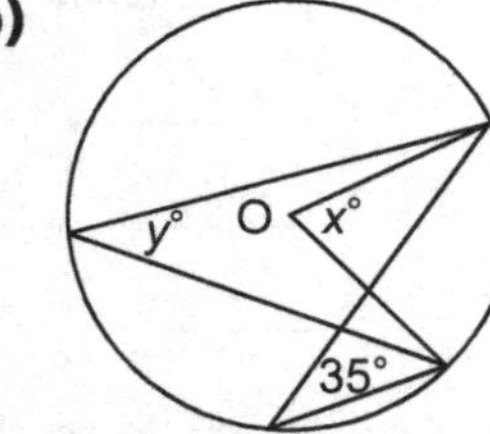

$x° = 2 \times$ ______

$=$ ______

$y° =$ ______

8. Find the value of $w°$, $x°$, $y°$, and $z°$.

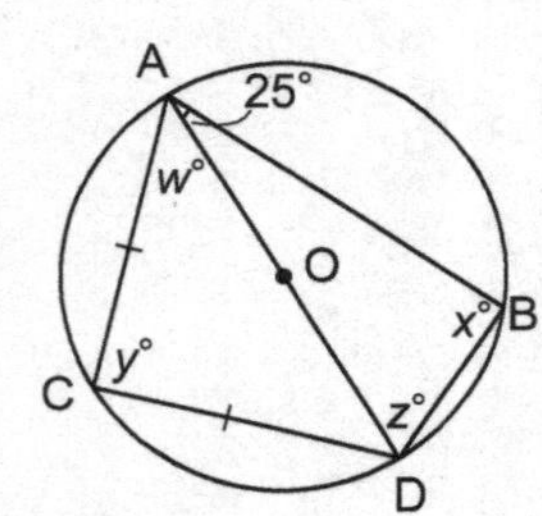

$x° = y° =$ ______

$z° =$ ______ $-$ ______ $-$ ______ By the angle sum property

____

$z° =$ ______

$\triangle ACD$ is ____________. So, $\angle CDA = \angle CAD = w°$

$w° + w° =$ ______ $-$ ______ By the angle sum in $\triangle ACD$

$2w° =$ ______

$w° = \frac{\text{\_\_\_\_}}{2}$

$w° =$ ______

UNIT 9

Probability and Statistics

What You'll Learn

- Understand how probability is used in everyday situations.
- Identify and solve problems related to data collection.
- Use either a population or a sample to answer a question.
- Develop and carry out a plan to collect, display, and analyze data.

Why It's Important

Probability is used by

- weather forecasters to more accurately predict the weather
- insurance companies to help determine insurance premiums

Key Words

experimental probability
theoretical probability
subjective judgment
assumptions
fair
biased question
census
population
sample
valid conclusions
simple random sampling
systematic sampling
cluster sampling
self-selected sampling
convenience sampling
stratified random sampling

9.1 Skill Builder

Relating Fractions, Decimals, and Percents

This hundredths chart represents one whole, 1, or 100%.
The shaded part of this grid can be described in 3 ways.

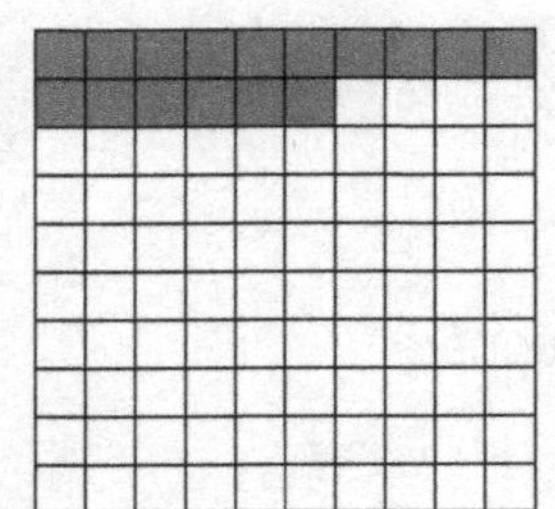

As a fraction: $\frac{16}{100}$

As a decimal: 0.16

As a percent: 16%

We can also read $\frac{16}{100}$ as 16 out of 100.

Write a fraction as a percent:

$$\frac{3}{4} \overset{\times 25}{=} \frac{75}{100} = 75\%$$

Write a percent as a fraction:

$$12\% = \frac{12}{100} \overset{\div 4}{=} \frac{3}{25}$$

We can think of a fraction as division.

Write a fraction as a decimal: $\frac{1}{2} = 1 \div 2 = 0.5$

Check

1. Write each fraction as a percent.

a) $\frac{35}{100} =$ ______

b) $\frac{2}{5} = \frac{2 \times \_\_}{5 \times 20}$
$= \frac{\_\_}{100}$
$=$ ______

c) $\frac{3}{20} =$ ______
$=$ ______
$=$ ______

2. Write each fraction as a decimal.

a) $\frac{17}{100} =$ ______

b) $\frac{13}{50} =$ ______
$=$ ______

c) $\frac{3}{8} =$ ______
$=$ ______

3. Write each percent as a fraction in simplest form.

a) $27\% = \frac{\_\_}{100}$

b) $25\% = \frac{\_\_}{100}$
$= \frac{\_\_}{100 \div 25}$
$=$ ______

c) $70\% = \frac{\_\_}{100}$
$=$ ______
$=$ ______

Theoretical Probability

The **theoretical probability** of an event occurring is:

$$\frac{\text{number of outcomes favourable to that event}}{\text{number of possible outcomes}}$$

Usually we refer to theoretical probability as the *probability*.

A die is labelled from 1 to 6.
The theoretical probability of rolling a number greater than 4 is:

When a die is rolled, there are 6 possible outcomes.
They are: 1, 2, 3, 4, 5, 6

A number greater than 4 is: 5 or 6
So, there are 2 favourable outcomes.

The probability of rolling a number greater than 4 is:

$$\frac{\text{number of outcomes favourable to that event}}{\text{number of possible outcomes}} = \frac{2}{6}, \text{ or } \frac{1}{3}$$

Check

1. Twenty marbles are placed in a bag: 9 red, 2 green, 5 purple, and 4 yellow
You take one marble from the bag without looking.
What is the theoretical probability of picking:

a) a green marble?

The outcomes are:

____

The favourable outcome is ______.

There are ____ green marbles.

Theoretical probability

$= \frac{\text{number of \_\_\_\_\_ marbles}}{\text{total number of marbles}}$

$= \frac{\_\_}{\_\_}$, or $\frac{\_\_}{\_\_}$

= ____________

b) a marble that is not red?

The favourable outcomes are:

____

Add to find the total number of favourable outcomes:

_____ + _____ + _____ = _____

There are _____ favourable outcomes.

Theoretical probability

$= \frac{\_\_\_\_\_\_\_\_\_\_\_\_\_\_\_\_}{\_\_\_\_\_\_\_\_\_\_\_\_\_\_\_\_}$

$= \frac{\_\_}{\_\_}$

= ____________

Experimental Probability

Experimental probability is the likelihood that something occurs based on the results of an experiment. The **experimental probability** of an event occurring is:

$$\frac{\text{number of times the outcome occurs}}{\text{number of times the experiment is conducted}}$$

It can be written as a fraction, decimal, or percent.

When a paper cup is slid off the edge of a table, it can land open end up, open end down, or on its side. Here are the results of 50 trials.

Outcome	Frequency
(on its side)	24
(open end up)	10
(open end down)	16

The experimental probability of landing on its side was:

$$\frac{\text{number of times cup landed on its side}}{\text{total number of trials}} = \frac{24}{50}$$

$$= 0.48\text{, or } 48\%$$

The experimental probability of landing open end down was:

$$\frac{\text{number of times cup landed open end down}}{\text{total number of trials}} = \frac{16}{50}$$

$$= 0.32\text{, or } 32\%$$

Check

1. A counter is randomly drawn from a bag of counters.
Here are the results of 25 draws:

Colour	Frequency
Red	7
Blue	4
Yellow	11
Green	3

What is the experimental probability of drawing:

a) a yellow counter?
Experimental probability

$= \frac{\text{number of times yellow was drawn}}{\text{total number of draws}}$

$= \frac{\_\_\_}{\_\_\_}$

$=$ ____________

b) a blue or green counter?
Experimental probability

$= \frac{\text{number of times \_\_\_\_\_\_\_\_\_\_ was drawn}}{\text{total number of draws}}$

$= \frac{\_\_\_}{\_\_\_}$

$= \frac{\_\_\_}{\_\_\_}$

$=$ ____________

9.1 Probability in Society

FOCUS **Explain how probability is used outside the classroom.**

We can use probability to help us make decisions.

Sometimes the decisions we make are influenced by our state of mind or by our gut feeling. When we make a decision in this way, we make a **subjective judgment.**

For example, I think I did well on the test because I wrote in red pencil and red is my favourite colour.

Example 1 Identifying Predictions Based on Probabilities and Judgments

Emily was the first to arrive at 9 of the last 10 volleyball practices. So, the coach thinks Emily will arrive first for today's practice.

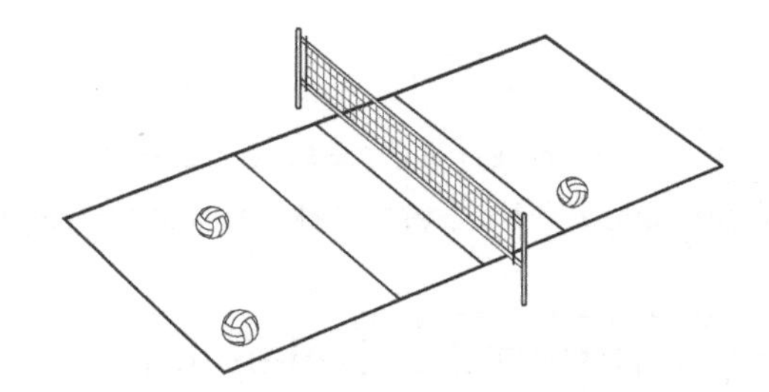

Is the coach's prediction based on theoretical probability, experimental probability, or subjective judgment? Explain.

Solution

The coach's decision is based on past experience. He has seen Emily arrive first at almost all the practices. This is an example of experimental probability.

Check

1. Explain how each decision is based on theoretical probability, experimental probability, or subjective judgment.

a) A school survey finds that 5 out of 6 students wear jeans to school on any given day. So, Felix decides that if he randomly selects 6 students at his school, 5 of them will be wearing jeans.

Circle the statement that best describes how Felix made his prediction.

Felix made his prediction based on the results of a survey or experiment.

Felix made his prediction based on theoretical probability.

Felix made his prediction based on his personal thoughts or feelings.

Explain your thinking.

____

____

b) Carlos and Natalie are playing a board game. Natalie needs to roll double ones on her next roll to win. The probability of rolling double ones with a standard pair of dice is 1 out of 36. So, Natalie predicts she will lose the game.

Circle the statement that best describes how Natalie made her prediction.

Natalie made her prediction based on the results of a survey or experiment.

Natalie made her prediction based on theoretical probability.

Natalie made her prediction based on her personal thoughts or feelings.

Explain your thinking.

____

____

Sometimes when we make a decision, we take certain things for granted.
When we do this, we make **assumptions.**

Example 2 Explaining How Assumptions Affect a Probability

A health magazine claims that people who exercise 3 times per week for an hour each time reduce their risk of developing heart disease by 40%. Monique plays soccer 3 times a week for an hour each time. So, Monique thinks she has reduced her risk of developing heart disease by 40%.

a) What assumptions is Monique making?

b) How might the outcome change if the assumptions change?

Solution

a) Monique assumes that no other factors increase the risk of developing heart disease. Other factors that should be considered are:
- eating habits
- family history
- age
- blood pressure
- exposure to smoking

b) If Monique eats a lot of junk food, or if she lives with people who smoke, Monique's risk of developing heart disease might increase.

If Monique's family has no history of heart disease, she lives with non-smokers, and she eats a healthy diet, her risk of developing heart disease might decrease.

Check

For each situation:

a) What assumptions are being made?

b) How might the outcome change if the assumptions change?

1. Sam thinks more people purchase Ultra White toothpaste than Shine.

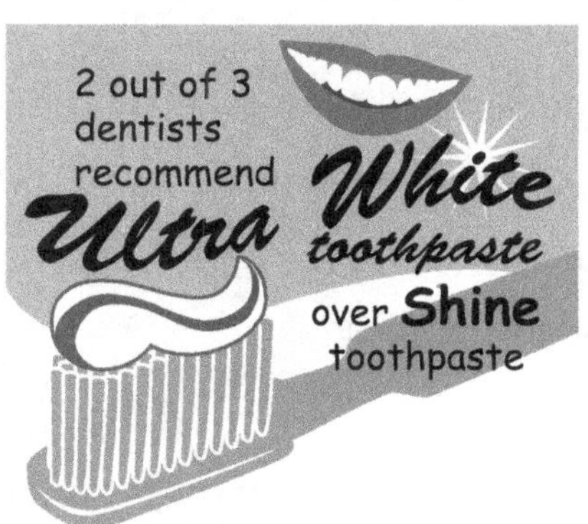

a) Do you listen to all of your dentist's suggestions? _____
Does the cost of an item help you decide whether to purchase it or not? ______
Would you make a trip to a specialty store to buy toothpaste?_____
What assumptions has Sam made?
– ____________________
– ____________________
– ____________________

b) Fewer people might purchase Ultra White toothpaste if:
– its cost is __________ than that of Shine
– it is only available in __________ stores
More people might purchase Ultra White toothpaste if:
– its cost is __________ than that of Shine
– it is available in __________ stores where toothpaste is sold

2. Salima has scored more than 10 points in her last 5 basketball games.
With 15 min left to play in the game, Salima has 4 points.
Salima has scored as often as she usually does so far in the game.
So, Salima will score at least 6 points in the remaining 15 min.

a) Assumptions: – ____________________
– ____________________

b) Salima may not score 10 points in the game if:
– the opposing team is __________ in the standings than Salima's team
– Salima is playing __________
Salima may score more than 10 points in the game if:
– the opposing team is __________ in the standings than Salima's team
– ____________________

Practice

1. Explain how each decision is based on theoretical probability, experimental probability, or subjective judgment.

a) Josh is given a bag that contains 5 red marbles and 5 blue marbles. He is to pick one marble from the bag without looking. He decides that his chance of picking a red marble is 1 out of 2, or 50%.

Circle the statement that best describes how Josh made his prediction.

Josh made his decision based on the results of a survey or experiment.

Josh made his decision based on theoretical probability.

Josh made his decision based on his personal thoughts or feelings.

Explain your thinking.

____

____

____

b) A quality control officer for a light bulb manufacturer tested 10 light bulbs. Nine of the bulbs burned for more than 1000 h. So, the manufacturer decides that 90% of the light bulbs will burn for more than 1000 h.

Circle the statement that best describes how the manufacturer made the decision.

The manufacturer made the decision based on the results of an experiment.

The manufacturer made the decision based on theoretical probability.

The manufacturer made the decision based on personal thoughts or feelings.

Explain your thinking.

____

____

____

c) A pair of concert tickets is hidden in an envelope. There are 3 envelopes to choose from: red, green, and blue. Desi chooses the green envelope because green is his favourite colour.

Circle the statement that best describes how Desi made his decision.

Desi made his decision based on the results of an experiment.

Desi made his decision based on theoretical probability.

Desi made his decision based on personal thoughts or feelings.

Explain your thinking.

____

____

____

2. What assumptions are being made in each situation?

a) Lin passed 6 of her last 7 math tests. So, Lin is sure she will pass tomorrow's test.

Assumptions: – ______________________

– ______________________

b) The first tunnel on a road through the mountains has height 4.5 m. A truck driver decides that her truck of height 4.3 m will pass safely through all tunnels on the road.

Assumption: – ______________________

c) The Tigers have won all of their home games to date. So, they will win tomorrow's home game.

Assumptions: – ______________________

– ______________________

3. For each situation:

i) What assumptions are being made?

ii) How might the outcome change if the assumptions change?

a) One hundred bottles of water were sold in the school cafeteria yesterday. So, the school orders 500 bottles of water for next week.

i) Assumptions: – ______________________

– ______________________

– ______________________

____

ii) Fewer bottles of water might be sold if:

– ______________________

– ______________________

More bottles of water might be sold if:

– ______________________

– ______________________

b) Marcel's dad leaves for work 5 min later when there is a school holiday because traffic is always lighter.

i) Assumptions: – ______________________

– ______________________

– ______________________

____

ii) ______________________

____

____

____

9.2 Skill Builder

Writing a Questionnaire

A questionnaire should contain questions that are **fair.**
A question should not influence a person's answer.
If it does, the question is a **biased question.**
Each person should also be able to answer the question.

To find out whether her classmates prefer dogs or cats, a student asks:
Do you like cute little puppies or mean hairy cats?

This question is worded in such a way that dogs are favoured over cats.
So, the question is biased. A better question would be:
Do you prefer dogs, cats, or neither?

Check

1. Which of these questions are biased? Insert *is* or *is not* each time.
For each biased question, write a better question.

a) Do you prefer to watch boring newscasts or exciting reality shows?
Is there a favoured answer? _____
So, the question ______ biased.
A better question is: ____
____

b) What is your favourite type of music: Rap _____ Pop _____ Classical _____
Hip Hop _____ Other _____?
Is there a favoured answer? _____
So, the question ______ biased.
A better question is: ____________

c) I love eating apple pie. What is your favourite dessert: Apple pie _____
Cheesecake _____ Ice Cream _____ Fruit _____ Other _____?
Is there a favoured answer? _____
So, the question ______ biased.
A better question is: ____
____

9.2 Potential Problems with Collecting Data

FOCUS **Describe how the collection of data may be unfavourably affected.**

A biased question could lead to a problem with data collection.
There are many other factors that should also be considered.

Potential Problem	Example
Timing – when data are collected could affect the results	The results of a survey on ski jackets may differ if the survey is conducted in the summer instead of the winter.
Privacy – people may not want to share personal information	Students may not want to share their school grades.
Cultural Sensitivity – the survey must not offend other cultures	A question about Christmas shopping may offend those who do not celebrate Christmas.
Cost – the cost of collecting data should be considered	The cost of mailing a survey to a large number of people may be too expensive.
Time – the amount of time needed to collect the data should be considered	A 30-min survey conducted over the lunch period may not interest students. Most would not want to give up 30 min of their lunch time.
Use of Language – the way a question is worded could lead people to answer in a certain way	This question may lead students to answer "Yes": Don't you think the price of a concert ticket is too high?

Example 1 Identifying Potential Problems

A survey is conducted to determine the favourite computer game among students aged 13 to 16. Students aged 13 to 16 are asked to participate in a 1-h survey. Identify the potential problem with this survey and how the problem could be avoided.

Solution

The potential problem is time. The survey takes too long to complete. Students of this age will not want to participate in a 1-h survey.

The amount of time that it takes to complete the survey should be reduced to no more than 15 min.

Check

1. For each survey question, identify the potential problem and how the problem could be avoided.

   a) A survey is conducted to find out which part of a chocolate Easter Bunny students eat first. The question asked was:

   Which part of a chocolate Easter bunny do you eat first:
   Ears _____ Tails _____ Leg _____ Nose _____ Other _____?

   Is there a problem with the survey question? _____
   Is there a problem with cost or time? _____
   Is there a problem with privacy or cultural sensitivity? _____
   Explain your thinking.

   ____
   ____

   How would you avoid the problem? ______________________
   ____
   ____
   ____

   b) A survey is conducted to find out which type of sunscreen people use. A sunscreen manufacturer conducts the survey in January because business is slow. The question asked was:

   What sun protection factor does your usual sunscreen have:
   Less than 30 _____ 30 _____ Greater than 30 _____ Do not use _____?

   Is there a problem with the survey question? _____
   Is there a problem with cost or timing? _____
   Is there a problem with privacy or cultural sensitivity? _____
   Explain your thinking.

   ____
   ____
   ____

   How would you avoid the problem? ______________________
   ____

Example 2 Analyzing Data Collection for Problems

Sarah wants to determine the average annual income of each household in her neighbourhood. She plans to go door-to-door in her neighbourhood to ask this question: What is your annual household income? Her brother thinks Sarah's plan is problematic. Explain why.

Solution

Sarah is asking her neighbours, many of whom she knows, to share personal information. This may make her neighbours uncomfortable. Sarah should consider an anonymous survey where the possible choices are given in ranges.

Check

Explain why each survey might be problematic.

1. A local theatre wants to survey the community to find out what types of events they might be interested in attending. The theatre plans to send out 5000 surveys, with a self-addressed stamped envelope in each survey.

Is there a problem with time? _____
Is there a problem with cost? _____
Is there a problem with cultural sensitivity? _____

Explain. ____

____

____

2. The local government wants to find out how satisfied its citizens are with their government's performance. They plan to conduct a random 30-min telephone survey 2 days after they raised taxes.

Is there a problem with time? _____
Is there a problem with timing? _____
Is there a problem with privacy? _____

Explain. ____

____

____

Practice

1. For each survey, identify a potential problem.

a) An on-line pet magazine surveys its readers by asking this question:

Don't you think people who don't walk their dogs at least once a day are being cruel to their dogs? Yes or No

Is there a problem with the survey question? _____
Is there a problem with cost or time? _____
Is there a problem with privacy or cultural sensitivity? _____
Explain your thinking.

____

____

____

b) The school cafeteria surveys all students to find out what meat dishes they would like to see served. They asked this question:

Which meat dish would you like to see served in the cafeteria:
Meatloaf _____ Roast Beef _____ Sausage _____ Chicken _____ Other _____?

Is there a problem with language? _____
Is there a problem with cost or time? _____
Is there a problem with privacy or cultural sensitivity? _____
Explain your thinking.

____

____

____

c) A French tutor asks all students in the school for their French marks so she can find out which students might need her help.

Is there a problem with cost or time? _____
Is there a problem with privacy or cultural sensitivity? _____
Explain your thinking.

____

____

____

2. For each scenario in question 1, describe how the problem could be avoided.

a) ______________________________

____

____

b) ______________________________

____

____

____

c) ______________________________

____

____

3. Explain why each survey might be problematic.

a) A cell-phone company wants to find out how satisfied customers are with the company and its product. They plan to conduct a 25-min telephone survey with 10 000 of its customers.
Is there a problem with time or cost? _____
Is there a problem with privacy? _____
Is there a problem with cultural sensitivity? _____
Explain. ______________________________

____

____

b) To find out which sport students in the school prefer, you plan to poll the students the day after the volleyball team wins the championship.
Is there a problem with timing or cost? _____
Is there a problem with privacy? _____
Is there a problem with cultural sensitivity? _____
Explain. ______________________________

____

____

____

4. For each scenario in question 3, describe how the problem could be avoided.

a) ______________________________

____

b) ______________________________

____

9.3 Using Samples and Populations to Collect Data

FOCUS **Select and defend the choice of using a population or a sample.**

When a **census** is conducted, data are collected from *all* people or items in the **population.** For example, to find the favourite movie of all Grade 9 students in a school, all Grade 9 students in the school are surveyed. Since all students in the population were surveyed, a census was conducted.

When the population is large, it is often too costly or time consuming to survey the entire population. So, we often collect data from a representative portion of the population, or a **sample.** For example, surveying 50 of the 300 Grade 9 students is a sample.

Example 1 Identifying the Population of a Survey

A local fitness club wants to find out how many members use the hot tub. What is the population of the survey?

Solution

The fitness club wants to find out how many members use the hot tub. So, the population is all members of the club.

Check

1. Identify the population of each survey.

a) A local restaurant owner wants to know which entree on the menu is preferred by the restaurant's customers.

____

____

b) A school principal wants to know how many of the teachers in the school assign homework on weekends.

____

c) A city planner wants to know which roads in Vancouver people think are in most need of repair.

____

Example 2 Deciding whether to Use a Census or a Sample

Winnipeg's city council wants to find out if its residents are satisfied with the garbage removal service. Should a sample or census be used to collect the data? Explain why.

Solution

A sample should be used. The city of Winnipeg is very large and it would be time consuming to survey every household in the city.

Check

In each case, should a sample or a census be used to collect the data?

1. A smoke detector manufacturer wants to test the product before it is packaged for sale.

Would it take a long time to test every smoke detector made? ______
Would it cost a lot of money to test every smoke detector? ______
Should a sample or a census be used? ___________

2. A teacher wants to find out whether her 30 students would prefer to have their math test on Friday or Monday.

Would it take a long time to ask each student? ______
Would it cost a lot of money to ask each student? ______
Should a sample or a census be used? ________

3. A radio station wants to poll its listeners to find the top 10 songs of the week.

Would it take a long time to poll every listener? ______
Would it cost a lot of money to poll every listener? ______
Should a sample or a census be used? ___________

The sample chosen should represent the population. For example, if the sample of Grade 9 students contains only girls, the results will not accurately represent the population. The sample should contain a proportional number of girls and boys to provide **valid conclusions.**

The size of the sample should also be considered. If the sample size is too small, the results may not represent the population. For example, a sample of ten Grade 9 students may not provide valid conclusions.

Example 3 Identifying and Critiquing the Use of Samples

The school council wants to know if students would like to have a dance to celebrate the end of the school year. The council surveys 2 girls from each grade to find out.

Was a sample or a census used to collect the data?
Do you think the conclusions would be valid? Explain.

Solution

Not all students in the school were surveyed. So, the data were collected from a sample.

All students surveyed were girls. And, 2 students from each grade is probably not enough to represent the population. So, the conclusions are probably not valid. More students should be surveyed, with a proportional number of boys and girls included.

Check

In each case, was a sample or a census used to collect the data?
Wherever a sample was used, do you think the conclusions would be valid?
Explain.

1. To find out what percent of the population of British Columbia has access to the Internet, all residents of Victoria were surveyed.

What is the population? ______________________________
Were all residents surveyed? ______
Was a sample or census used? _________

Do you think the results of the survey would represent the residents of a small northern mining community? ______
Do you think the conclusions would be valid? ______

2. To ensure safe and secure air travel from Edmonton International Airport, anyone boarding a plane at the airport goes through a security check.

What is the population? ____
____
Were all people boarding planes checked? ______
Was a sample or a census used? _________

3. To find out the favourite flavour of bubble tea, every fifth customer of the bubble tea café is surveyed.

What is the population? ______________________

Were all customers surveyed? ______

Was a sample or a census used? __________

Do you think the conclusions would be valid? ______ Why?

____

____

____

Practice

1. Identify the population of each survey.

a) The manager of an apartment building wants to find out which apartments in the building need new carpet.

____

b) The manager of a video game store in Grande Prairie wants to find out how many video games are owned by the average teenager in town.

____

c) A company that manufactures programmable thermostats wants to test its thermostats for defects.

____

2. In each case, describe why a sample was used instead of a census.

a) A car company in Canada wants to find out whether people plan to buy or lease their next vehicle.

____

____

b) A clothing manufacturer wants to make sure the guarantee that its clothing will not shrink when washed is upheld.

____

____

3. In each case, describe why a census was used instead of a sample.

a) The Canadian government wants to determine the population of the different towns and cities in Canada.

____

____

____

b) A school principal wants to find out how many students plan to take the Grade 10 Computer and Information Science course next year.

____

____

____

____

4. In each case, should a sample or a census be used to collect the data?

a) The school basketball teams are getting a new logo. There are 2 logos to choose from and 24 basketball team members in the school.

Would it take a long time to survey each team member? _____
Would it cost a lot of money to survey each team member? _____
Should a sample or census be used? ___________

b) The City of Regina wants to find out what percent of its residents use public transportation on a daily basis.

Would it take a long time to survey each resident of Regina? _____
Would it cost a lot of money to survey each resident? _____
Should a sample or a census be used? _________

5. In each case, a sample was used to collect data. Do you think the conclusions would be valid? Explain.

a) To find out if more daycare centres are needed in the city, all residents over the age of 50 were surveyed.

Do you think most residents over the age of 50 would have children in daycare? _____
Do you think the conclusions would be valid? _____

b) To find out what percent of the vehicles in a mall parking lot have a bumper sticker, Neil looks on the bumper of every fourth vehicle.

____

____

____

Can you ...

- Determine if a decision is based on theoretical probability, experimental probability, or subjective judgment?
- Identify assumptions associated with decisions based on probabilities?
- Identify potential problems with data collection?
- Determine if a sample or census should be used to collect data?
- Decide whether the data collection provides valid conclusions?

9.1 **1.** Explain how this decision is based on theoretical probability, experimental probability, or subjective judgment.

Peyton booked his piano exam for the 17th of the month. Peyton predicts he will do well because 17 is his lucky number.

Circle the statement that best describes how Peyton made his prediction.

Peyton made his decision based on the results of a survey or experiment.
Peyton made his decision based on theoretical probability.
Peyton made his decision based on his personal thoughts or feelings.

Explain your thinking.

____

____

____

2. Last week, Lori sat in the desk closest to the window. She was cold 4 out of 5 days. So, this week Lori decides to move away from the window so she will be more comfortable.

a) What assumptions is Lori making?

– ______________________________________

– ______________________________________

– ______________________________________

b) How might the outcome change if the assumptions change?
Lori might still be cold if:

– ______________________________________

– ______________________________________

Lori might be too hot if:

– ______________________________________

– ______________________________________

____

9.2 **3.** A survey is conducted at the local shopping mall over the Victoria Day weekend to find out how much money families usually spend on "back-to-school" shopping.

a) Identify potential problems with this survey.

Is there a problem with cost or timing? ______
Is there a problem with privacy or cultural sensitivity? ______
Explain your thinking.

____

____

____

____

____

b) How could the problem be avoided?

____

____

____

9.3 **4.** Identify the population of this survey. Should a sample or a census be used to collect the data?

The city of West Vancouver wants to know how many people plan on attending the fireworks display on Canada Day.

Population: ______________________________
Would it take a long time to survey each resident? _______
Would it cost a lot of money to survey each resident? _______
Should a sample or a census be used? ________

5. A sample was used to collect data. Do you think the conclusions would be valid? Explain.

To find out how many players on the hockey team use a wooden stick, the coach surveyed all defensemen.

____

____

____

____

9.4 Selecting a Sample

FOCUS **Understand and choose an appropriate sample.**

There are many different ways to select an appropriate sample.

Sampling Method	Example
Simple Random Sampling – each member of the population has an equal chance of being chosen	The name of each student in your class is put in a hat and 5 names are drawn.
Systematic Sampling – every *n*th member of the population is chosen	Every 10th name in the telephone directory is chosen.
Cluster Sampling – the population is divided into groups and every member of *one* group is chosen	A school has five Grade 9 classes. All students in one class are chosen.
Self-Selected Sampling – members of the population volunteer to be chosen	Forty students volunteer to take a survey about homework habits.
Convenience Sampling – members of the population who are convenient are chosen	The principal chooses the first 3 boys she sees to help her choose the word of the week.
Stratified Random Sampling – some members of each group of the population are randomly chosen	The school's population is divided into grades and 20 students from each grade are chosen.

Example 1 Identifying the Sampling Method

Five members of each team in the soccer league are surveyed to find out how satisfied they are with the practice facilities.
Which sampling method was used? Explain.

Solution

The league's population is divided into teams. Five members from each team are surveyed. So, stratified random sampling was used.

Check

1. Identify the sampling method used for each survey.

a) To find out the favourite newspaper of city residents, the first 50 people entering a shopping mall were surveyed.
Did each shopper have an equal chance of being chosen? _____
Was the population divided into groups? _____
Were the shoppers chosen for convenience? _____
Sampling method: ______________________

b) To find out if a fitness club should extend its hours, every fifth person entering the club was surveyed.
Was every *n*th person surveyed? _____
Did people volunteer to be surveyed? _____
Sampling method: ______________________

c) To find out what students think of the new menu in the cafeteria, volunteers are asked to complete an on-line survey.
Was every *n*th person surveyed? _____
Did people volunteer to be surveyed? _____
Was the population divided into groups? _____
Sampling method: ______________________

Example 2 Choosing the Better Sampling Method

Shania wants to find out how many students in the school plan on attending the spring dance. Which sampling method would you suggest she use? Why?
Method A: Survey the first 50 students to enter school through the main doors.
Method B: Randomly select 20 students from each grade.

Solution

Using Method A, only those students who enter through the main doors can be surveyed. It is also unlikely that each grade will be represented equally, especially if the main doors are closer to the school bus drop-off area than to the parking lot.

Using Method B, each grade level is represented equally and each student has an equal chance of being surveyed.

Method B is the better method.

Check

In each case, which sampling method do you think is better? Why?

1. Jerome wants to find out what percent of the residents of Edmonton enjoy watching hockey.

Method A: Survey every 20th person who enters the arena to watch the Edmonton Oilers game.

Method B: Survey every 500th person in the Edmonton phonebook.

Method A: Do most people who go to an Oiler game like hockey? _____
Does each resident of Edmonton have an equal chance of being surveyed? _____
Is the sample representative of the population of Edmonton? _____

Method B: Does each resident of Edmonton have an equal chance of being surveyed? _____
Is the sample representative of the population of Edmonton? _____

Better method: ___________

2. Ms. Cheung wants to find out students' opinions on the class trip to the symphony.

Method A: Put the names of each student in a hat and survey the first 6 names drawn.

Method B: Survey all students in the class who take music lessons.

Method A: Does each student have an equal chance of being surveyed? _____
Is the sample representative of the class population? _____

Method B: Do most students who take music lessons enjoy music? _____
Does each student have an equal chance of being surveyed? _____
Is the sample representative of the class population? _____

Better method: ___________

Example 3 Identifying Appropriate Samples

A math teacher wants to find out what percent of his students completed the assigned homework. He divides the class into boys and girls, then asks to see the homework of all boys in the class.

Which sampling method was used?
Is the sampling method appropriate? Explain.

Solution

The class was divided into 2 groups – girls and boys – and all boys were chosen. So, the sampling method is cluster sampling.

The sampling method is not appropriate because the teacher only looked at the boys' homework. It is possible that many more boys than girls completed the homework, or vice versa. For the sample to be representative of the class, the teacher should look at the homework of an equal number of girls and boys.

Check

In each case, which sampling method was used? Is the sampling method appropriate? Explain.

1. An animal rights group wants to find out how residents feel about the proposed location of the new dog park. Surveys were left in all pet stores in the area.

Sampling method: ____________________
Is the method appropriate? _____
Why? ____
____
____
____

2. A newspaper wants to find out if its customers are satisfied with the time at which the Friday paper is delivered. Every 5th person that subscribes to home delivery of the paper is called.

Sampling method: ____________________
Does everyone who subscribes to the paper take the Friday paper? _____
Should the people who only get the weekend papers be surveyed? _____
Is the method appropriate? _____

Practice

1. Identify the sampling method used for each survey.

a) The population is divided into age groups and those people in the 18–24 age group are surveyed.

____

____

b) The members of the population are given ID numbers. The numbers are entered into a computer. The computer randomly selects 50 numbers. The people whose numbers are selected complete the survey.

____

____

c) A market researcher stands beside the information booth of a mall and asks people who walk by to participate in a survey.

____

____

____

2. Identify a potential problem with each sampling method.

a) To find out which new books students would like to see in the library, the school librarian surveys all English teachers in the school.

____

____

____

____

b) To find out the favourite vacation destination of Canadians, listeners of a Regina radio station were asked to complete an on-line survey on the radio station's website.

____

____

____

3. Which sampling method do you think is better? Why?

a) Marco wants to find out whether students who purchase lunch at the cafeteria would like to have the lunch special served with salad or coleslaw.

Method A: Survey every 5th person who enters the cafeteria.

Method B: Survey every 5th person in the cafeteria line-up.

Method A: Does everyone who eats in the cafeteria buy their lunch? _____
Does the opinion of those who don't buy their lunch matter to the cafeteria staff? _____

Method B: Does each student who buys from the cafeteria have an equal chance of being surveyed? _____
Is the sample representative of the students who purchase from the cafeteria? _____

Better method: ___________

b) A local street has an apartment building on the north side and houses on the south side. Maya wants to find out whether residents would like to see a 24-h bus service on the street.

Method A: Survey the occupants of every apartment on the third floor.

Method B: Survey every 5th house on the street and every 5th name on the apartment building's directory.

Method A: Who do you think would make more use of public transit: the residents of the apartment building or the residents of the houses?
____
Is the sample representative of all the residents of the street? _____

Method B: Is the sample representative of all the residents of the street? _____

Better method: ___________

4. In each case, which sampling method was used? Is the sampling method appropriate? Explain.

a) To find out about the study habits of students in the school, all students on the honour roll were asked to complete the survey.

Sampling method: _______________
Is the method appropriate? _____
Why? ____
____
____

b) To find out the favourite toast topping of people in the restaurant, every 4th person who orders toast is surveyed.

Sampling method: _______________
Is the method appropriate? _____
Why? ____
____
____
____

9.5 Designing a Project Plan

FOCUS Develop a project plan for data collection.

To design a plan to collect data, choose a topic that interests you, then follow these 5 steps.

Step	What to think about
1. Prepare a question.	– the wording should not influence a person's answer – the question should be sensitive to different cultures – participants should be anonymous if the question is personal
2. Identify the population and choose a sample.	– decide if you will you use a sample or a census – choose a sampling method (consider time and cost) – be sure the sample is representative of the population
3. Collect the data.	– be sure the timing of the survey is appropriate
4. Analyse and display the data.	– display the data in a table or on an appropriate graph – draw conclusions from your data
5. Evaluate your plan.	– have you avoided all potential problems with data collection? – was your sample appropriate? – are the conclusions valid?

Topic

Design a survey to find out the favourite radio station of students in your school.

Step 1

Name 4 radio stations students might listen to.

____

____

____

____

Some students might not listen to these stations.
Others may not listen to the radio.
What other choices could we add? ________________________

Write a question:

____

____

Is there a possible answer for all students? _____
Is the question biased? _____
Is the question sensitive to all cultures? _____

Step 2
What is the population? ____________________
Would it take a lot of time to survey each student in the school? _____
Will you use a census or a sample? __________

Which sampling method will you use?

Method A: Survey all students in Grade 9.

Method B: Randomly survey 15 students from each grade level.

Method A: Do students in Grade 9 like the same radio stations as students in other grades? _____
Is the sample representative of all students in the school? _____

Method B: Does each student in the school have an equal chance of being surveyed? _____
Is the sample representative of all students in the school? _____

Which method will you use? __________

Step 3
When will you collect the data? ____________________

____

____

Collect the data. Record the data in the tally chart.

Choice	Frequency	Total
________	________	_____
________	________	_____
________	________	_____
________	________	_____
________	________	_____
________	________	_____

Step 4

What is the favourite radio station of students in your school? ____________________

How do you know? ____

Display the data on a bar graph.

How can you tell which radio station is the favourite by looking at the bar graph?

____

Step 5

Did you have any problems with data collection?

Was your sample appropriate?

Are your conclusions valid?

Suppose you displayed the data on a circle graph. How would you know which radio station is the favourite?

____

____

Unit 9 Puzzle

Data Wordsearch

S	Y	S	T	E	M	A	T	I	C	A	E	R	N	T
S	U	Q	W	E	C	N	E	I	N	E	V	N	O	C
E	B	E	C	H	E	M	I	T	A	P	O	N	I	V
C	V	R	X	T	U	R	E	T	S	U	L	C	T	L
S	A	B	N	P	E	I	P	L	R	O	T	E	A	S
N	L	Z	T	H	E	O	R	E	T	I	C	A	L	E
O	I	M	N	C	X	R	T	A	Y	E	J	K	U	N
I	D	A	E	R	Q	G	I	F	N	T	L	O	P	S
T	T	Y	U	L	T	N	E	M	G	D	U	J	O	I
P	K	T	C	E	P	M	R	A	E	A	O	N	P	T
M	E	J	E	A	O	M	U	S	I	N	T	M	S	I
U	L	O	N	E	V	J	A	K	I	K	T	L	F	V
S	W	E	S	S	T	I	M	S	I	A	M	A	L	I
S	G	E	U	P	B	A	R	E	P	O	I	E	L	T
A	C	Y	S	T	C	O	U	P	F	R	A	N	D	Y

Find each of these words in the grid above.

Assumptions	Experimental	Sample
Biased	Fair	Sensitivity
Census	Judgment	Systematic
Cluster	Population	Theoretical
Convenience	Privacy	Time
Cost	Random	Valid

Unit 9 Study Guide

Skill	Description	Example
Determine if a decision is based on theoretical probability, experimental probability, or subjective judgment.	Is the decision based on: – results of a survey or experiment, – theoretical probability, or – personal thoughts or feelings?	Bronwyn takes her umbrella to school today because, despite the sunny weather forecast, she is sure it is going to rain. This is a decision based on subjective judgment.
Identify potential problems with data collection.	Consider: – timing – privacy – cultural sensitivity – cost – time – use of language	Martin surveys the students in his class to find out what costume they will wear for Halloween. He conducts the survey in March. There are 2 potential problems: The survey is insensitive to those who do not celebrate Halloween. The timing of the survey is wrong. Halloween is in October.
Decide whether to use a census or survey to collect data.	When the population is large, it is often too costly or time consuming to survey the entire population (a census), so we collect data from a portion of the population (a sample).	To find out the favourite Olympic event of students in his class, Ali surveyed all of his classmates. Ali conducted a census because the population (Ali's classmates) is not large.
Decide if a sample would provide valid conclusions.	Ask: – is the sample size appropriate? – does the sample represent the population?	To find out the favourite NHL hockey team of Canadians, Jocelyn surveyed all residents of Calgary. Most residents of Calgary would choose the Calgary Flames. The sample does not represent all Canadians.
Identify which sampling method was used to collect data.	– simple random sampling – systematic sampling – cluster sampling – self-selected sampling – convenience sampling – stratified random sampling	Every 5th car entering the parking lot was stopped. (systematic sampling) The names of all team members were put in a hat and 5 names were drawn. (simple random sampling)

Unit 9 Review

9.1 **1.** Explain how this decision is based on theoretical probability, experimental probability, or subjective judgment.

Craig observed his hamster for a science project. His hamster ran on the wheel at 3:00 P.M. on each of the last four days. Craig predicts his hamster will run on the wheel at 3:00 P.M. on the fifth day.

Circle the statement that best describes how Craig made his prediction.

Craig made his prediction based on the results of a survey or experiment.

Craig made his prediction based on theoretical probability.

Craig made his prediction based on his personal thoughts or feelings.

Explain your thinking.

____

____

____

2. What assumptions are being made?

The mayor won each of the last two elections. So, Carrie is sure the mayor will win the election this year.

Assumptions: – ______________________________

– ______________________________

____

9.2 **3. a)** Identify a potential problem with this survey.

To find out how many students will need help completing their high school course-selection forms in February, the guidance department surveys all Grade 9 students in September.

Is there a problem with cost? _____

Is there a problem with timing? _____

Is there a problem with privacy or cultural sensitivity? _____

Explain your thinking.

____

____

b) Describe how the problem could be avoided.

____

____

4. Explain why this survey might be problematic.

a) A tourist attraction employs 500 students for the summer. They want to find out how much money students save from each pay cheque. They plan to survey each student as he or she is given the next pay cheque.
Is there a problem with time or cost? _____
Is there a problem with privacy? _____
Is there a problem with cultural sensitivity? _____
Explain. ____

____

b) Describe how the problem could be avoided.

____

____

9.3 **5.** Should a sample or a census be used to collect the data?

a) A car manufacturer is switching to a new battery provider. The manufacturer wants to test the new batteries to make sure they are reliable in very cold weather.
Would it take a long time to test each battery? _____
Would it cost a lot of money to test each battery? _____
Should a sample or a census be used? __________

b) The Canadian Standards Association wants to survey approximately 2000 people in Alberta who purchased a defective child car seat.
Would it take a long time to survey each person who purchased a defective car seat? _____
Could using a defective child car seat lead to a serious problem? _____
Is it important that each of the purchasers be notified? _____
Should a sample or a census be used? __________

6. A sample was used to collect these data. Do you think the conclusions would be valid? Explain.
To find out if a local swimming pool should extend its hours, all people who use the pool before noon on weekdays were surveyed.

Do you think most people who use the pool in the morning would care whether the pool was open later on a night? _____
Do you think the conclusions would be valid? _____

9.4 **7.** Identify the sampling method used for each survey.

a) Ten students from each class are chosen to participate in a survey about the eating habits of teenagers.

____

____

b) A market researcher wants to find out the favourite brand of sunglasses of beach goers. He surveys people who walk in front of his deck chair as he sits on the beach.

____

____

____

8. Identify a potential problem with this sampling method.
To find out whether more residents of Vancouver drive an American-made car or an import, every 5th person entering a Ford dealership was surveyed.

____

____

____

____

____

9. Which sampling method do you think is better? Why?
Martina wants to find out whether residents would like to see more parks built.

Method A: Survey all families with children under 10.

Method B: Randomly survey every 50th person in the local phone book.

Method A: Do families with young children often go to a park? _____
Who do you think is more likely to want more parks built: families with young children or people without children? ______________

____

Do people with pets often use a park? _____
Is the sample representative of all residents? _____

Method B: Does each resident have an equal chance of being surveyed? _____
Is the sample representative of all residents? _____

Better method: ____________